Global Environmental Ethics

LOUIS P. POJMAN
U.S. Military Academy

Mayfield Publishing Company
Mountain View, California
London • Toronto

Dedicated to

Cdt. Kirk Brinker • Cdt. Timothy Brower • Maj. Ben Danner • Cdt. Jolie Erickson • Cdt. Jordan Gros • Cdt. John Scalia • Cdt. Donald Sisson Cdt. Ian Tyson • Cdt. Robert Underwood • Cdt. Stephen Walker of the United States Military Academy

They give new significance to the green uniforms they wear.

Library of Congress Cataloging-in-Publication Data

Pojman, Louis P.
Global environmental ethics / Louis P. Pojman.
p. cm.
Includes index.
ISBN 0-55934-991-3
1. Environmental ethics. I. Title.
GE42.P64 1999
179.'.1—dc21 98-53448
CIP

Manufactured in the United States of America
10 9 8 7 6 5 4 3 2 1

Mayfield Publishing Company
1280 Villa Street
Mountain View, California 94041

Sponsoring editor, Ken King; production editor, Linda Ward; manuscript editor, Joan Pendleton; design manager, Jean Mailander; art editor, Amy Folden; text designer, Anne Flanagan; cover designer, Jean Mailander; manufacturing manager, Randy Hurst. The text was set in 10/13 Garamond Light by Archetype Book Composition and printed on 50# Butte des Morts by Banta Book Group.

Cover image: "The Hetch-Hetchy Valley, California," by Albert Bierstadt, ca. 1874–80, reproduced in its entirety. © Wadsworth Atheneum, Hartford. Bequest of Mrs. Theodore Lyman in memory of her husband. As discussed in Chapter 9, around 1900 a debate occurred over whether to dam up part of the Tuolumne River in Yosemite National Park into a reservoir to pipe its water to San Francisco. Environmentalists like John Muir opposed it, whereas the head of the National Forest service, Gifford Pinchot, favored it and was eventually successful in getting Congress to authorize damming up the Hetch Hetchy.

This book is printed on acid-free, recycled paper.

Preface

Traditionally, two basic skills were thought to be necessary for a minimally educated person: literacy and numeracy. We need to be able to read and use numbers in calculations. In 1985 one of the foremost American ecologists, Garrett Hardin, proposed a third requirement for a minimally educated person in today's society: *ecolacy,* the understanding of ecology or the natural environment and our relationship to it. The word *ecology* comes from the Greek words *oikos* ("home") and *logos* ("logic of"), so we might say that ecology is the study or logic of our natural home, our earthly habitat. In many religious scriptures, including the Book of Genesis, we read that God created our natural home, Earth, and pronounced it good. God created man and woman and gave them dominion over the Earth, to care for it as good stewards of a valuable treasure. For thousands of years humans wrestled with nature, partook of her blessings, but felt her wild and unpredictable blows: tornadoes, hurricanes, volcanic eruptions, earthquakes, floods, drought, disease, and death. The dominion of human beings over nature was limited and unsure. Nevertheless, it was in humankind's interest to be good to Earth, for she would then be more willing to yield up her fruit, her wheat, her bounteous harvests.

With the onset of the industrial and technological revolutions the careful and precarious relationship of humans to nature underwent a radical transformation. For the first time human beings had the power to wreak enormous havoc on nature—even as nature had wreaked havoc on humanity. Throughout recorded history most people lived in rural locales, hunting and gathering or working in agricultural societies. In the twentieth century the trend toward urbanization commenced, especially in the Western nations. For example, in 1800 about 2% of the world's population lived in urban areas. Today nearly 50% live there. In 1820 72% of Americans worked on farms. By 1990 only 2% did, while 4% of the largest farms produced more than 50% of the nation's food.

This vision of our country is different from that of our founders, who sought a middle way between the untamed wilderness and urbanicity: the domesticated rural setting, a blend of clean rivers, forests, small towns, and farms. For Thomas Jefferson agriculture, education, and democracy were intimately connected: "Cultivators of the earth are the most valuable citizens. They are the most vigorous, the most independent, and the most virtuous, and they are tied to their country, and wedded to its liberty and interests by the most lasting bonds." Farmers know and love the land. They enjoy a symbolic relationship with nature. They know what city folk are apt to forget—that milk comes from cows and potatoes from the good earth. Children brought up on farms, close to nature, appreciate the virtues of wresting a livelihood from the earth. They know that if you don't work, you don't eat. If you don't plan for the future, you won't have one.*

But contrary to Jefferson's vision, the flight from the small farm to the city was encouraged, as large corporations bought out small farms or small farms were pressured to borrow from banks until the roof almost literally caved in and farmers were forced to declare bankruptcy.

For a while the impersonal mechanisms of technological farming seemed to work. Thanks to modern machinery, petrochemical fertilizers, herbicides, pesticides, and the mining of underground water resources, more food was produced than ever before. But then in the 1980s the chicken began to come home to roost. Because of poor agricultural policies, especially the failure to rotate crops and rest the land, soil erosion in the United States escalated as never before.

Combine the loss of topsoil and the loss of the family farm with increasing overuse and pollution of water—half our streams are polluted and areas of the great Ogallala Aquifer are in danger of drying up—and air pollution, and one begins to see the makings of a national catastrophe. Link this information with the frightening fact that the population of the Earth is growing at a doubling time of about 41 years. In 1,850 years the population only tripled from an estimated 300 million at the time of Christ to one billion in the mid-nineteenth century. With advances in modern medicine and public health, the population began to skyrocket, reaching a second billion in 1930, then a third billion in 1960. At present the Earth's population is 6 billion. Where will it end? Large numbers of unemployed, unhoused people crowd the cities of our world, potentiating crime and conflict. Famines and armed conflicts are already devastating large sections of the world, and a fifth of the world lives in abject poverty.

Perhaps nothing is more indicative of our disregard of nature than our overuse of the automobile. While the car offers us freedom and mobility, it is also environmentally deleterious. Since the automobile was introduced, 3 million Americans have been killed in highway accidents, more than were killed in all the battlefields in all our wars. Two-thirds of Los Angeles's downtown area is devoted to roads, parking lots, and gas stations. Each day 5 million vehicles crowd its freeways. In 1907 the average speed of a horse-

---

* Do not misunderstand me. I am not suggesting that we can go back to an agricultural society. The city is here, at least for the foreseeable future, and we need to deal with the problem of sustainable urbanization. I deal with these issues in Chapters 17 and 18. I am also aware that the agricultural revolution has been blamed for some aspects of our environmental crisis.

drawn carriage through the streets of Manhattan was 11.5 miles per hour. Today's automobile, built to speed at 130 miles per hour, crawls through the city at 5 miles per hour—barely keeping up with a slow jogger who runs a 12-minute mile. Today's city commuter will spend an average of two years of his or her life in traffic jams. In addition the motor vehicle produces more than 50% of our nation's air pollution. Carbon dioxide emissions from automobiles contribute to the growing greenhouse effect, which is causing dangerous weather changes in the world. A few afternoons ago, just after a warning had been issued that the ozone level in the valley where I'm living had reached dangerous proportions, I was riding the bus. There was one other passenger. Cars were speeding by, as though impervious to the damage they were causing. Thirty-eight percent of the families in nearby Salt Lake City own three cars. No wonder public transportation is as uneconomical as it is desperately needed.

Our nation may be justly proud of its democratic institutions and general high quality of life. But from an ecological perspective we have become victims of our own success, becoming more wasteful consumers and polluters. The average American's negative impact on the environment is about 20 to 50 times that of a person in the third world. With 4.5% of the world's population, we use about 34% of its mineral resources and 25% of its nonrenewable energy, and we produce more than 33% of its pollution. We're number 1.

However, environmental concerns are not simply national. They are global. The ecological motto is "Think globally. Act locally." It is true that we must begin at home, where we are, but we should not end there. We must also act globally, difficult as that may be. The radiation given off in the nuclear plant accident at Chernobyl affected not just the area in its immediate vicinity, but also Sweden, Switzerland, France, and most of the rest of Europe. Acid rain from the Ohio valley affects Canadian lakes and forests. Water mining in New Mexico and north Texas causes depletion of the Ogallala Aquifer, which extends to Colorado and Nebraska. The carbon dioxide emitted into the atmosphere from urban driving heightens the greenhouse effect, which may affect climate and weather patterns all over the world. Chlorofluorocarbons from developed nations deplete the ozone layer, contributing to increased rates of cancer worldwide. The burning of the rain forest in countries along the equator, now occurring at unprecedented rates, robs the whole world of valuable resources. During the 1994–95 burning season, some 11,196 square miles of the Amazon rain forest were destroyed (an area larger than New Jersey). This rain forest contains the world's largest collection of plant and animal species—many of which have yet to be studied for their potential medicinal, nutritional, and ecological value. Not only is it true that "no man is an island, entire of itself," but it is also true that no nation is an island, entire of itself. Ecologically, every person and nation is "a piece of the continent, a part of the main" holistic ecological system that constitutes the whole Earth. We need a global understanding of our relationship to the environment, a global environmental ethic. Educated people must add a global ethics of ecolacy to their repertoire of basic skills. If we are to save the planet for future generations, we will need an ethics of the environment at least as much as the skills of numeracy and literacy, perhaps more so.

Environmental ethics is the fullest extension of objective ethics, extending the scope of moral thought beyond one's community and nation to include not only all people everywhere, but also animals and the whole of nature, the biosphere, both now and beyond the imminent future to include future generations. Global environmental ethics, though the logical extension of traditional ethics, is revolutionary in that it calls on us to think and act in ways we hardly imagined, subordinating our politics, economics, and technology to a holistic global understanding of how we function within the ecosphere. It calls for a new, deeper moral consciousness.

What should be done? What can be done? That is what this book is about: the ethics of ecolacy. It is a textbook in global environment ethics.

The book is divided into two parts. The first part deals with *theoretical* issues in environmental philosophy and the second part deals with *applied* issues, specific environmental concerns. The kinds of questions we are concerned with in Part One, Theory, include

1. What is ethics all about? Why do we need it?
2. Are there universally correct answers to moral questions, or is every answer relative to culture or individual perception? That is, are there or are there not universal moral truths?
3. What is the place of self-interest in ethics? Is the good always good for you? Is it rational to be egoists or altruists?
4. Which is the correct ethical theory? How do environmental concerns affect the way we think about ethics in general? Do those concerns change our perspective?
5. What are the historical roots of our ecological crisis? What caused us to become environmental degraders?
6. Can an anthropocentric philosophy, which values only humans, possibly be sufficient to save the environment? Is enlightened self-interest adequate to resolve our environmental needs?
7. What is our obligation to future and distant people with regard to the environment? Do future people have rights, so that we must be frugal (and even sacrifice) with regard to population growth, natural resources, and pollution? Or is the notion of rights an unwarranted moral

view? Or do we at most have a duty to leave the next generation a world not ruined by pollution?

8. What is our obligation to animals, especially higher animals? Should we be as concerned about their welfare as about human welfare? Do animals have rights—not to be unnecessarily harmed, experimented on, hunted, or eaten?
9. What is the correct attitude toward nature? Should we regard nature as spiritually empowered? as inherently valuable or only instrumentally valuable? as God indwelt or as simply a wild force to be subdued?
10. Do we need a deeper understanding of reality that centers in nature? one that is nonanthropocentric and even transsentient, that is biocentric?

In Chapters 2 through 4 I develop this idea of a core morality, which, I argue, encompasses the heart of mainstream utilitarian, contractual, and deonotological (including intuitionism and Kantianism) thought. I employ this model throughout the book, especially in Part Two.

In Part Two, Application, we deal with specific issues such as population growth, pollution, resource and energy consumption (including nuclear energy), economics and the environment, and political theory and the environment. We are concerned with sustainable living, living in such a way as to preserve and pass on to future generations a vibrant, resourceful Earth, one fit for human, animal, and plant habitation. We will present the best arguments of both the cornucopians and doomsdayers and positions in between. The questions we will address include

1. How severe a problem is population growth? Is the Earth's present population size already straining the Earth's resources, and is it responsible for environmental havoc—or can the Earth comfortably support six or seven times the current population, as the cornucopians allege? If population growth is a problem, how should we deal with it? How should we be limiting population growth? Should we require prospective parents to obtain licences to beget children? What sorts of immigration and migration policies are morally supportable?
2. How serious is technology as a factor in resource exploitation and the production of pollution? Is technology part of the problem, part of the answer, or both problem and answer in different ways? Should we abjure some of technology's benefits, deeming the costs too high? For example, what should we be doing about energy? Should we turn to ecologically more benign forms, such as solar and wind-generated energy production? Should we bet on nuclear energy, which, if carefully monitored, could produce enormous supplies of clean energy? Or should we scale back our lifestyles so that we live more simply—that others may simply live?

3. What should be done with regard to areas of the world where the natural resources are in short supply, areas where famine, drought, and disease inflict great suffering on people? What are our responsibilities to distant people?
4. Exactly what is pollution and how does it affect us? What is the state of the Earth's air, rivers, streams, lakes, aquifers, and oceans? What exactly is the greenhouse effect, and how is it potentially harmful to ecosystems and human beings? What is the truth about the depletion of the ozone layer, and how can the problem be remedied?
5. What is our obligation to preserve other species? Is there a sacred holism that suffers as species are forever destroyed? Is there a hierarchy of being that is preserved in the variety of species? How do we measure the short-term economic benefits to humans of cutting down the rain forests versus eliminating rare species?
6. What kinds of economic policy ought we adopt to promote a sustainable world of the future? Is neoclassical economics, with its emphasis on the GNP and cost-benefit analyses, adequate for a sustainable economic policy?
7. What kind of political strategies should we adopt? Since the world populations are gravitating to cities, how can we ensure that our cities are ecologically healthy?
8. What are the implications of environmentalism for private property? Will environmental ethics require a new understanding of our so-called "absolute right" to do whatever we like with our property? If some profit-making uses of property are ecologically damaging, may the state forbid them without compensation to the owners?
9. What are the implications of global environmental ethics for nation-states? Must we adopt a universal regulatory system, such as the United Nations, that can oversee environmental laws? How far should we take globalism in the quest for sustainable development?
10. Should we change our lifestyles to promote a greater ethical concern for the environment? For example, should every healthy person use a bicycle instead of an automobile wherever possible? Should we seek to replace the internal combustion engine and the whole motor vehicle kingdom that surrounds it?

These are the kinds of questions the second part of this book deals with. A large amount of discussion will be given to understanding the issues, since before we can apply our moral theories, we need to be clear on the facts. I have attempted to the best of my ability to present all the relevant positions as fairly as I can, though my own positions will not altogether be submerged. While I have deep commitments to the environmental movement in general and argue for a core objective morality that is universal (global) in scope, I recognize that many issues are still open to moral debate. The book features the important debate among enlightened (conservationist) anthropocentrism, which sees it in our self-interest to promote a sustainable development; sentientism (which extends moral considerability to animals); preservationism; biocentrism (for example, deep ecology); and ecocentrism (the land ethic), which sees all life tied together in a holistic web.

Review questions and a list of suggested readings conclude each chapter. A glossary is included at the end of the book.

Several people were instrumental in making this a better book. Dean Darby, Mane Hajdin, Stephen Kershnar, John Jagger, Stephanie Lofgren, Sterling Harwood, Bill Throop, and Michael Levin made helpful suggestions at various points. Tal Scriven, California Polytechnic State University, San Luis Obispo; Leslie P. Francis, University of Utah; Charles Taliaferro, St. Olaf College; Martin Schönfeld, University of South Florida; Dr. E. R. Klein, Flagler College; J. Steven Kramer, University of Colorado at Boulder; and Arthur Millman, University of Massachusetts at Boston, gave me excellent constructive criticisms, for which I am much obliged. I consider myself the most fortunate writer in the world to have received such penetrating comments and helpful criticisms on earlier drafts of this work and can only hope that it has resulted in a better book. Linda Ward and Joan Pendleton did a superb job in improving this work as they brought it through production.

I am grateful to my editor Ken King for his friendship and energetic encouragement during the writing of this book; to my wife, Trudy, who has been both my most ardent supporter and staunchest critic during a wonderful 36 years of married life and who made critical comments on every page of this book; and to Col. Peter Stromberg, my departmental head at the United States Military Academy at West Point, a marvelous administrator who has patiently nourished my research and provided me with a convivial work environment. This book is dedicated to the cadets in my 1998 spring semester environmental ethics class at West Point, during which time most of the book was written. They are among the finest, most environmentally alert students I have taught in my 20 years of college and university teaching. They give new significance to the green uniforms that they wear. I have learned from each of them. This book was finished during my stay at Brigham Young University in Provo, Utah, surrounded by the magnificent Wasatch Mountains.

Louis P. Pojman

Contents

CHAPTER ONE

The Environment: A Global Perspective

Numberless are the world's wonders, but none
More wonderful than man; the stormgray sea
Yields to his prows, the huge crests bear him high;
Earth, holy and inexhaustible, is graven
With shining furrows where his plows have gone
Year after year, the timeless labor of stallions.

The lightboned birds and beasts that cling to cover,
The lithe fish lighting their reaches of dim water,
All are taken, tamed in the net of his mind;
The lion on the hill, the wild horse windy-maned,
Resign to him; and his blunt yoke has broken
The sultry shoulders of the mountain bull.

Words also, and thought as rapid as air,
He fashions to his good use; statecraft is his.
The searching arrows of frost he need fear no more,
That under a starry sky are endured with pain.
He has made himself secure from all but one enemy—
In the late wind of death he cannot prevail.

—SOPHOCLES (ca. 497–405 B.C.), *Antigone*[1]

NEVER HAVE SO FEW DONE SO MUCH

Human beings have lived on Earth about 100,000 years, a very short time in relation to the age of the universe (15 billion years) or even to the life of our planet (4.6 billion years). Civilization developed only 10,000 years ago, and the wheel was invented 4,000 years ago. If we compacted the history of the Earth into a movie lasting one year, running 146 years per second, life would not appear until March, multicellular organisms not until November, dinosaurs not until December 13 (lasting until December 26), mammals not until December 15, *Homo sapiens* (our species) not until 11 minutes before midnight of December 31, and civilization not until one minute before the movie ended. Yet in a very short time, say less than 200 years, a mere 0.000002% of Earth's life, humans have become capable of seriously altering the entire biosphere. In some respects we have already altered it more profoundly than it has changed in the past *billion* years. Haze from fossil-fuel combustion far to the south pollutes the Arctic sky, and both poles received heightened levels of ultraviolet radiation through human-induced depletion of the stratospheric ozone layer. Since preagricultural times, humans have cleared a net area of forest of some 8 million square kilometers, about the size of the lower 48 United States. Human activities account for about 150 million tons per year of sulfur through the biosphere, approximately doubling the natural flow and creating pollution problems throughout the globe.[2] Human activities greatly increase the amount of carbon dioxide in the atmosphere, exacerbating the greenhouse effect, thus increasing the Earth's climate and altering weather patterns. To paraphrase Winston Churchill's remark about the British Royal Air Force during the Second World War, "Never have so few done so much in so short a time."

How did things get this way? How did the "environmental crisis" come about? Humankind is part of nature. Humans are animals who have evolved from simpler biological forms over hundreds of thousands, even millions, of years. As we noted earlier, human beings as we know them (*Homo sapiens sapiens*) have lived on Earth about 100,000 years. Human life originated in Africa, and migration to Europe and Asia took place some 40,000 years ago. As omnivores (feeding on flesh and plants), we naturally eat plants and kill other animals for food. We have sometimes killed other humans for food. All animals must eat other organisms to live. The big fish eats the smaller fish and so on down the **trophic pyramid** (the levels of the **food chain**). Death is natural and inevitable as well. Most seeds are wasted and many infants die. Excess seed then provides insurance for survival of the species. Species normally maintain a steady population size through competition. The rhythm of life and death continues.

Our zone of life, the biosphere, inhabits land and air up to an altitude of 5 miles, water down to about 5 miles. Proportionately, this zone is thinner than the skin of an apple, and it represents all the life we know of in the universe. We conclude that life, being rare and fragile, is precious. Human beings are the most intelligent beings we know of in the universe. We are the only ones who seem to systematically deliberate on our actions. We use language, store information, and through science and technology exercise a powerful influence over the Earth.

Although we humans are animals, determined like other animals by our genes, we have created culture and artifacts, sometimes called *memes*,[3] a second nature whereby we can alter our natural fate. Whereas in preagricultural times the physically strong would dominate the weaker parties, today, given our memetic devices, those who control the machines and intelligently use them dominate both weak and strong. Our genes may program us to aggression, but our memes, weapons like guns and bombs, enable us to multiply the effects of that aggressivity, killing at a distance. And the meme of education and socialization can channel that aggressivity for socially approved purposes. Our genes endow us with intelligence, but our memes in the form of books, libraries, theories, and institutions of learning enhance the power of what our genes afford us. Through the invention of medicine, memes protect us from nature's onslaughts by allowing us to maintain health and extend life expectancy. Our genetic endowment prevents us from outrunning the gazelle or leopard; but through the use of our memes, our inventions of the internal combustion engine and wheeled wagons, we can travel over land more than twice as fast as the swiftest animal. We cannot defy gravity and fly like the eagle; but through our meme, the theory of aerodynamics, we can launch supersonic planes into the skies and make the fastest birds resemble skydiving turtles by comparison. We cannot swim underwater for more than a few minutes, but though our meme—the theory of hydrophysics—we can construct submarines that enable us to traverse the oceans faster than any fish. And now through our memes we stand on the brink of asexual modes of reproduction, the cloning of our own kind, the exact reproduction of our genetic type. Genes produce memes, which, in turn, safeguard genes. Through memes nature is swamped by culture, and evolution is replaced by the logic of technology. Cities replace the forest, highways cut through the wild, and the human imprint is everywhere felt. Humanity replaces natural selection as the decisive force in the development of life on Earth. Sophocles, writing more than 400 years before Christ, was right:

> Numberless are the world's wonders, but none
> More wonderful than man.

For the first nine-tenths of our species' existence, humans lived as *hunter-gatherers* (males mostly doing the hunting and females mostly doing the gathering). In what is sometimes referred to as the Old Stone Age (Paleolithic period), our ancestors lived nomadic lives in small communities (fewer than 50 people), gathering edible wild plants, hunting wild animals (bison, deer, mammoths), and fishing. They had few tools, lived mainly by their own muscle power and foot speed, and probably had a short life expectancy. Their impact on the environment was, for the most part, slight.[4]

About 10,000 years ago the *agricultural revolution* (Mesolithic period) occurred, marked by a shift from nomadic life to agricultural communities with domesticated animals and cultivated plants. To obtain farmland, settlers used slash-and-burn techniques: burning sections of forests and using the ashes as fertilizer. When the land was overcultivated, the community would move on to a new area and clear it. Meanwhile, the farmers noticed that when left alone, the old land would rejuvenate. This led to the practice of rotation farming: letting the land lie fallow for a period between plantings. During this period, domesticated animals were used for plowing and hauling material. Food production was more plentiful and secure, so the population increased. In this condition people acquired material goods, stored food for winter, and bartered with others, especially craftspeople who produced pottery, tools, rugs, and the like. Eventually, towns and villages grew up and served as centers of trade, government, social entertainment, and religion. During this period, while life expectancy increased and the quality of life became increasingly more civilized and secure, significant environmental damage began to occur. Poor agricultural policies led to the spread of desertification. Cities were often cesspools of sewage, garbage, and disease.

The third great cultural development, the *industrial revolution*, began in England about 1760, spreading to the rest of Europe and the United States in the nineteenth century. It ushered in the transition from manual to machine techniques of production, giving rise to the steam engine, the spinning jenny, and the power loom. Having depleted the forests that once covered most of England and Ireland, the industrialists turned to coal as the nonrenewable energy resource that would sustain factories and steamships. Coal mining was a dangerous occupation and urban coal burning produced a dangerous smog, more dangerous to health than the pollution in contemporary American cities. Many miners and city-dwellers died of lung disease. The use of machines meant that fewer farmers were needed to produce food, so urban factory jobs expanded. The story of the industrial revolution is one of increased wealth, especially for successful entrepreneurs, of the end of feudalism and rural-based society, of the development of city life, and of the exponential growth and dominance of technology in our lives.

In the last 100 years or so, we have invented electricity, the lightbulb, the refrigerator, the telephone, cinema, radio, television, the automobile, the airplane, the spaceship, the submarine, the air conditioner, the skyscraper, antibiotics, heart and liver transplants, the birth control pill, safe abortion, the microwave oven, the atom bomb, nuclear energy, and the digital computer—including communication on the World Wide Web. We have discovered how to clone animals and are on the verge of full-scale human genetic engineering. Breakthroughs or cures for Alzheimer's disease, cancer, and other diseases are on the horizon. Through the miracles of science and technology, we have enabled millions of people over the face of the Earth to live longer with more freedom, power, and knowledge than our ancestors could dream of. Only in science fiction were these wonders of modern life even hinted at.

Yet with this new freedom, power, and knowledge has come a dark side. The automobile kills hundreds of thousands of people throughout the world each year (more Americans have died in automobile accidents than in all the battlefields of all the wars our nation has fought). It produces chemical pollution, which degrades the atmosphere, causing illness and helping to bring on dangerous global warming, the greenhouse effect. Refrigerators and air conditioners enable us to preserve food and live comfortably in hot seasons and climates, but they also use chlorofluorocarbons (CFCs), which rise into the stratosphere and deplete the thin ozone layer that protects us from harmful ultraviolet radiation. Many scientists believe that this depletion has contributed to an increase in skin cancer, especially deadly melanoma, and has harmful effects on plankton, which form the base of much of the food chain of marine animals. Nuclear power could provide safe, inexpensive energy to the world, but is also used to exterminate cities and threaten a global holocaust. Disasters like the nuclear plant steam explosion at Chernobyl in the former Soviet Union have spread harmful radiation over thousands of square miles, causing death and deformity. Such horrendous threats have replaced the public's fear of nuclear war as the greatest potential catastrophe, causing widespread distrust of the nuclear power industry. Nuclear waste piles up; safe depositories have been proposed, but no one wants it in his or her "backyard." But our modern way of life does require energy, lots of it. So we burn fossil fuels, especially coal, which, unbeknownst to the public at large, is probably more dangerous than nuclear energy; coal burning causes diseases such as cancer, pollutes the air with sulfur dioxide, and produces acid rain, which is destroying our rivers and lakes and killing trees. Medical science has found cures for tuberculosis and syphilis and has aided in greatly lowering infant mortality, but in the process it has allowed an exponential growth of population. This growth has produced crowded cities and put a strain on our resources. The more people, the more energy needed; the more energy produced, the more pollution; the more pollution, the more threat to life from disease. Add to all this the problem of a *just* distribution of resources: The increasing discrepancy of wealth between the rich and the poor, both within and between nations, portends a global tragedy, one that in sheer suffering could make previous socially destructive events seem miniscule by comparison.

And so the story goes. For each blessing of modern technology, a corresponding risk comes into being, as the tail of the same coin. With each new invention comes frightful responsibility. Technophiles put their money on technology's ability to meet all our basic needs and satisfy everyone. They reason, "We have put a man on the moon, learned how to increase food production as never before thought possible, and discovered how to put cancer into remission and how to clone animals and human beings. All we need to do is put our best brainpower to work at our problems and we can create the necessary technology to solve them." Technophobes see this faith in technology not only as naive, but also as a dangerous impediment to progress. In fact, they see technology (or the overuse of technology) as the main problem, undermining the human spirit, devastating community, and threatening to dictate our lives.

There was once a man who discovered his own shadow. He became so fascinated with it that he gave pride of place to it, treated it with deference and respect, until finally, neglecting his other duties, he became a shadow to his own shadow. Likewise, according to the technophobes, we may be in danger of being slaves to our own devices. The recent Unabomber's "Manifesto" calls for a revolution against our inexorable and headstrong plunge toward technocracy. The technophobes liken our society to an alcoholic with a lot of hard liquor in the refrigerator of technology, who is advised by doctors to drink moderately. The prescription offered by the technophobes is deadly: What we need is not moderate reform but a radical cure. According to the Unabomber's "Manifesto,"

> The industrial revolution and its consequences have been a disaster for the human race. They have greatly increased the life-expectancy of those of us who live in "advanced" countries, but they have destabilized society, have made life unfulfilling, have subjected human beings to indignities, have led to widespread psychological suffering (in the Third World to physical suffering as well) and have inflicted severe damage on the natural world. The continued development of technology will worsen the situation. It will certainly subject human beings to greater indignities and inflict greater damage on the natural world, it will probably lead to greater social disruption and psychological suffering, and it may lead to increased physical suffering even in "advanced" countries. . . . We therefore advocate a revolution against the industrial system. This revolution may or may not make use of violence: it may be sudden or it may be a relatively gradual process spanning a few decades. . . . [We] outline in a very general way the measures that those who hate the industrial system should take in order to prepare the way for a revolution against that form of society. This is not to be a POLITICAL revolution. Its object will be to overthrow not governments but the economic and technological basis of the present society.[5]

Even if this is an exaggeration, even if we want to dissociate ourselves from its author's violence, the manifesto echoes a horror of the possibility that our technology will usurp our autonomy. Even if we reject the technophobe's message as extreme, we must admit that it's hard to use technology wisely, hard to live moderately and frugally as good stewards of the Earth, hard to conserve our resources so that posterity will get a fair share, hard to foster social justice and global equity.

Environmental ethics concerns itself with these global concerns: humanity's relationship to the environment, its understanding of and responsibility to nature, and its obligations to leave some of nature's resources to posterity. Pollution, population control, resource use, food production and distribution, energy production and consumption, the preservation of the wilderness and of species diversity all fall under its purview. It asks comprehensive, global questions, develops metaphysical theories, and applies its principles to the daily lives of men and women all over the Earth.

These are the issues that will concern us in this book. We are concerned with a global ethic, one that will allow us to live morally in the midst of an ever-shrinking world, one that will enable us to survive and flourish in the light of terribly difficult problems. In this book we will develop a global ethic and attempt to apply it to the major problems of the environment, especially those related to population, resource consumption, and pollution. But first let us examine some of the basic ecological concepts that underlie environmental philosophy and science and that will be necessary for understanding the discussions in this book.

BASIC ECOLOGICAL CONCEPTS

Most discussion of environmental science and ethics revolves around three basic concepts: population, resources, and pollution. Two other concepts are related to these basic ones: exponential growth and the carrying capacity of an ecosystem. Let us briefly examine each of them.

Population and Exponential Growth

Imagine a pond that has a few lilies in it. Each day the number of lilies doubles until on the thirtieth day the pond is completely covered with lilies. On which day is the pond one-half full of lilies?[6]

There was once a knight who saved the kingdom of a king. The king asked him what he wanted as a reward. "Simply some grains," the knight said. He brought out a chess board with the normal 64 squares, put down a grain of wheat, and said, "Please double it every day for 63 more days."

"Is that all you want?" replied the startled king. "I would have given you a million dollars if you had asked, but all you want is some grain?"

So the next day the king put down 2 grains on the second square, and the following day 4 grains, then 8, then 16, then 32. By then he had to use bags for each day's grains, which were too large for the squares. On the seventh day he placed 64 grains in the bag for the seventh square, then 128 grains in the bag for the eighth square. The tenth day's bag consisted of 512 grains. By the fifteenth day the bag of grains consisted of 16,384 grains and by the twentieth day 524,288. By the twenty-fifth day the grains would have come to 16,777,216 if the king had been able to get that many, but he wasn't. By the thirty-first day he would have needed 1,073,000,000 grains, and he would not have accounted for even half the squares on the board. By the sixty-fourth day he would have needed 2^{63} grains, 500 times all the wheat harvested in the world last year and more than all the wheat that has ever been harvested in the world.

Exponential growth is growth by a fixed factor or percentage—that is, multiplicative (2, 4, 8, 16, . . .) growth, rather than additive or linear (1, 2, 3, 4, 5, 6, . . .). Instead of growing by additions of one, something that grows exponentially increases by multiplying its size. In environmental studies we are often concerned with the doubling time of a population. In figuring out the rate of increase we use what is called *the rule of 70.* If a sum increases at a 1% per annum rate, it will double in 70 years. For example, if you invest $100 in a bank at 1% interest, in 70 years you will have doubled your investment. If a population is growing by 1% per annum, it will take 70 years to double its size. So a community of 1,000 people growing at 1% per annum will reach 2,000 in 70 years. If we increase the growth rate to the present 1.7% rate of the U.S. population, the doubling time is about 41 years ($70/1.7 \approx 41$). Rwanda, before its present civil war, had the highest growth rate in the world at 8% per annum. How fast was it doubling its population? Answer: every 8½ years. What is likely to occur when populations double this rapidly?

At the time of Christ the population of the world is estimated to have been less than 300 million. It took 1,850 years for the global population to reach 1 billion, then only 80 more years to reach its second billion. Since 1930 the Earth's population has increased from 2 billion to its present 6 billion (Figure 1-1). In 1968 Paul Ehrlich wrote *The Population Bomb*,[7] warning that the population of the world (then 3.5 billion) was growing exponentially at a rate of 70 million per year and that if strong measures were not taken, we would likely experience increased famine, disease, and war. Critics labeled such grim forecasters as "doomsdayers," but some of his prophecies have been realized: Famines and civil wars in Rwanda, Zaire, Ethiopia, and Somalia have occurred as he and others predicted.

Crowded conditions prevail in many parts of the world. Famines have worsened in areas of Africa and Asia. Today the global population is growing not by 70 million per year but by 90 million (1995). The growth rate is about 1.7%, translating into a doubling time of 41 years, so that, if the present trend continues, it will reach about 12 billion in 2040, doubling again to 24 billion by 2081 (see Figure 1-1).

But it is hard to imagine this trend continuing. A *dieback* caused by famine, disease, war, or other catastrophes will probably cull the population—if policies of population control are not implemented. The United Nations' demographers have used three different models to project growth scenarios for 2050: a high, medium, and low population growth, which turn out to yield populations of 11.2 billion, 9.4 billion, and 7.7 billion, respectively. There have been some signs of a decrease in worldwide fertility rates, but the present growth rates of some third world countries are still very high. In 1994 some of the fastest growth rates were:

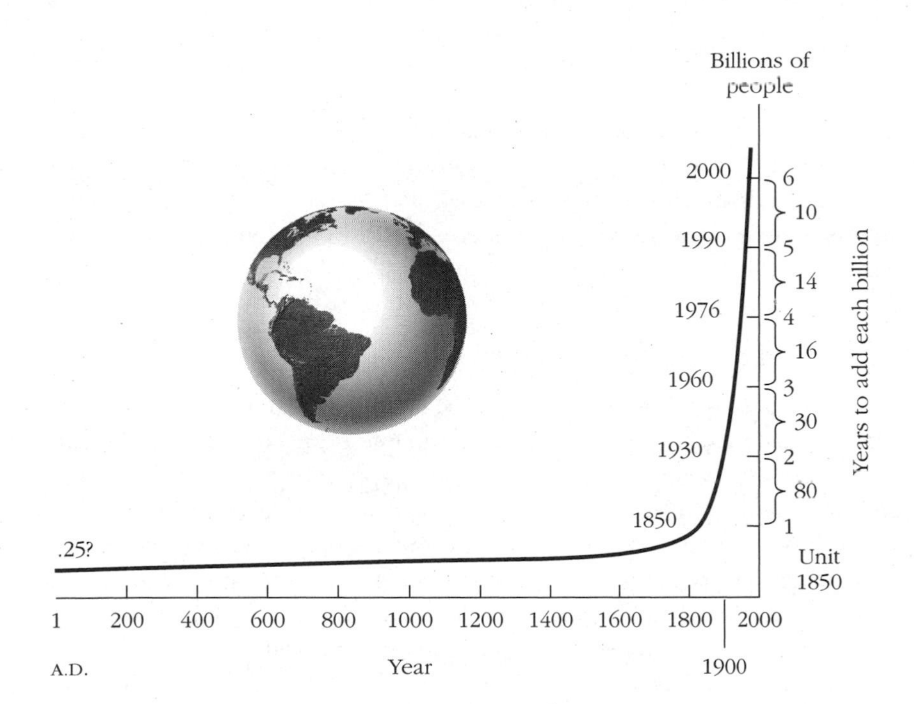

Figure 1-1 J curve of the world's population growth.

Jordan	5.2%	(doubling time of 13 years)
Malawi and Qatar	4.5%	
Oman	4.4%	(doubling time of 16 years)
Gambia and Yemen	4.1%	
Macao	3.7%	
Ivory Coast and Libya	3.6%	
Syria	3.5%	
Botswana, Iran, Zaire, and Zambia	3.3%	(doubling time of 21 years)

We can explain resource consumption and pollution growth in terms of exponential growth by factoring in population increase. If a population grows and does not reduce its resource consumption or pollution, resource consumption and pollution will exponentially increase. Our present environmental problems—population growth, resource use, extinction of species of plants and animals, destruction of ecosystems, depletion of the ozone layer, and pollution—are intertwined examples of exponential development. For example, if you have $1 million and invest it at 10% per year, you can spend up to $100,000 each year without depleting your capital; but if you spend more than your investment allows, your capital will eventually be gone, and you will become bankrupt. If you spend $200,000 per year, you will deplete your capital in less than seven years.

The Earth has many wonderful resources, our precious capital, but we are spending that capital at a dangerously fast rate, depleting our soil, aquifers, ozone layer, rivers, atmosphere, forests, petroleum deposits, and more.

The Idea of a Carrying Capacity

When populations grow, they tend to use up the available resources. Each land area has a certain carrying capacity, the rate of resource removal it can tolerate without degenerating. The carrying capacity can be somewhat enriched via artificial means (fertilizers, herbicides, pesticides, and irrigation) but still has limits. Because consumers use increasing resources and create larger levels of pollution, population growth strains an environment. Eventually, the increased exploitation of resources puts too great a strain on the environment and causes a negative reaction: The aquifer cannot replace the water as fast as it is mined; the land cannot replace the grass as fast as it is eaten; the forests cannot replace trees as fast as they are cut down; the ozone layer cannot replace ozone as fast as it is depleted; the oxygen in the air cannot be replaced as fast as it is depleted; species cannot survive the loss of their habitats. We call the increased population strain an "overshoot," which results in an environmental crash or dieback. Under these conditions the population is likely to face famine, disease, and war. Migration may occur and, if fortune prevails, the environment is gradually restored, though perhaps at a lower carrying capacity.

An example of such a population crash occurred in 1910 when 26 reindeer were introduced to an Aleutian island off the southwest coast of Alaska. When the reindeer were first introduced, the vegetation provided them an ample food supply. But no predators existed, and by 1935 the deer population had escalated to more than 2,000, overshooting the island's carrying capacity. A dieback occurred so that by 1950 only 8 deer remained. Ireland experienced a similar, but human, population crash in 1845 when the potato crop failed. An estimated one million people died and another three million emigrated, mostly to the United States.

Figure 1-2 diagrams the idea of carrying capacity. Note that when a dieback occurs, a rejuvenation of the ecosystem may follow, but it is likely to have a lower carrying capacity (CC2) due to the damage done in the dieback. One has to distinguish between *theoretic* and *actual* carrying capacity. The theoretic assumes equal distribution of resources, which seldom happens. So the actual carrying capacity is usually lower.

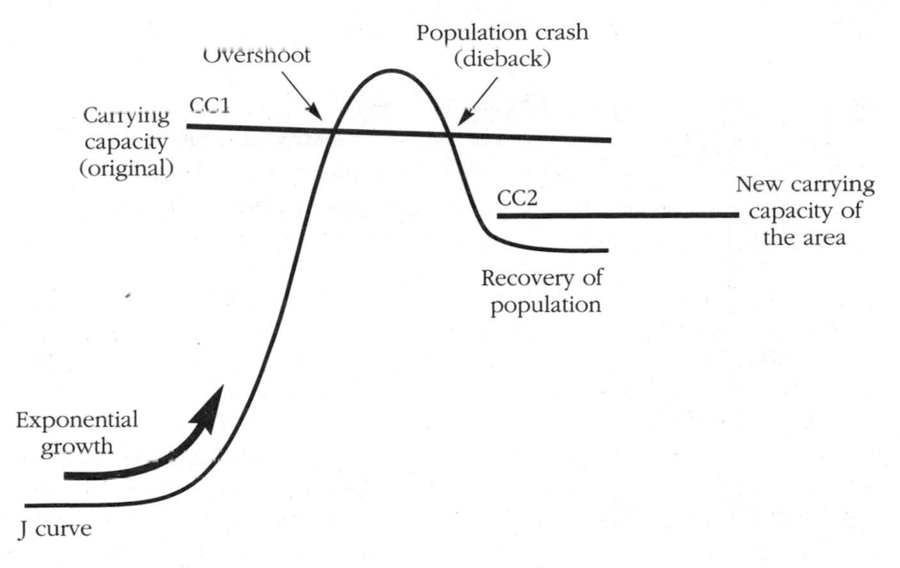

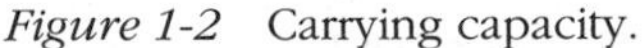

Figure 1-2 Carrying capacity.

What Is a Resource?

A resource is anything one gets from the environment to meet one's needs and desires. All living beings (organisms) need resources such as food, water, air, and shelter for health and survival. Natural resources include solar energy, fresh air, fresh water, fertile soil, and edible plants, which are directly available to us and other organisms. Other resources, such as petroleum, iron, copper, electricity, and modern crops, are available only through technology. Some resources are *renewable*; for example, solar energy is expected to last another 5 billion years. Other resources are relatively *nonrenewable*. When we eliminate a species; a rain forest; minerals such as copper, aluminum, and iron; or petroleum, the Earth may not produce any more of them. Other resources, such as fertile soil, clean underground water (aquifers), the ozone layer, and clean air, can be replenished slowly, but may be strained by our overuse. Some nonrenewable resources can be recycled to extend supplies.

What Is Pollution?

Pollution may be defined as unwanted substances, as contamination. The National Academy of Sciences defines it as "undesirable change in the physical, chemical, or biological characteristics of the air, water, or land that can harmfully affect health, survival, or activities of humans or other living organisms." A pollutant may result in a disruption of life-support systems for humans and other species, damage to wildlife, damage to human or animal health, and damage to property, as well as nuisances such as loud noise and unpleasant smells and sights.

Three factors determine the severity of a pollutant: (1) its chemical nature (how harmful it is to various types of living organisms), (2) its concentration (the amount per volume of air, water, soil, or body weight), and (3) its persistence (how long it remains in the air, water, soil, or body).

1. We think of carbon monoxide (CO), carbon dioxide (CO_2), nitrogen dioxide (NO_2), sulfur dioxide (SO_2), and methane (CH_4) fouling our atmosphere and water supply, and we are aware of noise pollution (screeching boom boxes, rock concerts, motorcycles, and supersonic jets). Carbon dioxide and methane gas are contributing to the greenhouse effect (see Chapter 13), threatening to melt the polar ice caps and change climate and weather patterns. Sulfur dioxide from coal-burning power plants is contributing to acid rain, which is killing our fish and forests. Nitrogen oxide, petrochemicals, and bacteria from animal feces are polluting our streams, rivers, and aquifers. CFCs (chlorofluorocarbons) used in aerosol cans, refrigerators, and air conditioners drift upward toward the ozone layer, transforming ozone (O_3) to oxygen (O_2), thus reducing our protection against harmful ultraviolet rays and threatening an increase in skin cancers, especially melanoma. In addition, radiation leaks from nuclear power plants may have far-reaching carcinogenic and mutagenic effects.

2. Ecologists measure pollution in terms of the concentration of chemicals in an organism or body of air or water, the amount per unit of volume or weight of air, water, or the body. A concentration of one part per million (1 ppm) equals one part pollutant per one million parts of the solid, liquid, or gas in which it is found. A measurement of 1 ppb refers to one part pollutant per billion of the medium, and 1 ppt refers to one part per trillion. These may seem very small percentages, but sometimes a very small part can have a gigantic influence, even resulting in cancer.

3. Finally, with regard to a pollutant's persistence, we may divide persistence into three types: degradable, slowly degradable, and nondegradable. Degradable pollutants, such as human sewage and soil, are usually broken down completely or reduced to acceptable levels by natural chemical processes. Slowly degradable pollutants, such as DDT, plastics, aluminum cans, and CFCs, often take decades to degrade to acceptable levels. Nondegradable pollutants, such as lead and mercury, are not broken down by natural processes.

We know little about the short- and long-range harmful potential of most of the more than 70,000 synthetic chemicals in commercial use on people and the environment. The Environmental Protection Agency (EPA) estimates that 80% of cancers are caused by pollution. We know that half of our air pollution is caused by the internal combustion engine of motor vehicles and that coal-burning stationary power plants produce unacceptable amounts of sulfur dioxidc (SO_2). Thc World IIcalth Organization (WIIO) cstimates that more than 1 billion urban people (nearly one-fifth of humanity) are being exposed to health hazards from air pollution and that emphysema, an incurable lung disease, is rampant in our cities. Studies tell us that smog (a mixture of smoke and fog) is hazardous to our health and that it has caused thousands of deaths in such cities as London, New York, and Los Angeles.

Now let us combine these concepts and relate them to the present state of the environment.

ENVIRONMENTAL IMPACT: THE PRESENT STATE OF THE WORLD

Former President Richard Nixon, upon being told that the United States uses 34% of the world's total mineral resources, 25% of its nonrenewable energy, and 30% of its food while producing over 33% of its pollution, is said to have exclaimed, "And so shall it ever be!" Nixon thought that these statistics reflected a high living standard, and, in a way, he was right. The United States is first in living high, but being first in these categories may not be a compliment; rather, it is perhaps tantamount to being the biggest pig on the block, a "petropig," as Amory Lovins calls the nation. If mineral and energy resources are limited and pollution is a threat to the globe, as most environmentalists believe, then with 4.5% of the world's population, the United States uses many times its reasonable share of mineral and energy resources and creates many times its fair share of pollution. Though in many ways more environmentally aware (especially in its laws requiring catalytic converters and lead-free gasoline and energy-efficient coal-burning plants) than most other nations, the United States still is the leading consumer of nonrenewable and scarce resources and the leading polluter in the world. It emits 15% of the world's SO_2 and 25% of NO_2, and it is the major manufacturer of CFCs, which damage the ozone layer. As of 1998 the United States had grown to 270 million people, with a growth rate of 1.7% per year—a doubling time of about 41 years. Though this is a lower rate than that of many countries, we create a greater stress on the environment than do most nations. Ecologists speak of a nation's *environmental impact*, meaning the product of its population *times* (×) its per capita resource use *times* its per capita pollution output. The formula is:

$$\text{Population} \times \begin{array}{l}\text{Resource use}\\\text{per person}\end{array} \times \begin{array}{l}\text{Pollution per}\\\text{unit of resource}\end{array} = \begin{array}{l}\text{Environmental}\\\text{impact}\end{array}$$

There are three ways to have an adverse environmental impact: (1) people overpopulation (PO), (2) consumption overpopulation (CO), and (3) both people and consumption overpopulation. People overpopulation occurs when the number of people in a geographical area results in diminished resources. Consumption overpopulation occurs when the people in an area consume more resources than are necessary for survival and produce large quantities of pollution. Many demographers believe that the world already has reached its population-carrying capacity and that we are experiencing people overpopulation in several areas of the world, especially in India, Bangladesh, China, and the Sahel in North Africa.

In the following formulas, let A = Population, B = Resource use per unit of resource, C = Pollution per unit of resource, and E = Environmental Impact:

$$A \times B \times C = E$$
$$\text{PO } 100 \times 1 \times 2 = 200 \text{ units}$$
$$\text{CO } 10 \times 10 \times 5 = 500 \text{ units}$$

Currently 90% of the world's births take place in the third world, so these countries have larger populations than do the Western industrialized nations and Japan, but industrialized nations have a far greater environmental impact (E) due to their consumption habits. In the example we see a community of 10 having an environmental impact 2½ times that of a poorer community of 100. Each member of a CO country has 25 times as great an impact as a member of a PO country. The total environmental stress on our planet is increasing 5.5% per year. As of 1990 it had increased to 450 times that of 1880, and it is still growing.

Consumption overpopulation occurs when people use high levels of nonrenewable and scarce natural resources and create high levels of pollution. The average U.S. citizen's negative impact on the environment is between 30 and 100 times that of an average person in a third world country like India, Bangladesh, or Rwanda. As already noted, with 4.5% of the world's population, we use about 34% of its mineral resources, 30% of its food, and 25% of its nonrenewable energy; and we create about 36% of its pollution. We also use 33% of the world's paper, cutting down 1,000 acres of forest each day to produce our paper. The Sunday edition of the *New York Times* requires about 150 acres of forest to be cut down.

Following this logic, it might not be a good thing to bring underdeveloped nations to our standard of living. To quote the eminent food scientist Jean Mayer:

> It might be bad in China with 700 million poor people but 700 million rich Chinese would wreck China in no time. It is the rich who wreck the environment, . . . occupy much more space, consume more of each natural resource, disturb ecology more, litter the landscape and create more pollution.[8]

We should also note that despite increased grain yields in recent years, our topsoil is being depleted at an alarming rate. Dust storms, poor tillage methods, erosion, and the use of pesticides and artificial fertilizers deplete the quality of the topsoil, resulting in the 3.2 million acres lost to the farm economy each year. It is estimated that the United States has lost two-thirds of its original topsoil. Some of our country's most productive agricultural lands, such as those in Iowa, have lost over half their topsoil. In California the erosion rate is 80 times faster than the rate of soil formation. Imagine a highway one-half mile wide stretching all the way from New York City to Los Angeles. That represents the amount of land taken out of the economy per year for shopping malls, highways, parking lots, and other automobile-related construction. Meanwhile, in our attempt to eradicate the original agricultural pests through the use of pesticides, new and more virulent "strain-resistant" pests and bacteria have evolved, which in turn require stronger pesticides, which in turn will result in more deadly pests and bacteria, and so forth. The costs of artificial fertilizer and pesticide are eating into the profits of agriculture and for the first time in years food production nationwide is on the downswing.

Fifty percent of our rivers and streams are polluted and our aquifers are being water-mined at a faster pace than they can be replenished by rain and runoff. The overuse of water due to population growth is a worldwide problem. For example, the oldest section of Mexico City has sunk 30 feet (at a rate of 19 inches per year) due to the fact that the water underneath the city is used to nourish its 18 million inhabitants. The result has been millions of dollars worth of damage to buildings in the city and a frenetic quest to locate new sources of fresh water.[9] Each year the United States produces 275 million tons of hazardous wastes. The EPA has identified 34,000 main hazardous waste sites, 1,211 of them so dangerous that they've been designated for cleanup, with the estimated cost of cleanup per site being $26 million. Acid rain is killing our forests and lakes. Add to this the heavy use of pesticides, fungicides, herbicides, and petrochemical fertilizers, all producing water and soil pollution, and you can begin to sense the magnitude of the problem.

Doomsdayers like Paul and Anne Ehrlich, Amory and Hunter Lovins, Helen Caldicott, and Garrett Hardin warn us to change our ways or encounter disaster. Biologist Gerald Durrell says, "At the present rate of 'progress,' and unless something is done quickly, disaster faces us in the face. We are bleeding our planet to death. We are led by sabre-rattling politicians who are ignorant of biology, and surrounded by powerful commercial interests whose only interest in nature is often to rape it."

On the other hand, *cornucopians* (environmental optimists) like Julian Simon, Greg Easterbrook, Michael Fumento, Stephen Budiansky, Rush Limbaugh, and Michael Zey believe that intelligent use of technology can keep disaster away and actually improve our lot. "Look," they point out, "never before have so many people lived so long and so happily. Be grateful and use the Earth wisely instead of playing Chicken Little. The sky isn't falling, but rather we're better off than ever. The earth's carrying capacity is probably over 40 billion, so we've a long way to go before we should take desperate measures to become ascetics." Still other views include those of the deep ecologists, who believe that we should reject both the doomsdayer and the cornucopian solutions. Salvation is in changing our anthropocentric ways of looking at the world—which both the doomsdayers and the cornucopians embody. Yet other ecologists, suspicious of technology, like Barry Commoner and Murray Bookchin, argue that not technology but social justice is the cure. Change the economic maldistribution and we can save our planet, they aver. The most radical of the technophobes call for an all-out revolution against our whole economic and technologically based way of life.

Who is right? The doomsdayers or the cornucopians or the deep ecologists or the social ecologists or the revolutionary technophobes? In the rest of this book, we shall encounter and analyze the arguments and evidence presented by these and other theorists on what are to my mind the most important ethical issues of our generation.

Review Questions

1. Discuss the idea of exponential growth. What is its significance for ecological studies and ethics?
2. Discuss the relationship of technology to environmental problems. How does technology help and/or hinder sound environmental policy? Can you think of a technological innovation that has no corresponding danger?
3. What is the difference between consumption overpopulation and people overpopulation?
4. It will be helpful for you, as you begin to read this book, to write a paragraph or two on your assessment of our relationship to the environment. Is there an environmental crisis? Explain your answer.

Suggested Readings

DesJardins, Joseph. *Environmental Ethics*. 2nd ed. Belmont, Calif.: Wadsworth, 1997.

Easterbrook, Gregg. *A Moment on the Earth*. New York: Viking, 1995.

Gruen, Lori, and Dale Jamieson, eds. *Reflecting on Nature*. New York: Oxford University Press, 1994.

Leopold, Aldo. *Sand County Almanac*. New York: Oxford University Press, 1949.

Miller, G. Tyler. *Living in the Environment*. 10th ed. Belmont, Calif.: Wadsworth, 1998.

Newton, Lisa H., and Catherine K. Dillingham. *Watersheds 2: Ten Cases in Environmental Ethics*. Belmont, Calif.: Wadsworth, 1997.

Passmore, John. *Man's Responsibility for Nature*. 2nd ed. London: Duckworth, 1980.

Pojman, Louis P., ed. *Environmental Ethics: Theory and Application.* 2nd ed. Belmont, Calif.: Wadsworth, 1998.

Regan, Tom, ed. *Earthbound: New Introductory Essays in Environmental Ethics.* New York: Random House, 1984.

Revelle, Penelope, and Charles Revelle. *The Environment: Issues and Choices for Society.* 3rd ed. Boston: Jones & Bartlett, 1988.

Rolston, Holmes, III. *Environmental Ethics.* Philadelphia: Temple University Press, 1988.

Simon, Julian. *The Ultimate Resource 2.* Princeton: Princeton University Press, 1996.

Sterba, James, ed. *Earth Ethics.* Upper Saddle River, N.J.: Prentice Hall, 1994.

VanDeVeer, Donald, and Christine Pierce, eds. *The Environmental Ethics and Policy Book.* Belmont, Calif.: Wadsworth, 1994.

Westra, Laura. *An Environmental Proposal for Ethics: The Principle of Integrity.* Lanham, Md.: Rowman & Littlefield, 1994.

Zimmerman, Michael, ed. *Environmental Philosophy.* Englewood Cliffs, N.J.: Prentice Hall, 1993.

Notes

1. Sophocles, *Antigone,* trans. Dudley Fitts and Robert Fitzgerald (New York: Harvest Books, 1939).
2. William B. Meyer, *Human Impact on the Earth* (New York: Cambridge University Press, 1996), p. 2.
3. *Memes* refers to human inventions, including cultural institutions and practices, that redirect natural behavioral patterns. Morality is a meme (it serves our purpose of promoting human flourishing). Law and political organization are other examples. Technology is also a meme. Art, music, science, education, and buildings all are memes. Both genes and memes are replicators, which reproduce themselves—one through the transference from body to body, the other through one mind to another. See Richard Dawkins, *The Selfish Gene* (New York: Oxford University Press, 1976) for a good discussion. Dawkins defines *meme* as "a unit of cultural inheritance, hypothesized as analogous to the particulate gene, and as naturally selected by virtue of its 'phenotypic' consequences on its own survival and replication in the cultural environment" (Dawkins, *The Extended Phenotype,* New York: Oxford University Press, 1982, p. 290).
4. Apparently, hunter-gatherers in Africa, Asia, and the Americas, including the Native Americans, used fire to turn forests into grasslands and savannahs. See Stephen Pyne, *Fire in America,* (Princeton: Princeton University Press, 1982), pp. 47, 71–80. See also William Cronon, *Changes in the Land* (New York: Hill & Wang, 1983).
5. "Unabomber Manifesto," *Washington Post,* 19 September 1997, 1 and 4.
6. The answer is the twenty-ninth day.
7. Paul Ehrlich, *The Population Bomb* (New York: Ballantine, 1968).
8. Jean Mayer, quoted in Lewis Moncrief, "The Cultural Basis of Our Environmental Crisis," *Science* 170 (30 October 1970). Reprinted in L. Pojman, ed., *Environmental Ethics,* 2nd ed. (Belmont, Calif.: Wadsworth, 1998).
9. Reported in the *New York Times,* February 2, 1998.

CHAPTER TWO

What Is Ethics?

We are discussing no small matter, but how we ought to live.
—SOCRATES, in *Plato's Republic*

What is it to be a moral person? What is the nature of morality, and why do we need it? What is the good and how shall I know it? Are moral principles absolute or simply relative to social groups or individual decisions? Is it in my interest to be moral? Is it sometimes in my best interest to act immorally? What is the relationship between morality and religion? What is the relationship between morality and law? What is the relationship between morality and etiquette?

These are some of the questions that we shall examine in this chapter. We want to understand the foundation and structure of morality. We want to know how we should live.

The terms *moral* and *ethics* come from Latin and Greek, respectively (*mores* and *ethos*), deriving their meaning from the idea of custom. Although philosophers sometimes distinguish these terms—*morality* referring to the customs, principles, and practices of a people or culture, and *ethics* referring to the whole domain of morality and moral philosophy—I shall use them interchangeably in this book, using the context to make any differences clear.

Moral philosophy refers to the systematic endeavor to understand moral concepts and justify moral principles and theories. It undertakes to analyze such concepts as "right," "wrong," "permissible," "ought," "good," and "evil" in their moral contexts. Moral philosophy seeks to establish principles of right behavior that may serve as action guides for individuals and groups. It investigates which values and virtues are paramount to the worthwhile life or society. It builds and scrutinizes arguments in ethical theories, and it seeks to discover valid principles (such as "Never kill innocent human beings") and the relationship between those principles (for example, does saving a life in some situations constitute a valid reason for breaking a promise?).

MORALITY COMPARED WITH OTHER NORMATIVE SUBJECTS

Moral precepts are concerned with norms—not with what is, but with what *ought* to be. How should I live my life? What is the right thing to do in this situation? Should one always tell the truth? Do I have a duty to report a student whom I have seen cheating in class or a coworker whom I have seen stealing office supplies? Should I tell my friend that her spouse is having an affair? Is premarital sex morally permissible? Ought a woman ever to have an abortion? Should we permit cloning of human beings? Morality has a distinct action-guiding or *normative* aspect,[1] an aspect it shares with other practical institutions such as religion, law, and etiquette.

Moral behavior, as defined by a given religion, is often held to be essential to the practice of that religion. But neither the practices nor precepts of morality should be identified with religion. The practice of morality need not be motivated by religious considerations. And moral precepts need not be grounded in revelation or divine authority—as religious teachings invariably are. The most salient characteristic of ethics—by which I mean both philosophical morality (or morality, as I will simply refer to it) and moral philosophy—is that it is grounded in reason and human experience.

To use a spatial metaphor, secular ethics is horizontal, omitting a vertical or transcendental dimension. Religious ethics has a vertical dimension, being grounded in revelation or divine authority, though generally using reason to supplement or complement revelation. These two differing orientations will often generate different moral principles and standards of evaluation, but they need not. Some versions of religious ethics, which posit God's revelation of the moral law in nature or conscience, hold that reason can discover what is right or wrong even apart from divine revelation.

Morality is also closely related to law, and some people equate the two practices. Many laws are instituted in order to promote well-being and resolve conflicts of interest and/or social harmony, just as morality does, but ethics may judge that some laws are immoral without denying that they are valid laws. For example, laws may permit slavery or irrelevant discrimination against people on the basis of race or sex. Catholics or anti-abortion advocates may believe that the laws permitting abortion are immoral.

In a recent television series, *Ethics in America* (PBS, 1989), James Neal, a trial lawyer, was asked what he would do if he discovered that his client had committed a murder some years back for which another man had been convicted and would soon be executed. Mr. Neal said that he had a legal obligation to keep this information confidential and that if he divulged it, he would be disbarred. It is arguable that he has a moral obligation which overrides his legal obligation and which demands that he take action to protect the innocent man from being executed.

Furthermore, some aspects of morality are not covered by law. For example, while it is generally agreed that lying is usually immoral, there is no general law against it (except under special conditions, such as in cases of perjury or falsifying income tax returns). Sometimes college newspapers publish advertisements for "research assistance," where it is known in advance that the companies will aid and abet plagiarism. The publishing of such ads is legal, but it is doubtful whether it is morally correct. In 1963, 39 people in Queens, New York, watched from their apartments for some 45 minutes as a man beat up a woman, Kitty Genovese; and they did nothing to intervene, not even call the police. These people broke no law, but they were very likely morally culpable for not calling the police or shouting at the assailant.

There is one other major difference between law and morality. In 1351 King Edward III of England promulgated a law against treason that made it a crime merely to think homicidal thoughts about the king. But, alas, the law could not be enforced, for no tribunal can search the heart and fathom the intentions of the mind. It is true that *intention,* such as malice aforethought, plays a role in the legal process in determining the legal character of an act, once the act has been committed. But preemptive punishment for people presumed to have bad intentions is illegal. If malicious intentions (called in law *mens rea*) were criminally illegal, would we not all deserve imprisonment? Even if it were possible to detect intentions, when should the punishment be administered? As soon as the subject has the intention? But how do we know that he will not change his mind? Furthermore, is there not a continuum between imagining some harm to X, wishing a harm to X, desiring a harm to X, and intending a harm to X?

While it is impractical to have laws against bad intentions, these intentions are still bad, still morally wrong. Suppose I plan to push Uncle Charlie off a 1,000-foot cliff when we next hike together in order to inherit his wealth, but never have a chance to do it (because, perhaps, Uncle Charlie breaks his leg and forswears hiking). While I have not committed a crime, I have committed a moral wrong. Law generally aims at setting an important but minimal framework in a society of plural values.

Finally, law differs from morality in that there are physical and financial sanctions (such as imprisonment and fines) enforcing the law but only the sanction of conscience and reputation enforcing morality.[2]

Morality also differs from etiquette, which concerns form and style rather than the essence of social existence. Etiquette determines what is polite behavior rather than what is right behavior in a deeper sense. It represents society's decision as to how we are to dress, greet one another, eat, celebrate festivals, dispose of the dead, express gratitude and appreciation, and, in general, carry out social transactions. Whether we greet each other with a handshake, a bow, a hug, or a kiss on the cheek or whether we uncover our heads in holy places (as males do in Christian churches) or cover them (as females do in Catholic churches and males do in synagogues), none of these rituals has any moral superiority.

People in Russia wear their wedding rings on the third finger of their right hand, whereas we wear them on our left hand. People in England hold their fork in their left hand when they eat, whereas people in other countries hold it in their right hand or in either hand according to preference; people in India typically eat without a fork at all, using the forefingers of their right hand for conveying food from their plate to their mouth.

Polite manners grace our social existence, but they are not what social existence is about. They help social transactions to flow smoothly, but are not the substance of those transactions.

At the same time, it can be immoral to disregard or flaunt etiquette. The decision whether to shake hands when greeting a person for the first time or to put one's hands together and forward as one bows, as people in India do, is a matter of cultural decision; but once the custom is adopted, the practice takes on the importance of a moral rule, subsumed under the wider principle of "show respect to people." Similarly, there is no moral necessity to wear clothes, although we have adopted the custom partly to keep us warm in colder climates and partly out of modesty. Nonetheless, there is nothing wrong with nudists who decide to live together naked in nudist colonies. But it may well be the case that running nude in classrooms, stores, and along the road would constitute behavior so offensive as to count as morally insensitive. Recently, there was a scandal on the beaches of South India, where American tourists swam in bikinis, shocking the more modest Indians. There was nothing immoral in itself about wearing bikinis, but given the cultural context, the Americans were guilty of moral impropriety in willfully violating etiquette.[3]

Law, etiquette, and religion are all important institutions, but each has limitations. The limitation of the law is that you can't have a law against every social malady nor can you enforce every desirable rule. The limitation of etiquette is that it doesn't get to the heart of what is of vital importance for personal and social existence. Whether or not one eats with one's fingers pales in significance compared with the importance of being honest or trustworthy or just. Etiquette is a cultural invention, but morality claims to be a discovery.

The limitation of the religious injunction is that it rests on authority, and we are not always sure of or in agreement about the credentials of the authority, nor on how the authority would rule in ambiguous or new cases. Because religion is not founded on reason but on revelation, you cannot use reason to convince someone who does not share your religious views that your view is the right one. I hasten to add that when moral differences are caused by fundamental moral principles, it is unlikely that philosophical reasoning will settle the matter. Often, however, our moral differences turn out to be rooted in world views, not moral principles. For example, the anti-abortionist and pro-choice advocate often agree that it is wrong to kill innocent persons, but differ on the facts. The anti-abortionist may hold a religious view that states that the fetus has an eternal soul and thus possesses

Table 2-1 Normative Institutions

Subject	*Normative Disjuncts*	*Sanctions*
Ethics	Right or wrong or permissible as defined by conscience or reason	Conscience—praise and blame; reputation
Religion	Right or wrong (sin) or permissible as defined by religious authority	Conscience—eternal; reward and punishment caused by a supernatural agent or force
Law	Legal or illegal as defined by a judicial body	Punishments determined by the legislative body
Etiquette	Proper or improper as defined by culture	Social disapprobation and approbation

a right to life, while the pro-choice advocate may deny that anyone has a soul and hold that only self-conscious, rational beings have rights to life.

Table 2-1 characterizes the relationship between ethics, religion, etiquette, and law. It describes the normative disjunctive statements ("or" statements) for each.

In summary, morality distinguishes itself from law and etiquette by going deeper into the essence of rational existence. It distinguishes itself from religion in that it seeks reasons, rather than authority, to justify its principles. The central purpose of moral philosophy is to secure valid principles of conduct and values that can be instrumental in guiding human actions and producing good character. As such, it is the most important activity known to humans, for it has to do with how we are to live.

DOMAINS OF ETHICAL ASSESSMENT

It might seem at this point that ethics concerns itself entirely with rules of conduct based solely on an evaluation of acts. However, the situation is more complicated than this. There are four domains of ethical assessment:

Domain	Evaluative Terms
1. Action, the act	right, wrong, obligatory, permissible
2. Consequences	good, bad, indifferent
3. Character	virtuous, vicious, neutral
4. Motive	good will, evil will, neutral

Let us examine each of these domains.

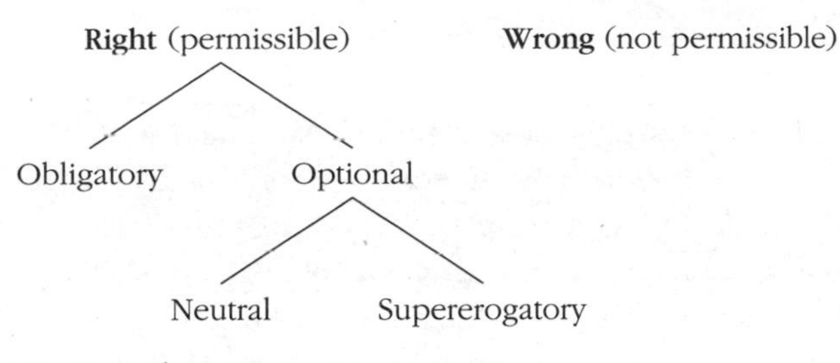

Figure 2-1 Action schema.

Types of Action

The most common distinction may be the classification of actions as right or wrong, but the term *right* is ambiguous. Sometimes it means "obligatory" (as in "*the right* act"), but sometimes it means "permissible" (as in "a right act"). Usually, philosophers define *right* as permissible, including under that category what is obligatory (Figure 2-1).

A *right act* is an act that is permissible for you to do. It may be either optional or obligatory. An *optional* act is an act that is neither obligatory nor wrong to do. It is not your duty to do it, nor is it your duty not to do it. Neither doing it nor not doing it would be wrong. An *obligatory* act is one morality requires you to do; it is not permissible for you to refrain from doing it. It would be wrong not to do it.

A *wrong act* is an act that you have an obligation, or a duty, to refrain from doing. It is an act you ought not to do. It is not permissible to do it.

Let us briefly illustrate these concepts. The act of lying is generally seen as a wrong type of act (prohibited), whereas telling the truth is generally seen as obligatory. But some acts do not seem to be either obligatory or wrong. Whether you decide to take a course in art history or English literature or whether you write your friend a letter with a pencil or a pen seems morally neutral. Either is permissible. Whether you listen to pop music or classical music is not usually considered morally significant. Listening to both is allowed, and neither is obligatory. A decision to marry or remain single is of great moral significance. (It is, after all, an important decision about how to live one's life.) The decision reached, however, is usually considered to be morally neutral or optional. Under most circumstances, to marry (or not to marry) is thought to be neither obligatory nor wrong, but permissible.

Within the range of permissible acts is the notion of **supererogatory,** or highly altruistic, acts. These acts are not required or obligatory, but exceed what morality requires, going "beyond the call of duty." You may have an obligation to give a donation to help people in dire need, but you are probably not obliged to sell your car, let alone become destitute, in order to help them.

Theories that place the emphasis on the nature of the act are called **deontological** (from the Greek word for duty). These theories hold that there is something inherently right or good about such acts as truth telling and promise keeping and something inherently wrong or bad about such acts as lying and promise breaking. Illustrations of deontological ethics include the Ten Commandments, found in the Bible (Exodus 20); natural law ethics, such as is found in the Roman Catholic Church; and Immanuel Kant's theory of the categorical imperative, which we discuss in Chapter 5.

Consequences

We said earlier that lying is generally seen as wrong and telling the truth is generally seen as right. But consider this situation. You are hiding in your home an innocent Jewish woman named Sarah, who is fleeing Nazi officers. When the Nazis knock on your door and when you open it, they ask if Sarah is in your house. What should you do? Should you tell the truth or lie? Those who say that morality has something to do with consequences of actions would prescribe lying as the morally right thing to do. Those who deny that we should look at the consequences when considering what to do when there is a clear and absolute rule of action will say that we should either keep silent or tell the truth. When no other rule is at stake, of course, the rule-oriented ethicist will allow the foreseeable consequences to determine a course of action. Theories that focus primarily on consequences in determining moral rightness and wrongness are called **teleological** ethical theories (from the Greek *telos*, meaning "goal-directed"). The most famous of these theories is utilitarianism, set forth by Jeremy Bentham (1748–1832) and John Stuart Mill (1806–73), which enjoins us to do the act that is most likely to have the best consequences: Do that act which will produce the greatest happiness for the greatest number. We will discuss utilitarianism in Chapter 5.

Character

While some ethical theories emphasize principles of action in themselves and some emphasize principles involving consequences of action, other theories, such as Aristotle's ethics, emphasize character or virtue. According to Aristotle, it is most important to develop virtuous character, for if and only if we have good people can we ensure habitual right action. Although it may be helpful to have action-guiding rules, what is vital is the empowerment of character to do good. Many people know that cheating or gossiping or overindulging in food or alcohol is wrong, but they are incapable of doing what is right. The virtuous person may not be consciously following the moral law when he or she does what is right and good. While the virtues are not central to other types of moral theories, most moral theories include the virtues as important. Most reasonable people, whatever their notions about ethics, would judge that the people who watched Kitty Genovese get assaulted lacked good character. Different moral systems emphasize different virtues and emphasize them to different degrees.

Motive

Finally, virtually all ethical systems, but especially Kant's system, accept the relevance of *motive*. It is important to the full assessment of any action that the intention of the agent be taken into account. Two acts may be identical, but one might be judged morally culpable and the other excusable. Consider John's pushing Joan off a ledge, causing her to break her leg. In situation A he is angry and intends to harm her, but in situation B he sees a knife flying in her direction and intends to save her life. In A what he did was clearly wrong, whereas in B he did the right thing. On the other hand, two acts may have opposite results, but the actions may be equally good judged on the basis of intention. For example, two soldiers may try to cross the enemy lines to communicate with an allied force, but one gets captured through no fault of his own and the other succeeds. In a full moral description of any act, motive will be taken into consideration as a relevant factor.

THE PURPOSES OF MORALITY

What is the role of morality in human existence? I believe that morality is necessary to stave off social chaos, what Thomas Hobbes called a "state of nature" wherein life becomes "solitary, poor, nasty, brutish and short," a "war of all against all," a lose-lose situation. It is a set of rules that, if followed by nearly everyone, will promote the flourishing of nearly everyone. These rules restrict our freedom but only in order to promote greater freedom and well-being. More specifically, morality seems to have these five purposes:

1. To keep society from falling apart
2. To ameliorate human suffering
3. To promote human flourishing
4. To resolve conflicts of interest in just and orderly ways
5. To assign praise and blame, reward and punishment, and guilt

None of these is the sole purpose of morality—but each of them is part of a comprehensive purpose that enables us to live a good life in a just society. The first purpose, keeping society from falling apart, is a necessary but not sufficient condition for a good society, for a tyrant could keep a society together through oppression. Likewise, the second purpose, ameliorating suffering, is a vital component—one of the reasons we cooperate with each other is to come to each other's aid in time of need—but this is not the sole aim of society or morality. The third goal, promoting human flourishing, means enabling people to reach their potential, to live happily; it may be as central to morality as any of the other four purposes, but even here, we would not want happiness at any price. Suppose the price we had to pay for our happiness was treating other people unjustly. One thinks of Ursula LeGuin's short story "The Ones Who Walk Away from Omelas," in which the flourishing of the community (Omelas) depends on the torturing of a child. People with a sensitive conscience walk away from such "happiness." They reject such a social arrangement as morally incomplete. Certainly, something is amiss. Hence, the fourth goal—justice. Yet, justice isn't the *whole* of morality either. Imagine two different social arrangements, equally fair, but one goes beyond mere justice and promotes high altruism, thereby making people more fulfilled. Finally, morality has the function of holding people responsible for their actions, of assigning praise and blame in accordance with the springs of our actions.

Imagine what society would be if everyone or nearly everyone did whatever he or she pleased without obeying moral rules. I would make a promise to you to help you with your philosophy homework tomorrow if you fixed my car today. You believe me. So you fix my car, but you are deeply angry when I laugh at you on the morrow as I drive away to the beach instead of helping you with your homework. Or you loan me money but I run off with it. Or I lie to you or harm you when it is in my interest or even kill you when I feel the urge.

Parents would abandon children and spouses betray each other whenever it was convenient. Under such circumstances society would break down. No one would have an incentive to help anyone else because reciprocity (a moral principle) would not be recognized. Great suffering would go largely unameliorated, and certainly people would not be very happy. We would not flourish or reach our highest potential.

Once while visiting a city in the former Soviet Union, I rented a fifth-floor apartment in which the stairwell lacked lighting. It was difficult navigating stairs at night in complete darkness. I asked why there were no lightbulbs in the stairwells, only to be told that the residents stole them, believing that if they did not take them, their neighbors would. Absent a dominant authority, the social contract has been eroded and everyone must struggle alone in the darkness.

We need moral rules to guide our actions in ways that light up our paths and prevent and reduce suffering, that enhance human well-being (and animal well-being?), that allow us to resolve our conflicts of interests according to recognizably fair rules, and that allow us to assign responsibility for actions so that we can praise and blame, reward and punish people according to how their actions reflect moral principles.

Even though these five purposes are related, they are not identical, and different moral theories emphasize different purposes and in different ways. Utilitarianism fastens on human flourishing and the amelioration of suffering, whereas contractual systems rooted in rational self-interest accent the role of resolving conflicts of interest. Contractualism, as set forth by Thomas Hobbes, emphasizes the first (survival purpose) and fourth purpose: to resolve conflicts of interest in just and orderly ways. A complete moral theory would include a place for each of the five purposes. Its goal would be to internalize the rules that promote these principles in each person's life, producing the virtuous person, someone who is "a jewel that shines in [morality's] own light," to paraphrase Kant. The goal of morality is to create happy and virtuous people, the kind that create flourishing communities. That's why it is the most important subject on Earth.

In this book we shall discuss how and to what extent this classical idea of morality is applicable to environmental concerns. Some philosophers believe that this historic pattern is too heavily anthropocentric and atomistic—they believe that a biocentric or sentient-based ethic must replace it if we are adequately to account for environmental concerns. A holistic ethic, like Leopold's land ethic or Gaian ethics, refocuses the center of moral concern from the individual to the ecological community (see Chapter 9). Biocentric egalitarianism extends our moral principles to every living object, plants and animals as well as humans. But other environmental ethicists believe that this classical account is adequate for environmental concerns. They believe that we must expand the circle of concern to include animals and our relationship with ecosystems, but that we need not change any fundamental principle. Integrity, justice, ameliorating of suffering, and promoting of happiness are universal principles, but they must be extended better to cover animals and our relationship to the environment. For these ethicists, environmentalism raises a higher consciousness, but it is not fundamentally discontinuous with sound classical ethical thinking.

Which of these views of classical ethics is correct will be a major question throughout this work. But there is one basic question to which we must attend before we can have any intelligent discussion of ethics at all—that is the question of ethical relativism. Are moral principles objectively and universally valid, or are they simply valid relative to culture or individual choice?

Review Questions

1. Illustrate the difference between ethics, law, religion, and etiquette. How are these concepts related? Do you think either law, religion, or

etiquette is more important than morality in guiding human action? Explain your answer.

2. Discuss the four domains of morality. Which domains are more crucial than others? Or do you think they are all equally important? Explain your answer.
3. What are the purposes of morality? Which ones seem more important than others? Do you agree that these are the central purposes? Are there others?
4. Based on what you know now, do you think that environmental concerns force us radically to revise our understanding of morality or merely extend it or neither?

Suggested Readings

Frankena, William. *Ethics*. Englewood Cliffs, N.J.: Prentice Hall, 1973.
Kagan, Shelly. *Normative Ethics*. Boulder, Colo.: Westview Press, 1997.
Mackie, J. L. *Ethics: Inventing Right and Wrong*. Harmondsworth, N.Y.: Penguin, 1976.
Pojman, Louis. *Ethics: Discovering Right and Wrong*. Belmont, Calif.: Wadsworth, 1998.
Pojman, Louis, ed. *Ethical Theory*. Belmont, Calif.: Wadsworth, 1998.
Rachels, James. *Elements of Moral Philosophy*. New York: McGraw-Hill, 1993.
Singer, Peter. *The Expanding Circle: Ethics and Sociobiology*. New York: Oxford University Press, 1983.
Taylor, Richard. *Good and Evil*. Buffalo, N.Y.: Prometheus, 1970.
Williams, Bernard. *Morality*. New York: Harper, 1972.
Wilson, James Q. *The Moral Sense*. New York: Free Press, 1993.

Notes

1. The term normative means seeking to make certain types of behavior a norm or standard in a society.
2. A sanction is a mechanism for social control, used to enforce society's standards. It may consist of rewards or punishment, praise or blame, approbation or disapprobation.
3. Although Americans pride themselves on tolerance, pluralism, and awareness of other cultures, custom and etiquette can be—even among people from similar backgrounds—a bone of contention. A friend of mine, John, tells of an experience early in his marriage. He and his wife, Gwen, were hosting their first Thanksgiving meal. He had been used to small celebrations with his immediate family, whereas his wife had been used to grand celebrations. He writes, "I had been asked to carve, something I had never done before, but I was willing. I put on an apron, entered the kitchen, and attacked the bird with as much artistry as I could muster. And what reward did I get? [My wife] burst into tears. In her family the turkey is brought to the *table,* laid before the [father], grace is said, and then he carves! "So I fail patriarchy," I hollered later. "What do you expect?"

CHAPTER THREE

Ethical Relativism: Who's to Judge What's Right and Wrong?

In the nineteenth century, Christian missionaries sometimes used coercion to change the customs of pagan tribal people in parts of Africa and the Pacific Islands. Appalled by the customs of public nakedness, polygamy, work on the Sabbath, and infanticide, they paternalistically went about reforming the pagans. They clothed them, separated wives from their husbands in order to create monogamous households, made the Sabbath a day of rest, and ended infanticide. In the process they sometimes created social malaise, causing the estranged women to despair and their children to be orphaned. The natives often did not understand the new religion, but accepted it in deference to the power of the white people, who had guns and medicine.

Since the nineteenth century, we've made progress in understanding cultural diversity and realize that the some of the social dissonance caused by do-gooders was a bad thing. In the last century or so, anthropology has exposed our penchant for *ethnocentrism,* the prejudicial view that interprets all of reality through the eyes of our cultural beliefs and values. We have come to see enormous variety in social practices throughout the world.

Eskimos allow their elderly to die by starvation, while we believe that this is morally wrong. The Spartans of ancient Greece and the Dobu of New Guinea believe that stealing is morally right, but we believe it is wrong. Many cultures, past and present, have practiced or still practice infanticide. A tribe in East Africa once threw deformed infants to the hippopotamus, but our society condemns infanticide. Anthropologist Ruth Benedict describes a tribe in Melanesia that views cooperation and kindness as vices, and British anthropologist Colin Turnbull has documented that the Ik in northern Uganda have no sense of duty toward their children or parents. Some societies make it a duty for children to kill (sometimes strangle) their aging parents. Sexual practices vary over time and clime. Some cultures permit, while others condemn, homosexual behavior. Some cultures, including Moslem societies, practice polygamy, while Christian cultures view it as immoral. Some cultures accept cannibalism, while we detest it; but while our culture has generally encouraged meat eating, not to mention the killing of animals for sport, Hinduism and Jainism (both religions in India) look upon the killing of animals as immoral and the eating of meat as wrong. **Cultural relativism** is well documented, and custom seems king o'er all.

Today we condemn ethnocentrism, the uncritical belief in the inherent superiority of one's own culture, as a variety of prejudice tantamount to racism and sexism. What is right in one culture may be wrong in another, what is good east of the river may be bad west of the same river, what is a virtue in one nation may be seen as a vice in another, and so it behooves us not to judge others but to be tolerant of diversity.

This rejection of ethnocentricism in the West has contributed to a general shift in public opinion about morality, so that for a growing number of Westerners, consciousness-raising about the validity of other ways of life has led to a gradual erosion of belief in moral **objectivism,** the view that there are universal moral principles, valid for all people at all times and climes. For example, in polls taken in my ethics and introductory philosophy classes over the past several years (in three different universities in three areas of the country), students by a 2-to-1 ratio affirmed a version of ethical relativism over moral absolutism—with hardly 3% seeing something in between these two polar opposites. Of course, I'm not suggesting that all of these students have a clear understanding of what relativism entails, for many who say that they are ethical relativists also state on the same questionnaire that "abortion except to save the mother's life is always wrong," that "capital punishment is always morally wrong," or that "suicide is never morally permissible." The apparent contradictions signal an apparent confusion on the matter.

In the first part of this chapter, I want to examine the central notions of ethical relativism and look at the implications that seem to follow from it. Later, I want to set forth the outlines of a very modest objectivism, which holds to the objective validity of moral principles but takes into account many of the insights of relativism.

AN ANALYSIS OF ETHICAL RELATIVISM

Ethical relativism is the theory that there are no universally valid moral principles, but that all moral principles are valid relative to culture or individual choice. It is to be distinguished from *moral skepticism,* the view that there are no valid moral principles at all (or at least that we cannot know whether there are any), and from all forms of moral objectivism or absolutism. The following statement by John Ladd is a typical characterization of the theory:

> Ethical relativism is the doctrine that the moral rightness and wrongness of actions varies from society to society and that there are no absolute universal

> moral standards binding on all men at all times. Accordingly, it holds that whether or not it is right for an individual to act in a certain way depends on or is relative to the society to which he belongs.[1]

If we analyze this passage, we come up with the following argument:

1. What is considered morally right and wrong varies from society to society, so that there are no moral principles accepted by all societies.
2. All moral principles derive their validity from cultural acceptance.
3. Therefore, there are no universally valid moral principles, objective standards that apply to all people everywhere and at all times.

The Diversity Thesis

The first premise of the argument, which may be called the *diversity thesis* and identified with *cultural relativism*, is simply an anthropological thesis, which registers the fact that moral rules differ from society to society. It is purely *descriptive*, having no prescriptive force. As we noted in the introduction to this chapter, there is enormous variety in what may count as a moral practice or principle in a given society. The human condition is malleable in the extreme, allowing any number of folkways or moral codes. As Ruth Benedict has written,

> The cultural pattern of any civilization makes use of a certain segment of the great arc of potential human purposes and motivations, just as we have seen in an earlier chapter that any culture makes use of certain selected material techniques or cultural traits. The great arc along which all the possible human behaviors are distributed is far too immense and too full of contradictions for any one culture to utilize even any considerable portion of it. Selection is the first requirement.[2]

It may or may not be the case that there is not a single moral principle held in common by every society, but if there are any, they seem to be few, at best. Certainly, it would be very hard to derive one single "true" morality on the basis of observation of various societies' moral standards.

The Dependency Thesis

The second premise, the *dependency thesis,* asserts that individual acts are right or wrong depending on the nature of the society from which they emanate. Morality does not occur in a vacuum, but what is considered morally right or wrong must be seen in a context, depending on the goals, wants, beliefs, history, and environment of the society in question. As anthropologist William Graham Sumner says, "We learn the [morals] as unconsciously as we learn to walk and hear and breathe, and they never know any reason why the [morals] are what they are. The justification of them is that when we wake to consciousness of life we find them facts which already hold us in the bonds of tradition, custom, and habit."[3] In other words, there is no independent justification for moral practices—they are just assumed as givens from the outset. Trying to see things from an independent, noncultural point of view would be like taking out our eyes in order to examine their contours and qualities. We are simply culturally determined beings.

We could, of course, distinguish a weak and a strong thesis of dependency. The nonrelativist can accept a certain relativity in the way moral principles are *applied* in various cultures, depending on beliefs, history, and environment. For example, Asians show respect by covering the head and uncovering the feet, whereas Westerners do the opposite, but both adhere to a principle of respect for deserving people. They just apply the principle of respect differently. Drivers in Great Britain, Southeast Asia, and Australia drive on the left side of the road, while those in the rest of Europe and the United States drive on the right side, but both adhere to a principle of orderly progression of traffic. The application of the rule is different, but the principle in question is the same principle in both cases. The ethical relativist, however, must maintain a stronger thesis, one that insists that the very validity of the principles is a product of the culture and that different cultures will invent different valid principles. The ethical relativist maintains that absolutely no justified morality exists apart from cultural acceptance. As Hegel said, "What is, is right." What culturally is approved is morally right.

In a sense, we all live in radically different worlds. Each person has a different set of beliefs and experiences, a particular perspective that colors all of his or her perceptions. Do the farmer, the real estate dealer, and the artist, all looking at the same spatiotemporal field, see the *same* field? Not likely. Their different orientations, values, and expectations govern their perceptions, so that different aspects of the field are highlighted and some features are missed. Even as our individual values arise from personal experience, so social values are grounded in the peculiar history of the community. Morality, then, is just the set of common rules, habits, and customs that have won social approval over time, so that they seem part of the nature of things, as facts. There is nothing mysterious or transcendent about these codes of behavior. They are the outcomes of our social history.

There is something conventional about *any* morality, so that every morality really depends on a level of social acceptance. Not only do various societies adhere to different moral systems, but in addition the very same society could (and often does) change its moral views over time and place. For example, the southern United States now views slavery as immoral, whereas just over one hundred years ago, it did not. We have greatly altered our views on abortion, divorce, and sexuality as well.

Classical Ethical Relativism

The conclusion that there are no absolute or objective moral standards binding on all people follows from the first two premises. Cultural relativism (the diversity thesis) plus the dependency thesis yields ethical relativism in its classic form. If there are different moral principles from culture to culture and if all morality is rooted in culture, then it follows that there are no universal moral principles valid for all cultures and people at all times.

SUBJECTIVE ETHICAL RELATIVISM (SUBJECTIVISM)

Some people think that even this conclusion is too tame and maintain that morality is not dependent on the society but on the individual him- or herself. As students sometimes maintain, "Morality is in the eye of the beholder." Ernest Hemingway wrote, "So far, about morals, I know only that what is moral is what you feel good after and what is immoral is what you feel bad after and judged by these moral standards, which I do not defend, the bullfight is very moral to me because I feel very fine while it is going on and have a feeling of life and death and mortality and immortality, and after it is over I feel very sad but very fine."[4]

This form of moral **subjectivism** has the sorry consequence that it makes morality a useless concept, for, on its premises, little or no interpersonal criticism or judgment is logically possible. Hemingway may feel good about killing bulls in a bullfight, while Albert Schweitzer or Mother Teresa may feel the opposite. No argument about the matter is possible. The only basis for judging Hemingway or anyone else wrong would be if he failed to live up to his own principles; but, of course, one of Hemingway's principles could be that hypocrisy is morally permissible (he feels good about it), so that it would be impossible for him to do wrong. For Hemingway hypocrisy and nonhypocrisy are both morally permissible. On the basis of subjectivism, it could very easily turn out that Adolf Hitler is as moral as Gandhi, as long as each believes he is living by his chosen principles. Notions of moral good and bad, right or wrong, cease to have interpersonal evaluative meaning. You may not like it when your teacher gives you an F on your test paper, while she gives your neighbor an A for one exactly alike, but there is no way to criticize her for injustice, since justice is not one of her elected principles.

Absurd consequences follow from subjective ethical relativism. If it is correct, then morality is reduced to aesthetic tastes over which there can be no argument nor interpersonal judgment. Although many students say that they hold this position, there seems to be a conflict between it and other of their moral views (for example, that Hitler is really morally bad or capital punishment is always wrong). There seems to be a contradiction between subjectivism and the very concept of morality, which it is supposed to characterize, for morality has to do with "proper" resolution of interpersonal conflict and the amelioration of the human predicament. Whatever else it does, it has a minimal aim of preventing a state of chaos where life is "solitary, poor, nasty, brutish, and short." But if so, subjectivism is no help at all in achieving this goal, for it doesn't rest on social *agreement* of principle (as the conventionalist maintains) or on an objectively independent set of norms that bind all people for the common good.

Subjectivism treats individuals as billiard balls on a societal pool table where they meet only in radical collisions, each aiming for its own goal and striving to do the other one in before being done in. This atomistic view of personality is belied by the fact that we develop in families and mutually dependent communities in which we share a common language, common institutions, and habits, and by the fact that we often feel each other's joys and sorrows. As John Donne said, "No man is an island, entire of itself; every man is a piece of the continent."

Subjective ethical relativism seems incoherent. If so, it follows that the only plausible view of ethical relativism must be one that grounds morality in the group or culture. This form of relativism is called "conventionalism," and to it we now turn.

CONVENTIONAL ETHICAL RELATIVISM (CONVENTIONALISM)

Conventional ethical relativism, the view that there are no objective moral principles but that all valid moral principles are justified by virtue of their cultural acceptance, recognizes the social nature of morality. That is precisely its power and virtue. It does not seem subject to the same absurd consequences that plague subjectivism. Recognizing the importance of our social environment in generating customs and beliefs, many people suppose that conventional ethical relativism is the correct ethical theory. Furthermore, they are drawn to it for its liberal philosophical stance. It seems to be an enlightened response to the sin of ethnocentricity, and it seems to entail or strongly imply an attitude of tolerance toward other cultures. As Benedict says, in recognizing ethical relativity "we shall arrive at a more realistic social faith, accepting as grounds of hope and as new bases for tolerance the coexisting and equally valid patterns of life which mankind has created for itself from the raw materials of existence."[5] The most famous of those holding this position is the anthropologist Melville Herskovits, who argues even more explicitly than Benedict that ethical relativism entails intercultural tolerance:

1. If morality is relative to its culture, then there is no independent basis for criticizing the morality of any other culture but one's own.
2. If there is no independent way of criticizing any other culture, we ought to be *tolerant* of the moralities of other cultures.
3. Morality is relative to its culture.
4. Therefore, we ought to be tolerant of the moralities of other cultures.[6]

Tolerance is certainly a virtue, but is this a good argument for it? I think not. If morality simply is relative to each culture and if the culture does not have a principle of tolerance, then its members have no obligation to be tolerant. Herskovits seems to be treating the *principle of tolerance* as the one exception to his relativism. He seems to be treating it as an absolute moral principle. But from a relativistic point of view, there is no more reason to be tolerant than to be intolerant; and neither stance is objectively morally better than the other.

Not only do relativists fail to offer a basis for criticizing those who are intolerant, but they also cannot rationally criticize anyone who espouses what they might regard as a heinous principle. If, as seems to be the case, valid criticism of morality supposes an objective or impartial standard, relativists cannot criticize anyone outside their own culture. Adolf Hitler's genocidal actions, as long as they are culturally accepted, are as morally legitimate as Mother Teresa's works of mercy. If conventional relativism is accepted, racism, genocide of unpopular minorities, oppression of the poor, slavery, and even the advocacy of war for its own sake are as equally moral as their opposites. And if a subculture decided that starting a nuclear war was somehow morally acceptable, we could not morally criticize these people.

For the relativist, any actual morality, whatever its content, is as valid as every other and more valid than ideal moralities—since the latter aren't adhered to by any culture.

There are other disturbing consequences of conventional ethical relativism. It seems to imply that reformers are always (morally) wrong since they go against the tide of cultural standards. William Wilberforce was wrong in the eighteenth century to oppose slavery; the British were immoral in opposing suttee in India (the burning of widows, which is now illegal in India). The early Christians were wrong in refusing to serve in the Roman army or bow down to Caesar, since the majority in the Roman Empire believed that these two acts were moral duties. In fact, Jesus himself was immoral in breaking the law of his day by healing on the Sabbath and by advocating the principles of the Sermon on the Mount, since it is clear that few in his time (or in ours) accepted them.

Yet we normally believe just the opposite, that the reformer is the courageous innovator who is right, who speaks the truth and takes a stand against the mindless majority. Sometimes the individual must stand alone with the truth, risking social censure and persecution. As Dr. Stockman says in Ibsen's *Enemy of the People*, after he loses the battle to declare his town's profitable but polluted tourist spa unsanitary, "The most dangerous enemy of the truth and freedom among us—is the compact majority. Yes, the damned, compact and liberal majority. The majority has *might*—unfortunately—but *right* it is not. Right—are I and a few others." Yet if conventional ethical relativism is correct, the opposite is necessarily the case. Truth is with the crowd and error with the individual.

Similarly, conventional ethical relativism entails disturbing judgments about the law. Our normal view is that we have a prima facie ("at first glance") duty to obey the law, because law in general promotes the human good. According to most objective systems, this prima facie obligation is not absolute but may be overridden in cases of significant injustice. Civil disobedience is warranted in some cases where the law seems to be in serious conflict with morality. However, if ethical relativism is true, then neither law nor civil disobedience has a firm foundation. On the one hand, from the side of the society at large, civil disobedience will be morally wrong, as long as the culture agrees with the law in question. On the other hand, if you belong to the relevant subculture that doesn't recognize the particular law in question, disobedience will be morally mandated. The Ku Klux Klan, which believes that Jews, Catholics, and blacks are evil or undeserving of high regard, are, given conventionalism, morally permitted or required to break the laws that protect these groups. Why should I obey a law that my group doesn't recognize as valid?

To sum up, unless we have an independent moral basis for law, it is hard to see why we have any general duty to obey it; and unless we recognize the priority of a universal moral law, we have no firm basis to justify our acts of civil disobedience against "unjust laws." Both the validity of law and morally motivated disobedience of unjust laws are annulled in favor of a power struggle.

There is an even more basic problem with the notion that morality is dependent on cultural acceptance for its validity. The problem is that the notion of a culture or society is notoriously difficult to define. This is especially so in a pluralistic society like our own where the notion seems to be vague and to have unclear boundary lines. One person may belong to several societies (subcultures) with different value emphases and arrangements of principles. John and Mary may belong to the nation as a single society with certain values of patriotism, honor, courage, obedience to laws (including some that are controversial but have majority acceptance, such as the law on abortion). But they may also belong to a church that opposes some of the laws of the state. They may also be integral members of a socially mixed community where different principles hold sway, and they may belong to clubs and a family where still other rules are adhered to. Relativism would seem to tell us that where they are members of societies with conflicting moralities, they must be judged both wrong and not-wrong whatever they do. For example, if Mary is a U.S. citizen and a member of the Roman Catholic Church, she is wrong (as Catholic) if she chooses to have an abortion and not-wrong (as citizen) if she acts against the teaching of the Church on abortion. As a member of a racist university fraternity, KKK, John has no obligation to treat his fellow black student as an equal, but as a member of the university community itself (where the principle of equal rights is accepted), he does have the obligation; as a member of the surrounding community (which may reject the principle of equal rights), he again has no such obligation; but then again as a member of the nation at large (which accepts the principle), he is obligated to treat his fellow student with respect. What is the morally right thing for John to do? The question no longer makes much sense in this moral Babel. It has lost its action-guiding function.

Perhaps the conventional ethical relativist would adhere to a principle which says that in such cases the individual may choose which group to belong to as primary. If Mary chooses to have an abortion, she is choosing to belong to the general society relative to that principle. And John must likewise choose between groups. The trouble with this option is that it seems to lead back to counterintuitive results. If Gangland Gus of Murder, Incorporated feels like killing Bank President Ortcutt and wants to feel good about it, he identifies with the Murder, Incorporated society rather than with the general public morality. Does this justify the killing? In fact, couldn't one justify anything simply by forming a small subculture that approved of it? Charles Manson would be morally pure in killing innocents simply by virtue of forming a little coterie. How large must the group be in order to be a legitimate subculture or society? Does it need ten or fifteen people? How about just three? Come to think of it, why can't my burglary partner and I found our own society with a morality of its own? Of course, if my partner dies, I could still claim that I was acting from an originally social set of norms. But why can't I dispense with the interpersonal agreements altogether and invent my own morality—since morality, according to this view, is only an invention anyway? Conventionalist relativism seems to reduce to subjectivism. And subjectivism leads, as we have seen, to the demise of morality altogether.

However, while we may fear the demise of morality, as we have known it, this in itself may not be a good reason for rejecting relativism—that is, for judging it false. Alas, truth may not always be edifying. But the consequences of this position are sufficiently alarming to prompt us to look carefully for some weakness in the relativist's argument.

So let us examine the premises and conclusion listed at the beginning of this chapter as the three theses of relativism.

1. *The diversity thesis:* What is considered morally right and wrong varies from society to society, so that there are no moral principles accepted by all societies.
2. *The dependency thesis:* All moral principles derive their validity from cultural acceptance.
3. *Ethical relativism:* Therefore, there are no universally valid moral principles, objective standards that apply to all people everywhere and at all times.

Does any one of these seem problematic? Let us consider the first thesis, the diversity thesis, which we have also called cultural relativism. Perhaps there is not as much diversity as anthropologists like Sumner and Benedict suppose. One can also see great similarities between the moral codes of various cultures. E. O. Wilson has identified more than a score of common features, and before him Clyde Kluckhohn had noted some significant common ground:

> Every culture has a concept of murder, distinguishing this from execution, killing in war, and other "justifiable homicides." The notions of incest and other regulations upon sexual behavior, the prohibitions upon untruth under defined circumstances, of restitution and reciprocity, of mutual obligations between parents and children—these and many other moral concepts are altogether universal.[7]

And Colin Turnbull, whose description of the sadistic, semidisplaced Ik in northern Uganda, was seen as evidence of a people without principles of kindness and cooperation, has produced evidence that underneath the surface of this dying society, there is a deeper moral code from a time when the tribe flourished—a code that occasionally surfaces and shows its nobler face.

On the other hand, there is enormous cultural diversity and many societies have radically different moral codes. Cultural relativism seems to be a fact; but, even if it is, it does not by itself establish the truth of ethical relativism. Cultural diversity in itself does not prove or disprove either ethical theory. The objectivist could concede complete cultural relativism but still defend a form of universalism; for he or she could argue that some cultures simply lack correct moral principles.

On the other hand, a denial of complete cultural relativism (that is, an admission of some universal principles) does not disprove ethical relativism. For even if we did find one or more universal principles, this would not prove that they had any objective status. We could still *imagine* a culture that was an exception to the rule and be unable to criticize it. So the first premise doesn't by itself imply ethical relativism, and its denial doesn't disprove ethical relativism.

We turn to the crucial second thesis, the dependency thesis, which asserts that morality does not occur in a vacuum, but what is considered morally right or wrong must be seen in a context, depending on the goals, wants, beliefs, history, and environment of the society in question. We distinguished a weak and a strong thesis of dependency. The weak thesis says that the application of principles (such as respect for the worthy) depends on the particular cultural predicament, whereas the strong thesis affirms that the principles themselves depend on that predicament (some cultures may have no concept of respect). The nonrelativist can accept a certain relativity in the way moral principles are applied in various cultures, depending on beliefs, history, and environment. For example, a raw environment with scarce natural resources may justify the Eskimos' brand of euthanasia to the objectivist, who in another environment would consistently reject that practice. The members of a tribe in the Sudan throw their deformed children into the river because of their belief that such infants *belong* to the hippopotamus, the god of the river. We believe that they have a false belief about this, but the point is that the same principles of respect for property and respect for human life are operative in these contrary practices. They differ with us only in belief, not in substantive moral principle. This is an illustration of how nonmoral beliefs (such as "deformed children belong to the hippopotamus") when applied to common moral principles (such as "give to each his due") generate different actions in different cultures. In our own culture the difference in the nonmoral belief about the status of a fetus generates opposite moral prescriptions. So the fact that moral principles are weakly dependent doesn't show that ethical relativism is valid. In spite of this weak dependency on nonmoral factors, there could still be a set of general moral norms applicable to all cultures and even recognized in most that are disregarded at a culture's own expense.

What the relativist needs is a strong thesis of dependency, a statement that somehow all principles are essentially cultural inventions. But why should we choose to view morality this way? Note that the ethical relativist nowhere gives an argument for the dependency thesis. He merely assumes it from the outset, as Sumner does when, in the passage quoted earlier, he writes, "The justification of them is that when we wake to consciousness of life we find them facts which already hold in the bonds of tradition, custom, and habit." But just because we begin with a set of beliefs or practices in no way implies that they are *true* or *the best* or *morally correct*. To assume otherwise is to beg the question.

Is there anything to recommend the strong thesis over the weak thesis of dependency? The relativist may argue that in fact we don't have an obvious impartial standard from which to judge. "Who's to say which culture is right and which is wrong?" But this assertion seems to be dubious. We can reason and perform thought experiments in order to make a case for one system over another. We may not be able to *know* with certainty that our moral beliefs are closer to the truth than those of another culture or those of others within our own culture, but we may be justified in *believing* that they are. If we can be closer to the truth regarding factual or scientific matters, why can't we be closer to the truth on moral matters? Why can't a culture simply be confused or wrong about its moral perceptions? Why can't we say that the society like the Ik that sees nothing wrong with enjoying watching its own children fall into fires is less moral in that regard than the culture that cherishes children and grants them protection and equal rights? To take such a stand is not to commit the fallacy of ethnocentrism, for we are seeking to derive principles through critical reason, not simply uncritical acceptance of one's own mores.

THE CASE FOR MORAL OBJECTIVISM

If nonrelativists are to make their case, they will have to offer a better explanation of cultural diversity and why we should nevertheless adhere to moral objectivism. One way of doing so is to appeal to a divine law and human sin, which causes deviation from that law. Although I think that human greed, selfishness, pride, self-deception, and other maladies have a great deal to do with moral differences and that religion may lend great support to morality, I don't think that a religious justification is necessary for the validity of moral principles. In any case, in this section, I shall outline a modest nonreligious objectivism first by appealing to our intuitions and second by giving a naturalist account of morality that transcends individual cultures.

First, I must make it clear that I am distinguishing moral **absolutism** from moral *objectivism*. The absolutist believes that there is at least one moral principle that ought never to be violated. Kant's system is a good example of this. One ought never break a promise or lie, no matter what. Act utilitarianism, the theory that says one ought always do the act that promises to maximize happiness, also seems absolutist, for its central principle—do that act that has the most promise of yielding the most happiness (or utility)—is nonoverridable. An objectivist need not posit any nonoverridable principles, at least not in unqualified general form, and so need not be an absolutist. As Renford Bambrough put it,

> To suggest that there is a *right* answer to a moral problem is at once to be accused of or credited with a belief in moral absolutes. But it is no more necessary to believe in moral absolutes in order to believe in moral objectivity than it is to believe in the existence of absolute space or absolute time in order to believe in the objectivity of temporal and spatial relations and of judgements about them.[8]

In the objectivist's view, moral principles are what W. D. Ross refers to as ***prima facie*** principles, valid rules of action that should generally be adhered to, but that may be overridden by another moral principle in cases of moral conflict. For example, while a principle of justice may generally outweigh a principle of benevolence, there are times when enormous good could be done by sacrificing a small amount of justice so that an objectivist would be inclined to act according to the principle of benevolence. There may be some absolute or non-overridable principles, but there need not be any or many for objectivism to be true.

If we can establish or show that it is reasonable to believe that there is at least one objective moral principle which is binding on all people everywhere in some ideal sense, we shall have shown that relativism is probably false and that a limited objectivism is true. Actually, I believe that there are many qualified general ethical principles that are binding on all rational beings, but one will suffice to refute relativism. The principle I've chosen is the following:

A: It is morally wrong to torture people for the fun of it.

I claim that this principle is binding on all rational agents, so that if some agent, S, rejects A, we should not let that affect our intuition that A is a true principle but rather try to explain S's behavior as perverse, ignorant, or irrational instead. For example, suppose Adolf Hitler doesn't accept A. Should that affect our confidence in the truth of A? Is it not more reasonable to infer that Adolf is morally deficient, morally blind, ignorant, or irrational than to suppose that his noncompliance is evidence against the truth of A?

Suppose further that there is a tribe of Hitlerites somewhere who enjoy torturing people. The whole culture accepts torturing others for the fun of it. Suppose that Mother Teresa or Gandhi tries unsuccessfully to convince them that they should stop torturing people altogether, and the Hitlerites respond by torturing them. Should this affect our confidence in A? Would it not be more reasonable to look for some explanation of Hitlerite behavior? For example, we might hypothesize that this tribe lacked the developed sense of sympathetic imagination that is necessary for the moral life. Or we might theorize that this tribe was on a lower evolutionary level than most *Homo sapiens*. Or we might simply conclude that the tribe was closer to a Hobbesian state of nature than were most societies and as such probably would not survive. But we need not know the reason why the tribe was in such bad shape in order to maintain our confidence in A as a moral principle. If A is a basic or core belief for us, we will be more likely to doubt the Hitlerites' sanity or ability to think morally than to doubt the validity of A.

We can perhaps produce other candidates for membership in our minimally basic objective moral set. For example:

1. Do not kill innocent people.
2. Do not cause unnecessary pain or suffering.
3. Do not cheat or steal.
4. Keep your promises and contracts.
5. Do not deprive another person of his or her freedom.
6. Do justice to others, treating equals equally and unequals unequally.
7. Tell the truth.
8. Help other people, at least when the cost to oneself is minimal.
9. Do good wherever feasible, at least when the cost to oneself is minimal.

These nine principles are examples of the *core morality,* principles necessary for the good life. They overlap but emphasize different aspects of morality. Perhaps some will argue that principle 7 should read "Don't lie" instead of the more positive "Tell the truth." I think we hold deeply to both principles—but as *prima facie*, not as absolute. Others will object that these principles are anthropocentric. Principles 2 and 9 cover animals. Principle 8 covers distant people. Which, if any, of these principles are applicable to future generations (see Chapter 5)? Principle 9 is a broad, general principle, identifying a willingness of the moral person to extend well-being further, as conscience becomes raised to new morally significant possibilities. How much further we should go with these principles to include species and ecosystems will be discussed later.

My main point is that these principles are not up for grabs or arbitrary, in the way a relativist holds, for we can give reasons why they are necessary to social cohesion and human flourishing. Principles like the Golden Rule as well as not killing innocent people, treating equals equally, truth telling, promise keeping, and the like are central to the fluid progression of social interaction and the resolution of conflicts that ethics is about (at least minimal morality is, even though there may be more to morality than simply these kinds of concerns). For example, language itself depends on a general and implicit commitment to the principle of truth telling. Accuracy of expression is a primitive form of truthfulness. Hence, every time we use words correctly we are telling the truth. Without this behavior, language wouldn't be possible. Likewise, without the recognition of a rule of promise keeping, contracts would be of no avail and cooperation would be less likely to occur. And without the protection of life and liberty, we could not secure our other goals.

A morality would be adequate if it contained a requisite set of these objective principles, of the core morality, but there could be more than one adequate morality that contained different rankings of these principles and other principles consistent with core morality. That is, there may be a certain relativity to secondary principles (whether to opt for monogamy rather than polygamy, whether to include a principle of high altruism in the set of moral duties, whether to allocate more resources to medical care than to environmental concerns, whether to institute a law to drive on the left side of the road or the right side of the road, and so forth), but in every morality a certain core will remain, though applied somewhat differently because of differences in environment, belief, tradition, and the like.

The core moral rules are analogous to the set of vitamins necessary for a healthy diet. We need an adequate amount of each vitamin—some humans more of one than another—but in prescribing a nutritious diet we don't have to set forth recipes, specific foods, place settings, or culinary habits. Gourmets will meet the requirements differently than ascetics and vegetarians, but the basic nutrients may be had by all without rigid regimentation or an absolute set of recipes.

Stated more positively, an objectivist who bases his or her moral system on a common human nature with common needs and desires might argue for objectivism somewhat in this manner:

1. Human nature is relatively similar in essential respects, having a common set of needs and interests.
2. Moral principles are functions of human needs and interests, instituted by reason in order to promote the most significant interests and needs of rational beings (and perhaps others).
3. Some moral principles will promote human interests and meet human needs better than others.
4. Those principles that will meet essential needs and promote the most significant interests of humans in optimal ways can be said to be objectively valid moral principles.
5. Therefore, since there is a common human nature, there is an objectively valid set of moral principles, applicable to all humanity.

If we leave out any reference to a common human nature, the argument would be even simpler:

1. Objectively valid moral principles are those that, if adhered to, meet the needs and promote the most significant interests of persons.
2. Some principles are such that adherence to them meets the needs and promotes the most significant interests of persons.
3. Therefore, there are some objectively valid moral principles.

Either argument would satisfy objectivism, but the former makes it clearer that it is our common human nature that generates the common principles. However, some philosophers might not like to be tied down to the concept of a common human nature, in which case the second version of the argument may be used. It has the advantage that even if it turned out that we did have somewhat different natures or that other creatures in the universe had somewhat different natures, some of the basic moral principles would still survive.

If this argument succeeds, there are ideal moralities (and not simply adequate ones). Some moralities are better than others. Of course, there could still be more than one ideal morality, which presumably an ideal observer would choose under optimal conditions. The ideal observer might conclude that out of an infinite set of moralities, two, three, or more combinations would tie for first place. One would expect that these would be similar, but there is every reason to believe that all of these would contain the set of core principles.

Of course, we don't know what an ideal observer would choose, but we can posit that such an observer would make choices impartially and with maximal knowledge about the consequences of action types because these conditions would ensure that agents have the best chance of making the best decisions. If this is so, then the more we learn to judge impartially and the more we know about possible forms of life, the better chance we have to approximate an ideal moral system. And if there is the possibility of approximating ideal moral systems with an objective core and other objective components, then ethical relativism is certainly false. We can confidently dismiss it as an aberration and get on with the job of working out better moral systems.

Let me make my appeal to your intuitions in another way to make the same point. Imagine that you have been miraculously transported to the dark kingdom of hell, and there you get a glimpse of the sufferings of the damned. What is their punishment? Well, they have eternal back itches that ebb and flow constantly. But they cannot scratch their backs, for their arms are paralyzed in a frontal position. And so they writhe with itchiness through eternity. But just as you are beginning to feel the itch in your own back, you are suddenly transported to heaven. What do you see in the kingdom of the blessed? Well, you see people with eternal back itches, who cannot scratch their own backs. But they are all smiling instead of writhing. Why? Because everyone's arms are stretched out to scratch someone else's back, and, so arranged in one big circle, a hell is turned into a heaven of ecstasy.

If ethical relativism is false and moral objectivism true, why is ethical relativism so popular?

Ethical relativism is appealing because we have become aware of the evils of ethnocentrism, which has plagued the relations of Europeans and Americans with other cultures. We are now more conscious of the frailty of many aspects of our moral repertoire so that there is a tendency to wonder, "Who's to judge what's really right or wrong?" However, the move from a reasonable cultural relativism, which rightly causes us to rethink our moral systems, to an ethical relativism, which causes us to give up the heart of morality altogether, is an instance of the fallacy of confusing factual or descriptive statements with normative ones. Cultural relativism doesn't entail ethical relativism. Our rationale for opposing ethnocentricism is the same as our rationale for supporting an objective moral system: Impartial reason draws us to it.

Philosophers from other nations have noted that ethical relativism is a particularly "American phenomenon" (Canadian as well, I'd say). Being a multicultural society, made of diverse ethnic, cultural, and religious groups, the United States "succeeds as a functional and democratic society because of its strong social pressures toward . . . tolerance. . . . That American college students tend to be ethical relativists . . . merely shows that they are well-adjusted citizens."[9] But tolerance can be confused with wishy-washy mindlessness, the acceptance of what is clearly evil. I may tolerate my neighbor's taking drugs because I value his or her freedom, but I do not think it's a good thing. I may tolerate an economic system that distributes wealth unjustly, but I may disapprove of it. There simply may not be anything I can do to improve the system that will be better than the present alternatives. As a vegetarian, I tolerate my friends and fellow citizens' consumption of meat, but I don't think it's right to kill animals for their meat when it's not necessary for our survival or health.

We may well agree that cultures differ and that we ought to be cautious in condemning what we don't understand, but this in no way need imply that there are not better and worse ways of living. We can understand and excuse, to some degree at least, those who differ from our best notions of morality without abdicating the notion that cultures without principles of justice or promise keeping or protection of the innocent are morally poorer for these omissions.

In sum, I have argued (1) that cultural relativism (the fact that there are cultural differences regarding moral principles) does not entail ethical relativism (the thesis that there are no objectively valid universal moral principles), (2) that the dependency thesis (which asserts that morality derives its legitimacy from individual cultural acceptance) is mistaken, and (3) that there are universal moral principles based on a common human nature and a need to solve conflicts of interest and flourish.

Note that if either subjective or conventional ethical relativism is correct, we will have no basis for arguing against anyone else's view of how he or she should act with regard to the environment. If subjective ethical relativism is correct, we each may individually choose our own moral perspective. If conventional ethical relativism is correct, all we have to do is identify with a specific culture to justify our treatment of the environment—no matter how degrading, no matter how much suffering it causes, no matter how selfish. But if the arguments in the last two chapters are correct, morality has a set of purposes that generate a set of principles to reach those purposes.

So "who's to judge what's right or wrong?" *We are*. We are to do so on the basis of the best reasoning we can bring forth and with sympathy and understanding.

Review Questions

1. Explain the difference between cultural and ethical relativism. Are there criteria by which we can say some cultures are morally better than others?
2. Explain the difference between subjectivism and conventionalism. Discuss their strengths and weaknesses. Why is relativism so appealing to many people?
3. Discuss the difference between absolutism and objectivism. What is the meaning of prima facie duty?
4. What are the implications of the debate over ethical relativism and objectivism for our treatment of the environment? Can one be a relativist and still promote global environmental ethics? That is, does a commitment to practices such as preserving species, reducing pollution, and reducing our consumptive patterns so that future generations may flourish demand holding that certain principles are universally valid? Explain your answer.

Suggested Readings

Bambrough, Renford. *Moral Skepticism and Moral Knowledge*. London: Routledge & Kegan Paul, 1979.

Benedict, Ruth. *Patterns of Culture*. New York: New American Library, 1934.

Brink, David. *Moral Realism and the Foundation of Ethics*. New York: Cambridge University Press, 1989.

Gert, Bernard. *Morality: A New Justification of the Moral Rules*. New York: Oxford University Press, 1988.

Harman, Gilbert, and Judith Jarvis Thomson. *Moral Relativism and Moral Objectivity*. London: Blackwell, 1996.

Mackie, J. L. *Ethics: Inventing Right and Wrong*. Harmondsworth, N.Y.: Penguin, 1977.

Pojman, Louis, ed. *Ethical Theory*. Belmonth, Calif.: Wadsworth, 1998.

Sumner, William Graham. *Folkways*. Boston: Ginn, 1906.

Notes

1. John Ladd, *Ethical Relativism* (Belmont, Calif.: Wadsworth, 1973).
2. Ruth Benedict, *Patterns of Culture* (New York: New American Library, 1934), p. 219.
3. William Graham Sumner, *Folkways* (Boston: Ginn, 1906), sec. 80. Ruth Benedict indicates the depth of our cultural conditioning this way: "The very eyes with which we see the problem are conditioned by the long traditional habits of our own society." "Anthropology and the Abnormal," *The Journal of General Psychology* (1934): 59–82.
4. Ernest Hemingway, *Death in the Afternoon* (New York: Scribner's, 1932), p. 4.

5. Benedict, *Patterns of Culture*, p. 257.
6. Melville Herskovits, *Cultural Relativism* (New York: Random House, 1972).
7. Clyde Kluckhohn, "Ethical Relativity: Sic et Non," *Journal of Philosophy* 52 (1955).
8. Renford Bambrough, *Moral Skepticism and Moral Knowledge* (London: Routledge & Kegan Paul, 1979), p. 33.
9. An anonymous reviewer made this point. I am indebted.

CHAPTER FOUR

Egoism, Self-Interest, and Altruism

Nice guys finish last.

—LEO DUROCHER

The achievement of his own happiness is man's highest moral purpose.

—AYN RAND

Universal ethical egoism is the theory that everyone ought always to serve his or her own self-interest. That is, one ought to do what will maximize one's own expected utility or bring about one's own greatest happiness, even if it requires harming others. Brian Medlin defines ethical egoism this way: "Everyone (including the speaker) ought to look after his own interest and to disregard those of other people except insofar as their interests contribute towards his own."[1] Jesse Kalin defines it thus: For everyone (x) and for every act (y), "x ought to do y if and only if y is in x's overall self-interest."[2] Ethical egoism is utilitarianism reduced to the pinpoint of the single individual ego. Instead of advocating the greatest happiness for the greatest number, as utilitarianism does, it advocates the greatest happiness for myself, whoever I may be. It is a self-preoccupied prudence, urging one to postpone enjoyment today for long-term benefits. In its more sophisticated form, it compares life to a competitive game, perhaps a war game, and urges each person to *try* to win in the game of life.

AN ANALYSIS OF ETHICAL EGOISM

We'll consider two arguments for ethical egoism: the economist argument and Ayn Rand's argument for the virtue of selfishness.

The Economist Argument

Economists, following Adam Smith (1723–90), often argue that in a state of competition in the marketplace individual self-interest produces a state of optimal goodness for society at large because the peculiar nature of self-interested competition causes each competitor to produce a better product and sell it at a lower price than others do. Enlightened self-interest leads, as by an invisible hand, to the best overall situation.

The *economist argument,* essentially not an argument for ethical egoism, is really an argument for utilitarianism (see Chapter 5), which makes use of self-interest to attain (paradoxically) the good of all. The argument's goal is social utility, but it relies on an invisible hand inherent in the free-enterprise system to guide enlightened self-interest to reach that goal. We might say that it is a two-tier system. On the highest level, it is utilitarian; but on a lower level of day-to-day action, it is practical **egoism.** It tells us to worry not about the social good but only about our own good and in that way we will attain the highest social good possible.

There may be some truth in such a two-tier system, but two objections arise. First, it is, at best, unclear whether we can transpose the methods of economics (which are debatable) into the realm of personal relations. Personal relations may have a logic different from that of economic relations. The best way to maximize utility in an ethical sense may be to give one's life for others rather than kill another person, as an egoist might enjoin. Second, it is not clear that classical laissez-faire capitalism works. Since the 1929 depression, most economists have altered their faith in classical capitalism, and Western nations have supplemented capitalism with some government intervention. Similarly, although self-interest may often lead to greater social utility, it may get out of hand and need to be supplemented by a concern for others. Just as classical capitalism has been altered to allow government intervention—resulting in a welfare system for the worst-off people, public education, Social Security, and Medicare—an adequate moral system may need to draw attention to the needs of others and direct us to meeting those needs even where we do not see such action to be in our immediate self-interest.

The Ayn Rand Argument

In her books *The Fountainhead, The Virtue of Selfishness,* and *Atlas Shrugged,* Ayn Rand argues for a radical version of ethical egoism, one that promotes selfishness. In *The Fountainhead* she paints Howard Roark, her hero, as an egoist dedicated to his own happiness, and Ellsworth Toohey, the altruist philanthropist, as a scoundrel. **Altruism,** in the hands of the likes of Toohey, Rand avers, calls on one to sacrifice his or her life, not to find happiness, which is the highest goal of life.

According to Rand, the perfection of one's abilities in a state of happiness is the highest goal for humans. We have a moral duty to attempt to reach this goal. Because the ethics of altruism prescribes that we sacrifice our interests and lives for the good of others, it is incompatible with the goal of happiness. But ethical egoism prescribes that we seek our own happiness exclusively, and as such it is consistent with the happiness goal. Therefore, for Rand, ethical egoism is the correct moral theory.

In *The Virtue of Selfishness* she argues that selfishness is a virtue and altruism a vice, a totally destructive idea that leads to the undermining of individual worth. She defines *altruism* as the view that

> any action taken for the benefit of others is good, and any action taken for one's own benefit is evil. Thus, the *beneficiary* of an action is the only criterion of moral value—and so long as the beneficiary is anybody other than oneself, anything goes.[3]

As such, altruism is suicidal:

> If a man accepts the ethics of altruism, his first concern is not how to live his life, but how to sacrifice it. . . . Altruism erodes men's capacity to grasp the value of an individual life; it reveals a mind from which the reality of a human being has been wiped out. . . . Altruism holds *death* as its ultimate goal and standard of value—and it is logical that renunciation, resignation, self-denial, and every other form of suffering, including self-destruction, are the virtues it advocates.[4]

Since finding happiness is the highest goal and good in life, altruism, which calls on us to sacrifice our happiness for the good of others, is contrary to our highest good.

Rand's argument seems to go like this:

1. The perfection of one's abilities in a state of happiness is the highest goal for humans. We have a moral duty to attempt to reach this goal.
2. The ethics of altruism prescribes that we sacrifice our interests and lives for the good of others.
3. Therefore, the ethics of altruism is incompatible with the goal of happiness.
4. Ethical egoism prescribes that we seek our own happiness exclusively, and as such it is consistent with the happiness goal.
5. Therefore, ethical egoism is the correct moral theory.

The *Ayn Rand argument* for the virtue of selfishness is flawed by the fallacy of a false dilemma. It simplistically assumes that absolute altruism or absolute, selfish egoism are the only alternatives, but this is an extreme view of the matter. There are plenty of options between these two positions. Even a predominant egoist would admit that (analogous to the *hedonistic paradox*) sometimes the best way to reach self-fulfillment is for us to forget about ourselves and strive to live for goals, causes, or other persons.*

* The hedonistic paradox states that the best way to be happy is to forget about happiness and devote yourself to worthwhile goals and people.

Even if altruism is not required as a duty, it may be permissible in many cases.

In addition, Rand conflates and confuses *selfishness* with *self-interest*. Actually, she slides back and forth between advocating each of these concepts, so that the perceptive reader is left wondering whether she knows the difference. *Selfishness* is the disposition to seek one's own good *without* regard to other people. *Self-interest* is simply a disposition to be concerned about one's interests. Virtually every moral theory allows, if not advocates, self-interest. Self-interest is compatible with concern for others. Even the Second Commandment, "Love your neighbor as yourself," set forth by Moses and Jesus, states not that one must always sacrifice oneself for the other person but that one ought to love one's neighbor *as* oneself (Lev. 19:18; Matt. 22:39). Self-interest and self-love are morally good things, but not at the expense of other people's legitimate interests, which is how they turn into selfishness.

Of course, there are moral conflicts of interest in which we are tempted to prefer our own interest without regard for others. But morality comes into the picture precisely to require us to allow a fair adjudication process to decide the issue.

ARGUMENTS AGAINST ETHICAL EGOISM

Not only do the arguments for ethical egoism have drawbacks, but there are also four arguments against this doctrine.

The Publicity Argument

On the one hand, it seems a necessary condition for something to be a moral theory that one publicize one's moral principles. Unless principles are put forth as universal prescriptions that are accessible to the public, they cannot serve as guides to action or as helps in resolving conflicts of interest. On the other hand, it is not in the egoist's self-interest to publicize his or her moral principles. Egoists would rather that the rest of us be altruists. (We might ask why Nietzsche and Rand wrote books announcing their positions—were the royalties taken in by announcing ethical egoism worth the price of letting the cat out of the bag?)

It would be a bad thing for the egoist to argue for his position and even worse for him to convince others of it! But it is perfectly possible to have a private morality that does not resolve conflicts of interest (for that, the egoist publicizes standard principles of traditional morality). So, if you're willing to pay the price, you can accept the solipsistic-directed norms of egoism. (**Solipsism** is the view that only I exist. No one else really does. They are mere projections of my imagination.)

If the egoist is prepared to pay the price, egoism can be a consistent system but have some limitations. Although the egoist can cooperate with others in limited ways and perhaps even have friends—as long as their interests don't conflict with his—he has to be very careful about preserving his isolation. The egoist can't give advice or argue about his position, not sincerely at least. He must act alone, atomistically or solipsistically, in moral isolation; for to announce his adherence to the principle of egoism would be dangerous to his project. He can't teach his children or justify himself to others or forgive others.

The Paradox of Egoism

The situation, however, may be even worse than the sophisticated, self-conscious egoist supposes. Could the egoist have friends? And if limited friendship is possible, could she ever be in love or experience deep friendship? Suppose the egoist discovers that in the pursuit of the happiness goal, deep friendship is in her best interest. Can she become a friend? What is necessary to deep friendship? A true friend is one who is not always preoccupied about her own interest in the relationship but who forgets about herself altogether, at least sometimes, in order to serve or enhance the other person's interest. "Love seeketh not its own" is an altruistic disposition, the very opposite of egoism. One could go on to argue that friendship is a necessary ingredient for psychological health. Since egoists cannot have true friends, they cannot attain psychological health. Since an egoist would agree that any theory that undermines psychological health is an inadequate moral theory, ethical egoism must be rejected. So the paradox of egoism is that in order to reach happiness, the goal of egoism, one must give up egoism and become (to some extent) an altruist, the very antithesis of egoism.

Does the egoist have a reply to this criticism? Perhaps she can construct a split-level egoism. On the *higher* level, I am committed to advancing my own good above all else, whatever the effect on others. But I may conclude that in order to maximize personal utility (happiness), I must have friends. Because having friends requires having altruistic dispositions, I must on a *lower* or *practical* level be selectively *altruistic,* rather than egoistic. But this move seems to abolish egoism. For the kind of altruism involved in friendship really to work, I need to forget, or, at least, leave aside, my original egoist motivation and act spontaneously. So even if egoism is the overall justification for friendship, even if it originally led me to have friends, it seems that once I adopt the policy of promoting friendship, I must cease to be an egoist. An area of my psyche has been claimed by altruism.

The Relevant Difference Argument

The relative difference argument, based on a concept of universalizability and developed by James Rachels, goes like this: All difference of treatment between people must be justified by some relevant difference in description of the people or their acts.[5] For example, I am justified in paying Mary twice as much as John because she is working twice as long and producing twice as many widgets, but I am not justified in paying Mary twice as much as Sam simply because Mary is an African American and Sam is an Asian American. Race is an irrelevant difference. Racism, sexism, and fanatical nationalism are all prejudices that violate the relevant difference principle. But this principle applies to egoism as well. For the question is, What makes you so different from everyone else that you will allow your preferences to count for more than those of other rational beings? It seems unjust.

Of course, the egoist will reject the relevant difference principle in his or her behavior and so allow racism and sexism and other forms of discrimination. I've heard egoists deny that they are racists, but the truth is that nothing prohibits them from being racists. If it's in my interest to be a racist, I *ought* to be one; and if it's not, then I *ought* not be one. It all depends on whether being one serves my interest. If it does, the version of egoism described here would require that we be racists.

The Argument from Counterintuitive Consequences

The final argument against ethical egoism is that it leads to consequences that seem abhorrent. If we followed its dictates, we would be prohibited from doing acts that seem obviously good. Egoism is an absolute ethics that not only permits egoistic behavior but also demands it. Helping others at one's own expense is not only not required but also morally wrong in the egoist's view. Whenever I do not have good evidence that my helping you will end up to my advantage, I must refrain from helping you. According to the egoist's viewpoint, the 38 people who watched Kitty Genovese being beaten and repeatedly stabbed to death in New York City (Queens) in March 1964 not only had *no* obligation to call for help but also would have been wrong to do so. If I can save the whole of Europe and Africa from destruction by pressing a little button, as long as there is nothing for me to gain by it, it is wrong for me to press that button. The Good Samaritan was, by this logic, a vicious man in helping the injured victim and not collecting on it. It is certainly hard to see why egoists should be concerned about environmental matters if they are profiting from polluting the environment (for example, if an egoist gains 40 hedons in producing *P*, which creates pollution that, in turn, causes others 1,000 dolors—units of suffering—but the egoist suffers only 10 of those dolors, by an agent-maximizing calculus, the egoist is morally obligated to produce *P*). There is certainly no obligation to preserve scarce natural resources for future generations. "Why should I do anything for posterity?" the egoist asks. "What has posterity ever done for me?"

In the preceding examples we have taken a strong version of ethical egoism. One could accuse me of attacking a straw man if it weren't that people like Rand, Kalin, and others defend this strong position, making serving one's self-interest a necessary and sufficient condition for moral obligation. But perhaps there is a weaker form of ethical egoism which states that it is always *permissible* to do whatever is in one's self-interest. I don't think this works. Why? Because it makes everything permissible! If it's always permissible to do what is in my self-interest, it is permissible to do what is not in my self-interest. If, given any act, it is always permissible to act egoistically, then I am not *required* to do anything. So what's the purpose of morality? Answer: it has none. So this version of permissible egoism is dead from the start.

EVOLUTION AND ALTRUISM

If sheer unadulterated egoism is an inadequate moral theory, does that mean that we ought to aim at complete altruism, total self-effacement for the sake of others? What is the role of self-love in morality? An interesting place to start is with the new field of sociobiology, which posits the theory that social structures and behavioral patterns, including morality, have a biological base, explained by evolutionary theory.

In the past, linking ethics to evolution meant justifying exploitation. Social Darwinism justified imperialism and the principle that "might makes right" by saying that the law of nature is the survival of the fittest. This philosophy lent itself to a promotion of ruthless egoism. This is nature's law, "nature red in tooth and claw." Against this view, ethologists (those who study animal behavior) like Robert Ardrey and Konrad Lorenz argue for a more benign view of the animal kingdom—reminiscent of Rudyard Kipling—where the animal kingdom survives by cooperation, which is at least as important as competition. In Ardrey and Lorenz's view, it is the group or species, not the individual, that is of primary importance.

With the onset of sociobiology in the work of E. O. Wilson, but particularly with the work of Robert Trivers, J. Maynard Smith, and Richard Dawkins, a theory has come to the fore that combines radical individualism with limited altruism. It is not the group or species that is of evolutionary importance but the gene or, more precisely, the gene type. Genes, the parts of the chromosomes that carry the blueprints for all our natural traits (such as height, hair color, skin color, intelligence), copy themselves as they divide and multiply. In conception they combine with the genes of the member of the opposite sex to form a new individual.

In his fascinating sociobiological study, Richard Dawkins describes human behavior as determined evolutionarily by stable strategies set to replicate the gene.[6] This is not done consciously, of course, but it's the invisible hand that drives the consciousness. We're essentially gene machines.

Morality—or successful morality—can be seen as an evolutionary strategy for gene replication. Here's an example. Birds are inflicted with life-endangering parasites. Because they cannot use limbs to pick them off their heads, they—like much of the animal world—depend on the ritual of mutual grooming. It turns out that nature has evolved two basic types of birds in this regard: those who are disposed to groom anyone (the nonprejudiced type?) and those who refuse to groom any but those who present themselves for grooming. The former type of bird Dawkins calls "Suckers" and the latter "Cheaters."

In a geographic area with harmful parasites where there are only Suckers or Cheaters, Suckers will do fairly well, but Cheaters will not survive for want of cooperation. But in a Sucker population where a mutant Cheater arises, he will prosper, and his gene type will multiply. As the Suckers are exploited, they will gradually die out. But if and when they become too few to groom the Cheaters, the Cheaters will start to die off too and eventually become extinct.

Why don't birds all die off, then? Well, somehow nature has come up with a third type—call them "Reciprocators." Reciprocators groom only those who reciprocate in grooming them. They groom one another and Suckers, but not Cheaters. In fact, once a Cheater is caught, he is marked forever. There is no forgiveness. It turns out then that unless there are a lot of Suckers around, Cheaters have a hard time of it—harder even than Suckers! But it is the Reciprocators that prosper. Unlike Suckers, they don't waste their time messing with unappreciative Cheaters, and so they are not exploited and have ample energy to gather food and build better nests for their loved ones.

J. L. Mackie argues that the real name for Suckers is "Christian," one who believes in complete altruism, even turning the other cheek to one's assailant and loving one's enemy. Cheaters are ruthless egoists who can survive only if there are enough naive altruists around. Whereas Reciprocators are *reciprocal* altruists who have a rational morality based on cooperative self-interest, Suckers like Socrates and Jesus advocate "turning the other cheek and repaying evil with good."[7] Instead of a rule of reciprocity, "I'll scratch your back if you'll scratch mine," the extreme altruist substitutes the Golden Rule, "If you'd like the other fellow to scratch your back, you scratch his—even if he won't reciprocate."

The moral of the story is this: The altruist's morality (so interpreted) is only rational given the payoff of eternal life (with a scorekeeper, as Woody Allen says). Take that away, and it looks like a Sucker system. What replaces the "Christian" vision of submission and saintliness is the reciprocal altruist with his tit-for-tat morality, one who is willing to share with those willing to cooperate.

Mackie may caricature the position of the religious altruist, but he misses the subtleties of wisdom involved (Jesus said, "Be as wise as serpents but as harmless as doves"). Nevertheless, he does remind us that there is a difference between core morality and complete altruism. We have duties to cooperate and reciprocate, but no duty to serve those who manipulate us and no obvious duty to sacrifice ourselves for people outside our domain of special responsibility. We have a special duty of high altruism toward those in the close circle of our concern—namely, our family and friends.

Martin Luther once said that humanity is like a man mounting a horse who always falls off on the opposite side, especially when he tries to overcompensate for his previous exaggerations. So it is with ethical egoism. Trying to compensate for an irrational, guilt-ridden, complete Sucker altruism, the morality of self-effacement, it falls off the horse on the other side, embracing a Cheater's preoccupation with self-exaltation that robs the self of the deepest joys in life. Only the person who mounts properly, avoiding both extremes, is likely to ride the horse of happiness to its goal.

Let us apply these thoughts about egoism and altruism to environmental ethics, beginning with the story of the *tragedy of the commons.*

THE TRAGEDY OF THE COMMONS

In the nineteenth century, English villages usually had a commons, land that was available to all the citizens of a community to use for grazing. When the commons of a village was used judiciously, individuals could gradually increase their wealth. But as the community grew, the temptation to overgraze was strong. The story now known as the *tragedy of the commons*, first documented by William Lloyd in the middle of the nineteenth century and utilized in the services of environmental ethics and population policy by Garrett Hardin in the 1960s and 1970s, goes like this:[8]

Imagine an unmanaged village commons in which ten villagers pasture their cattle, each having 8 cattle. Since the carrying capacity of the commons is sufficient for 100 cattle, every additional cow put onto the commons increases the individual wealth of the farmer without harming anyone else. The commons is a source of increased wealth, a blessing to anyone able to utilize its resources. Gradually the farmers grow in wealth, so that each has 10 cows grazing the commons, each producing 1 unit of utility (*utiles*), totaling 100 units of utiles.

Now the land has reached its carrying capacity, and it no longer is in the interest of the village to add further cows to the commons. In fact, adding another cow to the land will actually decrease the total utiles, for increased grazing tends to eliminate the sweet green grass on which the cattle feed and ruin the soil by constant trampling. But since the commons is unmanaged, it is in each individual farmer's interest to add an additional cow. The entrepreneurial farmer, call him Farmer Brown, who myopically sees short-term advantage for himself but not the long-term welfare of the community, reasons that "It is in my interest to add one more cow beyond the carrying capacity, because if I do this I will be obtaining one more unit of good but paying only a fraction of the negative impact on the land." If the carrying capacity has reached 100 utiles, by adding one more cow he will gain 1 more utile while diminishing the total utiles of the common by 1. The land is worth only 99 utiles. But he gains 1 whole utile while he shares the 1 negative unit with 9 other farmers, ending up with just under 11 utiles, compared with his neighbors who, instead of the original 10 utiles each, now have 9.9 utiles each.

But, of course, Farmer Jones sees what Brown has done, and he applies the same logic. He adds another cow to the commons, further diminishing the value by 2 units, the loss of which he shares with the other nine farmers while reaping a positive utile.

As Garrett Hardin writes of this situation,

> Adding together the component partial utilities, the rational herdsman concludes that the only sensible course for him to pursue is to add another animal to his herd. And another; and another. . . . But this is the conclusion reached by each and every rational herdsman sharing the commons. Each man is *locked* in to a system that compels him to increase his herd without limit . . . in a world that is limited. Ruin is the destination toward which all men rush, each pursuing his own best interest that believes in the freedom of the commons. *Freedom in a commons brings ruin to all.*[9]

This tragedy can be illustrated in many other ways, of course. A few farmers dumping their refuse into a river probably won't affect the river's purity very much, but when many companies see it as a "cheap" way of refuse disposal, the whole region suffers, while the company bears only a fraction of the disadvantage, not enough to offset the money saved by dumping waste into the river.

Twice I lived in a crowded neighborhood where store owners burned their refuse rather than having it hauled away. They saved a considerable sum of money, but since the city's air was already poor, the additional pollution affected the health of the whole community. In both cases the store owners lived in the suburbs, so they suffered the consequences of polluted air only during the daytime; and from a cost-benefit point of view, they profited.

Of course, the tragedy of the commons is the story of our global environmental catastrophe. A factory saves money by burning a fuel with high sulfur content, but the sulfur dioxide it spews out causes acid rain that kills a forest or pollutes a lake 1,000 miles away—perhaps in a different country. The history of imprudent agricultural practices in North Africa (the Sahel) and Spain are responsible for the deserts presently there. England and Ireland were once lands lush with forests, but over the centuries people unwisely denuded them, and now these countries import their wood from Sweden and other countries.

Hardin's solution is for us to develop *mutually agreed upon, mutually coercive rules* in order to save us all from destruction. With growing populations who all want to make use of the commons of life, we need to limit our consumption of resources and make sure that cheaters don't get away with their theft.

But the lesson of the tragedy of the commons has wider implications for society. It shows that individual rational decisions add up to collective disorder and illustrates the role of moral rules in general. Cheating, for example, adds one unit of utility to the cheat but degrades the system so that everyone's work is worth less. Seeing that cheating pays, a second person, who might have been deterred by seeing the first cheater punished, adds his input to the commons, and so the tragedy begins and ruins a good system. Or one man bribes an official in order to get special treatment. A second man notices that bribing pays and follows suit. A third infers that the second profited from bribing and so proceeds to follow his example, and so on until bribing becomes a part of the culture (a kind of anticipatory tipping) and corruption covers the community.

CONTRACTUAL ETHICS

If the lesson of the tragedy of commons is valid, the purpose of morality is to inaugurate a contract, a set of mutually coercive, mutually beneficial rules, which if followed will prevent us from harm and preserve those values that make life worth living. That is, it is in my egoistic self-interest to support the core morality (discussed in Chapter 3), even though the result will be a limiting of my self-interest in favor of a moderate altruism.

Some philosophers would argue that the term *ethical egoism* is an oxymoron, a contradiction in terms. But if our discussion has been correct, ethical egoism *may* paradoxically have all the resources necessary for a fully developed morality. Suppose I want very much to be happy, and I reflect on what will make me happy. I conclude that I need friends and love to be happy. But how do I attain these things? Observation and reason tell me that I attain these goods by loving and helping others—and doing so in self-forgetting ways. So, paradoxically, I discover that the best way to fulfill my self is to deny it and forget it and dedicate myself to helping others. Egoism turns out to be altruism! On the whole and for the most part, we discover that the best way to be happy is to dedicate ourselves to making others happy.

Similarly, we find it in our long-term self-interest to discipline ourselves and follow the moral rules, which—if followed by almost everyone almost all of the time—will benefit almost everyone almost all of the time. These are the rules we would agree to under a contractual situation, and they include the general rule to keep the results of the contract even when no one is looking.

But what about those not members of our group, not parties to the contract? What obligations do we have to them? Well, we have no positive duties toward them and only the weakest negative ones. We don't even have a duty to keep our promises or to honor any initial contract made with them, since we can back out if it's not in our interest to continue the contract. How exactly one establishes the contract so that it becomes rational to adhere to it is a difficult question, there being no definite point between a nonbinding contract and a binding one—except that gradually the force of the contract masters us, becoming a prime motivator for action. At some point we perceive that the contract is in our long-term interest.

In a system of contractual ethics, we have no obligations to the stranger or outsider. We might agree that there is an objective morality that is universal in scope (the core morality), but it simply doesn't *apply* to situations outside the contract—any more than it would apply to our treatment of rocks or other nonsentient objects. Outside an established contract, not the Golden Rule, but the Bronze Rule applies: Do it to the other person before he gets a chance to do it to you. All is fair in love and war, and we are at war—a cold war in which the outsider is seen as the enemy.

Of course, even within the social contract, degrees of obligation exist. I have a deeper contractual relationship with my loved ones and family than with the members of my community, and a deeper contractual commitment to those in my community than to members of my society at large. Furthermore, I have a deeper commitment to citizens of my country than to citizens of other countries with whom we share a common implicit contract. When I am in Canada, I recognize mutual respect between the Canadians and myself that I may not recognize in a strange culture which practices cannibalism or witchcraft. But I might not treat a Canadian stranger with the same regard as I would an American. For example, if each of our nations were at war, I would choose to defend my people, not the Canadians.

As I have no positive obligations to rational beings in Outer Galactica or Mars (if there were any), I have no positive obligations to rational beings in other countries or communities where no contract exists. So I have no strong contract with people in other countries or continents—in Asia, Africa,

South America or even Europe, where I do have some deep friendships. Even though the plight of these people in these places might be far worse than that of my neighbors, I have no sufficient reason to come to their aid. Of course, I may build *bridges* to these other cultures and establish contractual relationships with them—in which case I do take on positive duties—but I need not do this. I violate no rights by just leaving them alone.

Suppose I live on island A with my family and friends, and another island B, unbeknownst to me, exists 50 miles to the south. We don't affect each other and have no interaction. I have no obligations to the people on B. Even if I discovered B's existence and heard that great suffering was taking place on B, I would have no obligation to come to B's aid. But suppose that an earthquake occurs on the shallow ocean floor, so that B is now joined to my island. The situation has changed. I and my fellow citizens must decide how to cope with the new situation. We can go to war or we can try to strike a mutual-recognition pact. Well, wars are dangerous, so unless one side thinks it can win easily, we will try to settle the matter through a contract, agreeing to live by certain common rules. Perhaps we will continue to maintain our independence as two distinct cultures on separate ends of the island, but we will agree to cooperate on certain issues, such as mutual defense when attacked, trade in each other's goods, provision of aid in times of dire economic need, and so forth. But our relationship will be formal and legalistic, rather than spontaneous and trusting. Still, the young will probably begin to interact, and soon friendships and marriages will take place so that the two communities may eventually merge into one. Human beings have a way of merging. The circle expands, the contract binds more and more within its ambit.

Now suppose that we discover some toxic gases wafting to us from the west. An exploration party is sent westward by boat to explore the situation. The party discovers island C thirty miles away, where the inhabitants are heating their homes by foolishly burning dangerous chemicals, which release into the atmosphere poisons that are drifting westward toward us. What should we do? We could launch a war on island C, but that would be costly and dangerous. So we send out a negotiating party and communicate with C, trying to convince its islanders to stop burning dangerous chemicals. In order to make this advantageous to the people on C, we promise to establish trade with them, selling them safer fuels and lush fruit for their bounteous wheat. The people on C, being rational, see this offer as a good deal and gradually alter their heating policy so that they burn more benign fuels—and reap the benefits of cleaner air and better nutrition at the same time. It's a win-win situation.

Now imagine that the globe is covered with thousands of such islands, A1 through Z1000, and that we have the power to help or harm one another by the way we live. Though we start out with no contractual relationships, it is in our interest to create them, at least minimal ones. Here is where the logic of the tragedy of the commons comes in. The greater our use of technology, the more quickly we are able to use up the globe's limited resources and turn them into harmful pollution. C's air pollution harms A, but Z's water pollution harms C, and A's overconsumption tends to deplete the fish population upon which Z is dependent. So, as the world becomes more interdependent, a vast commons, it becomes rational to expand the circle of considerability to include more and more people in more and more countries. A local egoistic contractual ethic, by its own logic and in its own interest, expands to devise a globally applicable moral contract, with mutually coercive, mutually agreed upon rules that can turn a tragedy into a peaceable kingdom of prosperity for all.

Given limited resources and the ability to do mutual harm, rational people will see a global moral contract as the only rational solution to the mutual destruction that awaits us all if we do not cooperate. Unless we all hang together, we will each hang—alone. This is the message of the tragedy of the commons: to teach us to establish an environmental ethic sufficient to save our planet and all of us with it. Many environmentalists deplore contractualism because it seems so thoroughly anthropocentric. But, in its favor, it does give a rational explanation, as well as justification, for morality, even a rational objective morality that involves making global contracts. In the long run, morality is in our interest. In popular lingo—it's a global win-win situation.

So far, so good. But contractualism has two additional problems that are very difficult for it to solve: (1) we seem to have obligations to nonrational animals who cannot enter into contracts—and perhaps to species and ecosystems; (2) we seem to have obligations to future generations. These two problems indicate problems with contractualism as a complete ethical theory. At best, it needs to be supplemented. We will consider the problem of future generations at the end of the next chapter. We will consider the first problem in Chapters 7 through 10.

Review Questions

1. Discuss the two arguments in favor of ethical egoism and the four against it. Which side has the best arguments? Why?
2. What is the relationship between ethics and evolution? How does this relationship throw light on egoism? What is the significance of reciprocity for ethics?
3. Can an ethical egoist have friends? Some philosophers, beginning with Plato, have argued that ethical egoism is irrational, since it precludes psychological health. In an article entitled "Ethical Egoism and Psychological Dispositions" (*American Philosophical Quarterly* 17, 1980), Laurence Thomas sets forth the following argument:

1. A true friend could never, as a matter of course, be disposed to harm or to exploit a friend [definition of a friend].
2. An egoist could never be a true friend to anyone [for the egoist must be ready to exploit others whenever it is in his or her interest].
3. Only someone with an unhealthy personality could never be a true friend to anyone [definition of a healthy personality; that is, friendship is a necessary condition for a healthy personality].
4. Ethical egoism requires that we have a kind of disposition that is incompatible with our having a healthy personality [from 1–3].

Conclusion: Therefore, from the standpoint of our psychological makeup, ethical egoism is unacceptable as a moral theory.

Do you agree with Thomas? Explain your answer:

4. In this chapter we have argued against various forms of egoism. Can you think of a version of ethical egoism that can meet the objections presented?
5. Discuss the story of the tragedy of the commons. How does it throw light on ethics and especially environmental ethics?
6. Go over the nature of contractual ethics. What is its relationship to egoism?

Suggested Readings

Baier, Kurt. *The Moral Point of View.* Ithaca, N.Y.: Cornell University Press, 1958.

Gauthier, David. *Morality by Agreement.* Oxford: Clarendon Press, 1986.

Pojman, Louis, ed. *Ethical Theory: Classical and Contemporary Readings.* 3rd ed. Belmont, Calif.: Wadsworth, 1998.

Rachels, James. *The Elements of Moral Philosophy.* New York: Random House, 1986.

Rand, Ayn. *The Virtue of Selfishness.* New York: New American Library, 1964.

Rogers, Kelly, ed. *Self-Interest: An Anthology of Philosophical Perspectives.* New York: Routledge, 1997.

Singer, Peter. *The Expanding Circle: Ethics and Sociobiology.* Oxford: Oxford University Press, 1983.

Notes

1. Brian Medlin, "Ultimate Principles and Ethical Egoism," *Australasian Journal of Philosophy* (1957).
2. Jesse Kalin, "In Defense of Ethical Egoism," *Philosophical Review* (1968).
3. Ayn Rand, *The Virtue of Selfishness* (New York: New American Library, 1964), pp. vii and 27–32; 80ff.
4. Ibid.
5. James Rachels, *The Elements of Moral Philosophy* (New York: Random House, 1986), chap. 6.
6. Richard Dawkins, *The Selfish Gene* (Oxford: Oxford University Press, 1976), chap. 10.
7. J. L. Mackie, "The Law of the Jungle: Moral Alternatives and Principles of Evolution," *Philosophy* 53 (1978).
8. Garrett Hardin, "The Tragedy of the Commons," *Science* 162 (1968).
9. Ibid.

CHAPTER FIVE

Classical Ethical Theories and the Problem of Future Generations

Suppose you are on an island with a dying millionaire. As he lies dying, he entreats you for one final favor: "I've dedicated my whole life to football and have gotten endless pleasure, and some pain, rooting for the last 50 years. Now that I am dying, I want to give all my assets, $5 million, to the Dallas Cowboys. Would you take this money [he indicates a box containing the money in large bills] back to Dallas and give it to the Cowboy's owner, so that he can buy better players?" You agree to carry out his wish, at which point a huge smile of relief and gratitude breaks out on his face as he expires in your arms. After returning to Dallas, you see a newspaper advertisement placed by the World Hunger Relief Organization (whose integrity you do not doubt) pleading for $5 million to be used to save 100,000 people dying of starvation in East Africa. Not only will the $5 million save their lives, but it will also be used to purchase small technology and the kinds of fertilizers necessary to build a sustainable economy. You reconsider your promise to the dying Cowboy fan in light of this consideration. What should you do with the money?

Or suppose two men are starving to death on a raft floating in the Pacific Ocean. One day they discover some food in an inner compartment of a box on the raft. They have reason to believe that the food will be sufficient to keep one of them alive until the raft reaches a certain island where help is available, but that if they share the food both of them will most likely die. One man is a brilliant scientist who holds in his mind the cure for cancer. The other man is undistinguished. Otherwise, there is no relevant difference between the two men. What is the morally right thing to do? Share the food and hope against the odds for a miracle? Flip a coin to see which man gets the food? Give the food to the scientist?

What is the right thing to do in these kinds of situations?

If you decide to act on the principle of promise keeping or not stealing in the case of the millionaire's money or if you decide to share the food in the case of the two men on the raft on the basis of the principle of fairness or equal justice, then you adhere to a type of moral theory called **deontology** (from the Greek *deon,* meaning "duty," and *logos,* meaning "logic").

If, on the other hand, you decide to give the money to the World Hunger Relief Organization in order to save an enormous number of lives and restore economic solvency to the region, you side with a type of theory called **teleology,** or **consequentialist ethics** (*teleology* comes from *teleos,* meaning "having reached one's end" or "finished"). Similarly, if you decide to give the food to the scientist because he would probably do more good with his life, you side with the teleologist.

Traditionally, these two major types of ethical systems have dominated the field: the deontological, in which the locus of value is the act or kind of act, and the teleological, in which the locus of value is the outcome or consequences of the act. Whereas teleological systems see the ultimate criterion of morality in some nonmoral value that results from acts, deontological systems see certain features in the act itself as having intrinsic value. For example, a teleologist would judge whether lying was morally right or wrong by the consequences it produced, but a deontologist would see something intrinsically wrong in the very act of lying.

As we mentioned earlier, a teleologist is a person whose ethical decision making aims solely at maximizing such goods as pleasure, happiness, welfare, and the amelioration of suffering. That is, the standard of right or wrong action for the teleologist is the comparative consequences of the available actions. The act that produces the best consequences is right. Whereas the deontologist is concerned only with the rightness of the act itself, the teleologist asserts that acts do not have intrinsic worth. Whereas for the deontologist there is something intrinsically bad about lying, for the teleologist the only thing wrong with lying is the bad consequences it produces. If you can reasonably calculate that a lie will do even slightly more good than would telling the truth, you have an obligation to lie. In the next section we consider the dominant version of teleological ethics, utilitarianism. Later in the chapter, we examine deontological ethics, especially Immanuel Kant's ethics as the major form of deontological ethics.

WHAT IS UTILITARIANISM?

The greatest happiness for the greatest number

—FRANCIS HUTCHESON

As a moral philosophy, **utilitarianism** begins with the work of Scottish philosopher Francis Hutcheson (1694–1746) and comes into its classical stage in the persons of English social reformers Jeremy Bentham (1748–1832) and John Stuart Mill (1806–73). They were the nonreligious ancestors of the twentieth-century secular humanists, optimistic about human nature and our ability to solve our problems without recourse to providential grace. Engaged in a struggle for legal as well as moral reform, they were impatient with the rule-bound character of law and morality in eighteenth-

and nineteenth-century Great Britain and tried to make the law serve human needs and interests.

Bentham's concerns were mostly practical rather than theoretical. He worked for a thorough reform of what he regarded as an irrational and outmoded legal system. To paraphrase Jesus, he might have made his motto "Morality and Law were made for man, not man for Morality and Law." What good was adherence to outworn deontological rules that served no useful purpose? that kept the poor from enjoying a better life? What good were punitive codes that served only to satisfy sadistic lust for vengeance?

The changes the utilitarians proposed were not done in the name of justice; rather, they believed that even justice must serve the human good. The poor were to be helped, women were to be liberated, and the criminal rehabilitated if possible, not in the name of justice but because by so doing society could bring about more utility: the amelioration of suffering and promotion of more pleasure or happiness.

The utilitarian view of punishment is a case in point. Whereas deontologists believe in retribution—that all the guilty should be punished in proportion to the gravity of their crime—the utilitarians' motto is "Don't cry over spilt milk!" They believe that the guilty should be punished only if the punishment would serve some deterrent (or preventive) purpose. Rather than punish John in exact proportion to the heinousness of his deed, we ought to find the punishment that will serve as the optimum deterrent.

The proper amount of punishment to be inflicted upon the offender is the amount that will do the most good (or least harm) to all those who will be affected by it. The measure of harm inflicted on John should be preferable to the measured harm that would have been done to society if the penalty had been lower. If punishing John will do no good (because John is not likely to commit the crime again and no one will be deterred by the punishment), John should go free.

It is the *threat* of punishment that is the important thing! Every act of punishment is to that extent an admission of the failure of the threat. If the threat were successful, there would be no punishment to justify. Of course, utilitarians believe that, given human failing, punishment is vitally necessary as a deterrent, and so the guilty will seldom if ever be allowed to go free.

There are two main features of utilitarianism: the consequentialist principle (or its teleological aspect) and the utility principle (or its hedonic aspect). The *consequentialist principle* states that the rightness or wrongness of an act is determined by the goodness or badness of the results that flow from it. It is the end, not the means, that counts. The end justifies the means. The *utility principle* states that the only thing that is good in itself is some specific type of state (such as pleasure, happiness, welfare). Hedonistic utilitarianism views pleasure as the sole good and pain as the only evil. To quote Bentham, the first one to systematize classical utilitarianism, "Nature has placed mankind under the governance of two sovereign masters, pain and pleasure. It is for them alone to point out what we ought to do, as well as what we shall do."[1] An act is right if it promotes a balance of pleasure over pain or prevents pain, and an act is wrong if it brings about more pain than pleasure or prevents pleasure from occurring.

Although applying his theory mainly to humanity, Bentham included animals within the scope of moral considerability:

> The day may come when the rest of the animal creation may acquire those rights which never could have been witholden from them but by the hand of tyranny. The French have already discovered that the blackness of the skin is no reason why a human being should be abandoned without redress to the caprice of a tormentor. It may one day come to be recognized that the number of legs, the [type] of the skin . . . are insufficient for abandoning a sensitive being to the same fate. What else is it that should trace the insuperable line? Is it the faculty of reason, or perhaps the faculty of discourse? But a full-grown horse or dog is beyond comparison a more rational, as well as a more conversable animal, than an infant of a day, or a week, or even a month, old. But suppose they were otherwise, what would it avail? The question is not, Can they reason? Nor Can they *talk,* but, *Can they suffer?*[2]

For Bentham, sentience is the sole criterion for moral considerability. If an animal can suffer, can experience pleasure and pain, it is worthy of our concern. At least we may not do it harm without cause. Bentham's utilitarianism is not really a *human-based* ethical system (anthropocentric) but a *sentience-centric* system. This will be an important consideration when we discuss animal rights in Chapter 7.

Bentham invented a scheme for measuring pleasure and pain, which he called the *hedonic calculus.* The quantitative score for any pleasure or pain experience comes about by giving sums to seven aspects of an experience in terms of pleasure and pain. The seven aspects of a pleasurable or painful experience are its intensity, duration, certainty, nearness, fruitfulness, purity, and extent. By adding up the sums of each possible act in terms of pleasure and pain and comparing them, we would be able to decide on which act to perform. With regard to our example of deciding between giving the dying man's money to the Cowboys or giving it to the starvation victims, we should add up the likely pleasures to all involved in terms of these seven qualities. Suppose that we find that by giving the money to the East African famine victims we will cause at least 3 million **hedons** (units of happiness), but by giving the money to the Cowboys, we will probably cause less than 1,000 hedons. So we would have an obligation to give the money to the famine victims.

There is something appealing about Bentham's utilitarianism. It is simple in that there is only one principle to apply: Maximize pleasure and minimize suffering. It is commonsensical in that we think that morality really is about ameliorating suffering and promoting benevolence. It is scientific: Simply make quantitative measurements and apply the principle impartially, giving no special treatment to yourself or to anyone else because of race, gender, or religion.

However, Bentham's philosophy may be too simplistic in one way and too complicated in another. It may be too simplistic in that there are values other than pleasure, such as freedom and wisdom; and it seems too complicated in that its hedonic calculus is encumbered with too many variables and problems in attempting to give scores to the variables. What score does one give a cool drink on a hot day or a warm shower on a cool day? How do you compare a 5-year-old's delight over a new toy with a 50-year-old's delight with a new lover? Can I take your second car from you and give it to Beggar Bob who does not own a car and would enjoy it more than you? And if it's simply the overall benefits of pleasure that we are measuring, might it not turn out that if Jack or Jill would be "happier" in the Pleasure Machine, which provided constant pleasure at the cost of completely deluding the occupier, or if they would be happier on drugs than in the real world, then we would have an obligation to provide them with a Pleasure Machine or drugs? Because of these considerations, Bentham's version of utilitarianism was even in his own day referred to as the "pig philosophy," since a pig enjoying his life would exist in a higher moral state than would a slightly dissatisfied Socrates.

It was to meet these sorts of objections and save utilitarianism from the charge of being a pig philosophy that Bentham's brilliant successor John Stuart Mill sought to distinguish happiness from mere sensual pleasure. His version of utilitarianism, *eudaimonistic* (from the Greek *eudaimona,* meaning "happiness") *utilitarianism,* defines happiness in terms of certain types of higher-order pleasures, or satisfactions, such as intellectual, aesthetic, and social enjoyments, as well as minimal suffering. In other words, there are two types of pleasures: the lower or elementary (such as food, drink, sexuality, rest, and sensuous titillation) and the higher (such as intellectual, creative, and spiritual activities). Although the lower pleasures are perhaps more intensely gratifying, they also lead to pain when overindulged in. The pleasures of achievement and spirituality tend to be more protracted, continuous, and gradual.

Mill argues that the higher or more refined pleasures are superior to the lower ones: "A being of higher faculties requires more to make him happy, is capable probably of more acute suffering, and certainly accessible to it at more points, than one of an inferior type," but still he is qualitatively better off than the person without these higher faculties. "It is better to be a human being dissatisfied than a pig satisfied; better to be Socrates dissatisfied than a fool satisfied."[3]

Humans are the kind of creatures who require higher pleasures to be truly happy. We want the lower pleasures but also deep friendship, intellectual ability, culture, the ability to create and appreciate art, knowledge, and wisdom. But, one may object, how do we know that it really is better to have these higher pleasures? Here Mill imagines a panel of experts and says that of those who have had wide experience of pleasures of both kinds, almost all give a decided preference to the higher type. Since Mill was an empiricist, one who believed all knowledge and justified belief was based in our experience, he had no recourse but to rely on the composite consensus of human history. People who experience both rock music and classical music will, if they appreciate both, prefer Bach and Beethoven to the Rolling Stones or Dancing Demons. We generally move up from appreciating simple things (for example, from nursery rhymes to more complex poetry rather than the other way around).

Mill has been criticized for not giving a better reply, for being an elitist and unduly favoring the intellectual over the sensual. But he has a point. Don't we generally agree, if we have experienced both the lower and the higher types of pleasure, that while a full life would include both, a life with only the former is inadequate for human beings? Isn't it better to be Socrates dissatisfied than the pig satisfied? and better still to be Socrates satisfied?

The point is not merely that humans would not be satisfied with what satisfies a pig but also that somehow the quality of the higher pleasures is *better*. But what does it mean to speak of better pleasure? Is Mill unwittingly assuming some nonhedonic notion of intrinsic value to make this distinction? Is he assuming that knowledge, intelligence, freedom, friendship, love, health, and so forth are good things in their own right? Or is Mill simply saying that the lives of humans are generally such that we can predict that they will be happier with the more developed, refined, spiritual values? Which thesis would you be inclined to defend?

The Strengths and Weaknesses of Utilitarianism

Utilitarianism does have two very positive features. The first attraction or strength is that it is a system with a single absolute principle, with a potential answer for every situation. Do what will promote the most utility! It's good to have a simple action-guiding principle, applicable to every occasion—even if it may be difficult to apply (life is not simple). The second strength is that utilitarianism seems to get to the substance of morality. It is not merely a *formal system,* offering only formal principles (that is, broad guidelines for choosing substantive principles, such as the Golden Rule, the rule to "Let your conscience be your guide," or "Never do what you cannot will to make a universal law"), but also has a *material* core: promoting human and animal happiness and ameliorating suffering. The first virtue gives one a clear decision-making procedure when considering what to do. The second virtue appeals to our sense that morality is made for humans (and other animals) and that morality is not so much about rules as about helping people and alleviating the suffering in the world.

This seems commonsensical. Utilitarianism gives us clear and reasonable guidance in everyday matters. We should try to make our colleges, our towns, our families, as well as our nation and world, better places than they are. We should help people, ameliorate their suffering, whenever it does not cost us unduly. In the case of deciding what to do with the $5 million of the dead millionaire, something in us says that it is absurd to keep a promise to a dead man when it means allowing hundreds of thousands of famine victims to die. (How would we like it if we were in their shoes?) Far more good can be accomplished by helping the needy than by giving the money to the Cowboys!

But we need to address certain problems with utilitarianism before we can give it a clean bill of health.

Problem 1: How Can We Know the Consequences of Actions?

Sometimes utilitarians are accused of playing God. They seem to hold to an ethical theory that demands godlike powers—that is, knowledge of the future. Of course, we normally do not know the long-term consequences of our actions, for life is too complex and the consequences go on into the indefinite future. One action causes one state of affairs that, in turn, causes another state of affairs indefinitely, so that calculation becomes impossible. Recall the nursery rhyme:

For want of a nail
The shoe was lost;
For want of a shoe
The horse was lost;
For want of a horse
The rider was lost;
For want of a rider
The battle was lost;
For want of a battle
The kingdom was lost;
And all for the want
Of a horseshoe nail.

Poor, unfortunate blacksmith! What utilitarian guilt he must bear all the rest of his days.

But it is ridiculous to blame the loss of one's kingdom on the poor unsuccessful blacksmith, and utilitarians are not so foolish as to hold him responsible for the bad situation. Instead, following C. I. Lewis, they distinguish three different kinds of consequences: actual consequences of an act, consequences that could reasonably have been expected to occur, and intended consequences.[4] An act is *absolutely* right if it has the best actual consequences. An act is *objectively* right if it is reasonable to expect that it will have the best consequences. An act is *subjectively* right if its agent intends or actually expects it to have the best consequences. It is the second kind of rightness (*objective rightness*), based on reasonable expectations, that is central here, for only the subsequent observer of the consequences is in a position to determine the actual results. The most that the agent can do is use the best information available and do what a reasonable person would expect to have the best overall results. Suppose, for example, that while Stalin's aunt was carrying little baby Josef up the stairs to her home, she slipped and had to choose between dropping infant Josef, allowing him to be fatally injured, or breaking her arm. According to the formula just given, it would have been absolutely right for her to let him be killed, but it would not have been within her power to know that. She did what any reasonable person would do—saved the baby's life at the risk of some injury to herself. She did what was objectively right. The utilitarian theory is that, generally, by doing what reason judges to be the best act based on likely consequences, we will, in general, actually promote the best consequences.

Problem 2: The No-Rest Objection

According to utilitarianism, one should always do the act that promises to promote the most utility. But there is usually an indefinite set of possible acts from which to choose, and even if I can be excused from considering all of them, I can be fairly sure that there is often something preferable (in society's terms) that I could be doing. For example, when I am about to go to the movies with a friend, I should ask myself if helping the homeless in my community wouldn't promote more utility. When I am about to go to sleep, I should ask myself whether I could at this moment be doing something to help save the ozone layer. And why not simply give all my assets (beyond what is absolutely necessary to keep me alive) to environmental organizations in order to promote greater utility? How would a sophisticated utilitarian respond to this criticism?

Problem 3: The Absurd Implications Objection

W. D. Ross argued that utilitarianism is to be rejected because it is counterintuitive. If we accepted it, we would have to accept an absurd implication.

Consider two acts, *A* and *B,* that will both result in 100 hedons (units of pleasure of utility). The only difference is that *A* involves telling a lie and *B* involves telling the truth. The utilitarian must maintain that the two acts are of equal value, but this seems counterintuitive—at least, at first glance. Most of us think that telling the truth is an intrinsically good thing. Who is right here?

Similarly, in Arthur Koestler's *Darkness at Noon,* Rubashov writes of the Communist philosophy in the Soviet Republic:

> History has taught us that often lies serve her better than the truth; for man is sluggish and has to be led through the desert for forty years before each step in his development. And he has to be driven through the desert with threats and promises, by imaginary terrors and imaginary consolations, so that he should not sit down prematurely to rest and divert himself by worshipping golden calves.[5]

According to this interpretation, orthodox Soviet Communism justifies its lies and atrocities by utilitarian ideas. Something in us revolts at this kind of value system. Truth is sacred and must not be sacrificed on the altar of expediency.

Problem 4: The Justice Objection

Suppose that in a racially volatile community a rape and murder is committed. You are the sheriff who has spent a lifetime working for racial harmony. Now, just when your goal is about to be realized, this incident occurs. The crime is thought to be racially motivated, and a riot is about to break out, which will very likely result in the death of several people and create long-lasting racial antagonism. You are able to frame a tramp for the crime, so that a trial will show that he is guilty. He will then be executed. There is every reason to believe that a speedy trial and execution will head off the riot and save the community. Only you (and the real criminal, who will keep quiet about it) will know that an innocent man has been tried and executed. What is the morally right thing to do? The utilitarian seems committed to framing the tramp, but many would find this solution appalling.

Or consider this hypothetical situation. You are a utilitarian physician who has five patients under your care. One needs to have a heart transplant, one needs two lungs, one needs a liver, and the last two need kidneys. Now into your office comes a healthy bachelor needing a flu shot. You judge him to be a perfect sacrifice for your five patients. Doing a utilitarian calculus, there is no doubt in your mind that you could do more good by injecting the healthy man with a sleep-inducing drug and using his organs to save your five patients.[6]

This cavalier view of justice offends us. The very fact that utilitarians even countenance such actions, that they would misuse the legal system or the medical system to carry out their schemes seems frightening. It reminds us of the medieval Roman Catholic bishop's justification for heresy hunts, inquisitions, and religious wars:

> When the existence of the Church is threatened, she is released from the commandments of morality. With unity as the end, the use of every means is sanctified, even cunning, treachery, violence, simony, prison, death. For all order is for the sake of the community, and the individual must be sacrificed to the common good.[7]

Utilitarian Responses to Standard Objections

These objections are weighty and too complicated to attempt to refute here, but we can allow the utilitarian to make an initial defense. What sorts of responses are open to utilitarians? A sophisticated version of utilitarianism seems to be able to offset at least some of the force of these criticisms. The sophisticated utilitarian can use the *multilevel strategy,* which goes like this: We must split considerations of utility into two levels, the lower level dealing with a set of rules that we judge to be most likely to bring about the best consequences most of the time. We'll call this the *rule-utility* feature of utilitarianism. Normally, we have to live by the best rules our system can devise, and rules of honesty, promise keeping, and justice will be among them.

But sometimes the rules conflict or clearly will not yield the best consequences. In these infrequent cases, we will need to suspend or override the rule in favor of the better consequences. We call this the *act-utility* feature of utilitarianism. It constitutes the second level of consideration and is referred to only when there is dissatisfaction with the rule-utility feature. An example of this might be the rule against breaking a promise. Normally, the most utility will occur through keeping one's promises, but consider this situation: I have promised to meet you at the movies tonight at 7:00. Unbeknown to you, on the way to our rendezvous I come across an accident and am able to render great service to the injured parties. Unfortunately, I cannot contact you, and you are inconvenienced as you wait patiently in front of the theater for an hour. I have broken a utility rule in order to maximize utility and am justified in so doing.

Here is another example set forth by Judith Jarvis Thomson: You are a trolley-car driver who sees five workers on the track before you. You suddenly realize that the brakes have failed. Fortunately, the track has a spur leading off to the right, and you can turn the trolley onto it. Unfortunately, there is one person on the right-hand track. You can turn the trolley to the right, killing one person, or you can refrain from turning the trolley, in which case five people will die.[8] Under traditional views, a distinction exists between killing and letting die, between actively killing and passively allowing death, but the utilitarian rejects this distinction. You should turn the trolley, causing the lesser evil, for the only relevant issue is expected utility. So the normal rule against actively causing an innocent to die is suspended in favor of the utility principle.

This is the kind of defense the sophisticated utilitarian is likely to lodge against all of the preceding criticisms. The utilitarian responds to problem 2, the no-rest objection, by insisting that a rule prescribing rest and entertainment is actually the kind of rule that would have a place in a utility-maximizing set of rules. The agent should aim at maximizing his or her own happiness as well as other people's happiness. For the same reason, it is best not to worry overly much about the needs of those not in one's primary circle. Although one should be concerned about the needs of future and distant people, it actually would promote disutility for the average person to become preoccupied with these concerns. But, the utilitarian would remind us, we can surely do a lot more for suffering humanity than we now are doing.

With regard to problem 3, Ross's absurd implications objection, the utilitarian can agree that there is something counterintuitive in equating an act based on a lie with an act based on honesty; but, he argues, we must be ready to change our culture-induced moral biases. What is so important about truth telling or so bad about lying? If it turned out that lying really promoted human welfare, we'd have to accept it. But that's not likely. Our happiness is tied up with a need for reliable information (that is, truth) on how to achieve our ends. So truthfulness will be a member of the rule-utility set. But where lying will clearly promote utility without undermining the general adherence to the rule, we simply ought to lie. Don't we already accept lying to a gangster or telling "white" lies to spare people's feelings?

With regard to Rubashov's utilitarian defense of Communism and its inhumanity or the medieval defense of the Inquisition, the utilitarian replies that this abuse of utilitarianism only illustrates how dangerous the doctrine can be in the hands of self-serving bureaucrats. Any theory can be misused in this way.

We turn to the most difficult objection, the claim that utilitarianism permits injustice, as seen in the example of the sheriff framing the innocent derelict. The utilitarian counters that justice is not an absolute—mercy, benevolence, and the good of the whole society sometimes should override it; but, the sophisticated utilitarian insists, it makes good utilitarian sense to have a principle of justice that is generally obeyed. It is not clear what the sheriff should do in the racially torn community. More needs to be said, but if we could be certain that it would not start a precedent of sacrificing innocent people, it may be right to sacrifice one person for the good of the whole. Wouldn't we all agree, the utilitarian continues, that it is sometimes preferable to harm an innocent person in order to prevent great evil? The trolley-car case is one example. Here is another.

Virtually all standard moral systems have a rule against torturing innocent people. But suppose a maniac is about to set off a nuclear bomb that will destroy New York City. He is scheduled to detonate the bomb in one hour. His psychiatrist knows the lunatic well and assures us that there is one way to stop him: Torture his 10-year-old daughter and show it on television. Suppose, for the sake of the argument, there is no way to simulate the torture. Would you not consider torturing the child in this situation? (Just in case you don't think New York City is worth saving, imagine that the lunatic has a lethal gas that will spread throughout the globe and wipe out *all* life within a few weeks.)

Is it not right to sacrifice one innocent person to stop a war or save the human race from destruction? We seem to proceed on this assumption in wartime, in every bombing raid, especially in the dropping of the atomic bomb on Hiroshima and Nagasaki. We seem to be following this rule in our decision to drive automobiles and trucks even though we are fairly certain that the practice will result in the death of thousands of innocent people each year.

On the other hand, the sophisticated utilitarian may argue that in the case of the sheriff framing the innocent derelict, justice should not be overridden by current utility concerns, for human rights themselves are outcomes of utility consideration and should not lightly be violated. That is, because we tend to subconsciously favor our own interests and biases, we institute the principle of rights to protect ourselves and others from capricious and biased acts that would in the long run have great disutility. So we must not undermine institutional rights too easily—we should not kill the bachelor in order to provide a heart, two lungs, a liver, and two kidneys to the five patients—at least not at this present time, given people's expectations of what will happen to them when they enter hospitals. But neither should we worship rights! They are to be taken seriously but not given ultimate authority. The utilitarian cannot foreclose the possibility of sacrificing innocent people for the greater good of humanity. If slavery could be humane and resulted in great overall utility, utilitarians would accept it.

We see then that sophisticated, multileveled utilitarianism has responses to all the criticisms leveled at it. For most people most of the time, the ordinary moral principles should be followed, for they actually maximize utility in the long run. But we should not be tied down to this rule: "Morality was made for man, not man for morality." The purpose of morality is to promote flourishing and ameliorate suffering, and where these can be done by sacrificing a rule, we should sacrifice them. Whether this is an adequate defense, I must leave you to decide.

WHAT ARE DEONTOLOGICAL ETHICS?

Act only on that maxim whereby thou canst at the same time will that it would become a universal law.

—IMMANUEL KANT

What makes a right act right? The teleological answer to this question is that it is the good consequences that make it right. Moral rightness and wrongness are determined by such values as happiness or utility. To this extent, the end justifies the means. The deontological answer to this question is quite the opposite. The end *never* justifies the means. Indeed, you must do your duty whatever the consequences, simply because it is your duty. You must do your duty disinterestedly, as though it were the last act of your life, simply because it is your duty. The Danish philosopher Søren Kierkegaard (1813–55) described his childhood experience of sensing his duty to learn his first-grade grammar lesson thusly: "It was as if heaven and earth might collapse if I did not learn my lesson, and on the other hand as if, even if heaven and earth were to collapse, this would not exempt me from doing what was assigned to me."[9]

For deontologists, the consequences of an act do not determine the rightness or wrongness of an act, but certain features in the act itself do. For example, there is something right about truth telling and promise keeping even when acting thusly may bring about some harm, and there is something wrong about lying and promise breaking even when acting thusly may bring about good consequences. Acting unjustly is wrong even if it will maximize expected utility. Referring to our examples in the introduction of this chapter, as a deontologist you would very likely keep your promise and give the $5 million to the Cowboys and share or flip a coin for the food on the raft.

There are several different deontological ethical systems. Religious ethics that are based on God's commands are deontological systems. The Ten Commandments in the Hebrew Bible (Old Testament) are an example. Ethical **intuitionism** is another example. The intuitionist claims that we can consult our hearts and consciences in order to discern correct moral rules. The Oxford University philosopher, W. D. Ross (1877–1971) held that moral principles are self-evident upon reflection to any normal person, but they may not be absolute. He calls them *prima facie* (Latin for "at first glance"), or conditional duties. Ross contrasts *prima facie* duties with *actual* duties. He listed seven prima facie moral principles: (1) promise keeping, (2) fidelity, (3) gratitude for favors, (4) beneficence, (5) justice, (6) self-improvement, and (7) nonmaleficence. If we make a promise, for example, we put ourselves into a situation in which a duty to keep promises is a moral consideration. It has presumptive force; and if there is no conflicting prima facie duty that is relevant, then it becomes an actual duty.

What about situations of conflict? For an absolutist an adequate moral system can never produce moral conflict, nor can a basic moral principle be overridden by another moral principle. But Ross is no absolutist. He allows for overridability of principles. For example, suppose you have promised your friend that you will help her with her ethics homework at 3:00 P.M. As you are going to meet her, you encounter a lost, crying child. There is no one else around to help the little boy, so you help him find his way home. But in doing so, you miss your appointment. Have you done the morally right thing? Yes. You have broken your promise, but with good cause. You had an overriding duty to help the lost child. Whether intuitionism is a satisfactory system is a difficult question. One problem with it is that people with different upbringings and cultures will have different intuitions, so it may not be as reliable a guide as one would like. But intuitionism will prove important for environmental ethics. Systems like Albert Schweitzer's *reverence for life* and Holmes Rolston's *intrinsic value in nature* seem to be intuitionist.

Immanuel Kant's Rationalist Deontological System

The most famous deontological ethical system is Immanuel Kant's *categorical imperative*. Immanuel Kant (1724–1804), the greatest philosopher of the German Enlightenment and one of the most important philosophers of all time, was both an *absolutist* and a *deontological rationalist*. He believed that we could use reason to work out an absolute (that is, nonoverridable) consistent set of moral principles.

Kant was born in Konigsberg, Germany, in 1724 and died there in 1804, never having left the surroundings of the city. His father was a saddle maker. His parents were Pietists in the Lutheran church. The Pietists were a sect in the church, much like present-day Quakers, who emphasized sincerity, deep feeling, and the moral life rather than theological doctrine or orthodox belief. Pietism is a religion of the heart, not the head, of the spirit rather than ritual. However, Kant, as an intellectual, emphasized the head as much as the heart, but it was a head that was concerned about the moral life, especially good will.

The Good Will

The only thing that is intrinsically good, good in itself and without qualification, is the *good will*. All other virtues, both intellectual and moral, can serve the vicious will and thus contribute to evil. None of these are good in themselves but are good only for a further purpose. They can be united in themselves but only for further purposes. They are valuable only if accompanied by good will. Even success and happiness are not good in themselves. Honor can lead to pride. Happiness without good will is not worthwhile. Is honor with deceit worth attaining? No. Nor is utilitarianism plausible, for if we have a quantity of happiness to distribute, is it just to distribute it equally regardless of virtue? Should we not distribute it discriminately, according to moral goodness? Happiness should be distributed in proportion to one's moral worth, and happiness without moral worth is not inherently valuable.

How good is Kant's argument for the good will? Could we imagine a world where nonmoral virtues were always and necessarily put to good use, where it was simply impossible to use a virtue like intelligence for evil? Is happiness any less good simply because it can be distributed incorrectly? Can't the good will itself be put to bad use, as with the misguided do-gooder? As the aphorism goes, "The road to hell is paved with good intentions." Could Adolf Hitler have had good intentions in carrying out his dastardly programs? Can't the good will have bad effects?

We may agree that the good will is a great good, but it is not obvious in Kant's account that it is the *only* inherently good thing, for even as intelligence, courage, and happiness can be put to bad uses, so can the good will; and even as it does not seem to count against the good will that it can be put to a bad use, so it shouldn't count against the other virtues that they can be put to bad uses. The good will may be a necessary element to any morally good action, but it's another question whether it is also a *sufficient* condition to moral goodness.

But perhaps we can reinterpret Kant in such a way as to preserve his central insight. There does seem to be something morally valuable about the good will apart from any consequences. Consider this illustration. Two soldiers volunteer to cross the enemy lines to make contact with their allies on the other side. They both start off and do their best to make their way through the enemy's area. One succeeds, but the other doesn't and is captured. But aren't they both morally praiseworthy? The success of one in no way detracts from the goodness of the other. Judged from a commonsense moral point of view, their actions are equally good; judged from a utilitarian or consequentialist view, the successful act is far more valuable than the unsuccessful one. Here one can distinguish the agent's worth from the value of the consequences and make two separate, nonconflicting judgments.

Duty and the Moral Law

Kant wants to remove moral truth from the zone of contingency and empirical observation and place it securely in the area of *necessary truth*—that is, truth that is absolute and universal. The value of morality is not based on sentiment or intuition (as Ross believes), nor on the fact that it has *instrumental value,* that it often secures nonmoral goods such as happiness, but, rather, morality is valuable in its own right:

> Even if it should happen that, owing to special disfavor of fortune, or the stingy provision of a step-motherly nature, this [good] will should wholly lack power to accomplish its purpose, if with its greatest efforts it should yet achieve nothing, and there should remain only the good will . . . , then, like a jewel, it would still shine by its own light, as a thing which has its whole value in itself. Its usefulness or fruitfulness can neither add to nor take away anything from this value.[10]

All mention of duties (or obligations) can be translated into the language of imperatives or commands. As such, moral duties can be said to have imperative force. Kant distinguishes two kinds of imperatives: hypothetical and categorical. The formula for a hypothetical injunction is

If you want to *A,* then do *B.*

Two examples are "If you want to get a good job, get a good education," and "If you want to stay in good shape, jog in the mountains." The formula for a categorical injunction is simply

Do *B*!

That is, do what reason discloses to be the intrinsically right thing to do—for example, "Tell the truth!" **Hypothetical imperatives,** or means-ends imperatives, are not the kind of imperatives that characterize moral actions. Categorical or unqualified imperatives are the right kind of imperatives, for they show proper recognition of the imperial status of moral obligations. This imperative is an intuitive, immediate, and absolute injunction that all rational agents understand by virtue of their rationality.

Moral duty must be done solely for its own sake ("duty for duty's sake"). Some people conform to the moral law because they deem it in their own enlightened self-interest to be moral, but they are in fact *not* moral because they do not act for the sake of the moral law. For example, merchants may believe that "honesty is the best policy." That is, they may judge that it is conducive to good business to give their customers correct change and good-quality products; but unless they do these acts *because* they are a matter of duty, they are not acting morally, even though their acts are the same ones that they would be if the merchants were acting morally.

The kind of imperative that fits Kant's scheme as a product of reason is one that universalizes principles of conduct. He calls it the **categorical imperative:** "Act only on that maxim whereby thou canst at the same time

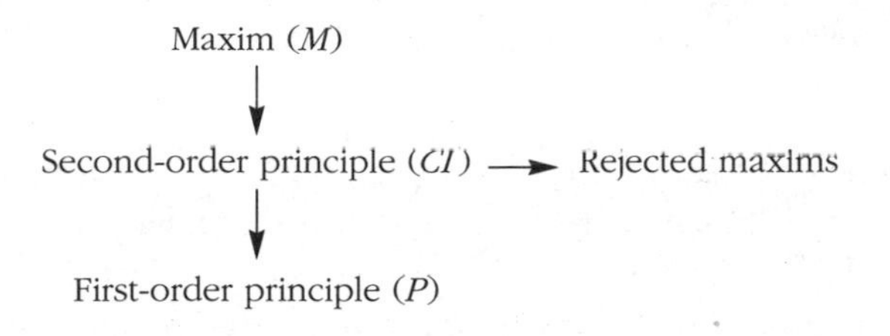

Figure 5-1 Test of universalization.

will that it would become a universal law of nature." This is given as the criterion (or second-order principle) by which to judge all other principles.

By *maxim,* Kant means the general rule in accordance with which the agent intends to act, and by *law,* he means an objective principle, a maxim that passes the test of universalization. The categorical imperative is the way to apply the universalization test. It enables us to stand outside our personal maxims and impartially and impersonally estimate whether they are suitable as principles for all of us to live by. If you could consistently will that everyone would do some type of action, then there is an application of the categorical imperative enjoining that type of action. If you cannot consistently will that everyone would do some type of action, then that type of action is morally wrong. The maxim must be rejected as self-defeated. The formula is shown in Figure 5-1.

To take one of Kant's examples, suppose I need some money and am considering whether it would be moral to borrow the money from you and promise to repay it without ever intending to do so.

> *M: Whenever I need money, I should make a lying promise while borrowing the money.*

Can I universalize the maxim of my act?

> *P: Whenever anyone needs money, that person should make a lying promise while borrowing the money.*

But something has gone wrong, for if I universalize this principle of making promises without intending to keep them, I would be involved in a contradiction. The resulting state of affairs would be self-defeating, for no sane person would take promises as promises unless there was the expectation of fulfillment. So the maxim of the lying promise fails the **universalizability** criterion. Hence, it is immoral. Now I universalize the opposite:

> M_1: *Whenever I need money, I should make a sincere promise while borrowing it.*

Can I universalize this maxim?

> P_1: *Whenever anyone needs money, that person should make a sincere promise while borrowing it.*

Yes, I can universalize M_1, for there is nothing self-defeating or contradictory in this. So it follows that making sincere promises is moral. We can make the maxim of promise keeping into a universal law.

Some of Kant's illustrations do not fare as well as the duty to promise keeping. For instance, he argues that the categorical imperative would prohibit suicide, for the principle

> *P: Whenever it looks as if one will experience more pain than pleasure, one ought to kill oneself*

is, according to Kant, a self-contradiction in that it would go against the very *principle of survival* on which it is based. But whatever the merit of the form of this argument, we could modify the principle to read

> P_1: *Whenever the pain or suffering of existence erodes the quality of life in such a way as to make nonexistence a preference to suffering existence, one is permitted to commit suicide.*

Why couldn't this (or something close to it) be universalized? It would not oppose the general principle of survival itself but cover rare instances where no hope is in sight for terminally ill patients and victims of torture or deep depression. It would not cover the normal kinds of suffering and depression that most of us experience in the normal course of life. Kant seems unduly absolutist in his prohibition of suicide.

Kant thought that he could generate an entire moral law from his categorical imperative. It seems to work with such principles as promise keeping, as we noted, and truth telling (try universalizing the principle of telling falsehoods—what do you come up with?) and a few other maxims, but it doesn't seem to give us all that Kant wanted. Kant's categorical imperative has been characterized as both *too wide* and *too unqualified,* leading to horrendous possibilities.

The charge that it is too wide is based on the perception that it seems to justify some actions that we would think to be trivial and even immoral. Consider, for example, principle *P:*

> *P: Everyone should always tie one's right shoe before one's left shoe.*

Can we universalize *P* without contradiction? Why not? Just as we universalize that people should drive cars on the right side of the street rather than the left, we could make it a law that everyone would tie the right shoe before the left shoe. It seems obvious that there would be no point to such a law; it would be trivial. It is justified, however, by the categorical imperative.

It may be objected that all this counterexample shows is that it may be permissible to live by the principle of tying the right shoe before the left, for we could also universalize the opposite maxim (tying the left before the right) without contradiction. That seems correct.

A more serious objection is the charge that the categorical imperative seems to justify acts that we judge to be horrendously immoral. Consider P_1:

> P_1: *Always kill blue-eyed children.*

Is there anything contradictory in this injunction? Could we make it into a universal law? Why not? Blue-eyed children might not like it (and might even be required to cooperate or commit suicide), but there is no logical contradiction involved in such a principle. Had I been a blue-eyed child when this command was in effect, I would not be around to write this book, but the world would have survived my loss without too much inconvenience.

Of course, it would be possible to universalize the opposite: No one should kill innocent people, but that shows only that either type of action is permissible.

It may be objected that Kant presupposed that only rational acts could be universalized, but this won't work, for the categorical imperative is supposed to be the criterion for rational action.

It may be that when we come to Kant's second formulation of the categorical imperative, he will have more ammunition with which to defeat P_1.

Finally, Kant thought that the categorical imperative yielded unqualified absolutes. The rules that the categorical imperative generates are universal and exceptionless. He illustrates this point with regard to truth telling. Suppose that an innocent man comes to your door, begging for asylum, for a group of gangsters is hunting him down in order to kill him. You take the man in and hide him in your third-floor attic. Moments later the gangsters arrive and inquire after the innocent man. "Is he in your house?" they inquire. What should you do? Kant's advice is to tell them the truth: "Yes, he's in my house."[11]

What is Kant's reasoning here? It is simply that the moral law is sacrosanct and exceptionless. It is your duty to obey its commands, not to reason about the likely consequences. You have done your duty: hidden an innocent man and told the truth when asked a straightforward question. You are absolved of any responsibility for the harm that comes to the innocent man. It's not your fault that there are gangsters in the world.

To many of us, this kind of absolutism seems counterintuitive. There are two ways in which we might alter Kant here. First, simply write in qualifications to the universal principles, changing the sweeping generalization "Never lie" to the more modest "Never lie except to save an innocent person's life." The trouble with this way of solving the problem is that there seem to be no limits on the qualifications that would need to be attached to the original generalization: for example, "Never lie *except* to save an innocent person's life (except when trying to save the innocent person's life will undermine the entire social fabric)" or "Never lie except when lying will spare people great anguish" (for example, telling a cancer patient the truth about her condition). And so on. The process seems infinite and time-consuming and thus impractical.

A second way of qualifying the counterintuitive results of the Kantian program is to follow W. D. Ross and distinguish between *actual* and *prima facie* duties. As we noted earlier, a prima facie duty that wins out in the comparison is called the *actual duty* or the *all-things-considered duty.* We can apply this distinction to Kant's example of the innocent man. First, we have the principle

L: Never lie.

Next, we ask whether any other principle is relevant in this situation, and we discover that the principle

P: Always protect innocent life.

also applies. But we cannot obey both *L* and *P* (we assume for the moment that silence will be a giveaway). We have two general principles, but neither is to be seen as absolute or nonoverridable, but rather as prima facie. We have to decide which of the two overrides the other, which has greater moral force. This is left up to our considered judgment (or the considered judgment of the reflective moral community). Presumably, we will opt for *P* over *L,* so that lying to the gangsters becomes our actual duty.

Will this maneuver save the Kantian system? Well, it changes it in a way that Kant might not have liked, but it seems to make sense. It transforms Kant's absolutism into an objectivist system. But now we need to have a separate criterion to adjudicate the conflict between two objective principles.

I conclude then that the categorical imperative is an important criterion for evaluating moral principles, but it needs supplementation. In itself it is purely formal and leaves out any understanding about the content or material aspect of morality. The categorical imperative with its universalizability test constitutes a necessary condition for being a valid moral principle, but it does not provide us with a sufficiency criterion. That is, if any principle is to count as rational or moral, it must be universalizable. It must apply to everyone and every case that is relevantly similar. If I believe that it's wrong for others to cheat on exams, then unless I can find a reason to believe that I am relevantly different from others, it is also wrong for me to cheat on exams. If premarital heterosexual coitus is prohibited for women, then it must also be prohibited for men. (Otherwise, with whom would the unmarried men have sex? other men's wives?) But this formal consistency does not tell us whether cheating itself is right or wrong or whether premarital sex is right or wrong. Those questions have to do with the substantive content of morality, and other considerations must help us decide the answers.

Kant's Second Formulation of the Categorical Imperative

Kant offered a second formulation of the categorical imperative, which has been referred to as the *principle of ends:* "So act as to treat humanity, whether in your own person or in that of any other, in every case as an end and never as merely a means only." Each person as a rational being has dignity and profound worth entailing that he or she must never be exploited or manipulated or used merely as a means to our idea of what is for the general good or to any other end.

What is Kant's argument for viewing rational beings as having ultimate value? It goes like this: In valuing anything, I endow it with value. It has no value apart from someone's valuing it. As a valued object, it has *conditional* worth, derived from my valuation. On the other hand, the person who values the object is the ultimate source of the object's value and, as such, belongs to a different sphere of beings. We, as valuers, must conceive of ourselves as having *unconditional* worth. We cannot think of our personhood as a mere thing, for then we would have to judge it to be without any value except that given to it by the estimation of other people. But then that person would be the source of value, and there is no reason to suppose that one person should have unconditional worth and not another who is relevantly similar. Therefore, we are not mere objects. We have unconditional worth and so must treat all such value givers as valuable in themselves, as ends, not merely means. I leave it to you to evaluate the validity of this argument, but most of us do hold that there is something exceedingly valuable about human life.

Kant thought that this formulation, the principle of ends, was substantively identical with his first formulation of the categorical imperative, but most scholars disagree with him. It seems better to treat this principle as a supplement to the first, adding content to the purely formal categorical imperative. In this way, Kant would limit the kinds of maxims that could be universalized. Egoism and the principle P_1, enjoining killing blue-eyed children, would be ruled out at the very outset since they involve a violation of the dignity of rational persons. The process would be as follows:

1. Maxim (*M*) formulated.
2. Ends test: does the maxim involve violating the dignity of rational beings?

3. Categorical imperative: can the maxim be universalized?
4. Successful moral principles survive both tests.

Does the principle of treating persons as ends in themselves fare better than the original version of the categorical imperative? Three problems soon emerge. The first problem has to do with Kant's setting such a high value on rationality. Why does reason and only reason have intrinsic worth? Who gives this value to rational beings, and how do we know that they have this value? What if we believe that reason has only instrumental value?

Kant's notion of the high inherent value of reason will be more plausible to those who believe that humans are made in the image of God and interpret that belief, as the mainstream of the Judeo-Christian tradition has—as entailing that our rational capabilities are the essence of being created in God's image. We have value because God created us with worth—that is, with reason. Kant doesn't use such an argument. Instead, he thinks that we must necessarily value rational nature, since we, as rational beings, must value ourselves and so, by the principle of consistency, anyone rational like us.

Kant seems to many to be correct in valuing rationality (the essence of our rational nature). It does enable us to engage in deliberate and moral reasoning and lifts us above lower animals. Kant holds that our rational nature endows us with free will, absent in all other creatures. Whether he is correct about this is a difficult matter, having to do with the question of whether the doctrine of freedom of the will is true. But even if reason gives us free will, Kant's theory is even more controversial in neglecting other values or possible criteria for being morally considerable. For example, he believed that we have no obligations to animals because they are not rational. Many of us believe (with Jeremy Bentham and Peter Singer) that the fact that animals can suffer should constrain us in our behavior toward them. We ought not cause unnecessary harm. Perhaps Kantians can supplement their system to accommodate this objection. Perhaps a principle, "Do no unnecessary harm" will enable them to reach a minimal animal-rights ethic. We will treat this subject in Chapter 7.

This brings us to our second problem with Kant's formulation. If we agree that reason (or rational nature) is an intrinsic value, then does it not follow that those who have more of this quality should be respected and honored more than those who have less? Doesn't *more* mean *better* here?

Following Kant's logic, we should treat people in exact proportion to their ability to reason. Thus, geniuses and intellectuals should be given privileged status in society (as Plato and Aristotle might argue). Kant could deny the second premise and argue that rationality is a threshold quality, that anyone having a sufficient quantity of it grants one equal worth. The question is whether Kant or Kantians have good (nonreligious) reasons to accept the egalitarian premise that all those who have rational nature have equal worth. One should note that Kant uses the dignity principle to establish conditions for desert and punishment. He tells us that we can forfeit our worth through immoral acts and that we can deserve happiness or punishment based on our moral actions.

There is a third problem with Kant's view of the dignity of rational beings. Even if we should respect them and treat them as ends, this principle does not tell us very much. It may tell us not to enslave them or act cruelly toward them without a good reason, but it doesn't tell us what to do in conflict situations. For example, what does it tell us to do about a terminally ill patient who wants us to help her die? What does it tell us to do in a war when we are about to aim our gun at an enemy soldier? Aren't we treating the soldier merely as a means?

Furthermore, what does it mean to treat this rational being as an end? What does it tell us to do with regard to the innocent victim and the gangsters who have just asked us about the whereabouts of the victim? What does it tell us about whether we should steal from the pharmacy in order to procure medicine that we can't afford to buy in order to bring healing to a loved one?

These are difficult questions. Depending on how we answer them, they will generate different moral strategies.

APPLICATION OF UTILITARIANISM AND DEONTOLOGY TO ENVIRONMENTAL ETHICS

In the last chapter we noted that contractual ethics was primarily anthropocentric, but we attempted to include environmental concerns within its scope via the metaphor of the tragedy of the commons. What kind of environmental ethics do utilitarianism and deontology generate?

Utilitarianism and deontological ethics can each be divided into two different sub-theories with regard to environmental ethics. Some utilitarians are anthropocentric or humanists, confining their maximizing principle to human beings alone. Most economic applications of utilitarianism are examples of this. Other utilitarians, such as Bentham and Peter Singer, include animals within the scope of moral consideration. Each sentient being deserves *equal consideration* as having needs, interests, or preferences of its own. That is, we are not to discriminate against sentient beings on the basis of rationality or species membership, but we are to consider their needs and interests as equally valid. We will examine Singer's view in Chapter 7. But utilitarianism rejects biocentric ethics, which posits intrinsic value in plants and nonsentient animals. Since sentience is a condition for moral considerability, these nonsentient entities fail the basic test and so fall outside the scope of moral consideration. They are *resources* for humans and sentient animals.

Deontological ethics also may be divided into two subtheories with regard to environmental ethics. Traditional deontological ethics, following Kant, excludes all but rational beings from the scope of morality. Kant generally has human beings in mind; however, the focus is not on the species *Homo sapiens*, but on the quality of rationality which characterizes us. If some animals, say chimpanzees and dolphins, turn out to be rational, they too have moral rights. But some humans, those who are severely retarded and those not yet rational (such as fetuses), will not count as moral agents and will be undeserving of full rights. Most animals will fail to qualify as rational beings and so will not be morally considerable. If we apply Kantianism to environmental ethics, we find ourselves with a *rational-agent-centered* morality. Nevertheless, the categorical imperative has a broad environmental scope—we are to will that the maxims of our actions be made universal laws "of nature."

A second type of deontological ethics rejects rationality as the sole criterion for moral worth and locates it instead in the quality of life itself. Every living entity has a *right to life*, simply in virtue of having a good—the developing of that life form into its biological destiny (for example, an acorn's destiny or telos is to become an oak tree). Albert Schweitzer's *reverence for life*, Holmes Rolston's notion of *intrinsic value in nature*, and Paul Taylor's *biocentric egalitarianism* are examples of this type of deontological ethics and will be examined in Chapters 8 and 10. A middle position that holds that only higher animals, mammals, have moral rights is espoused by Tom Regan (see Chapter 7).

Much of this book will consist of a debate among anthropocentrists, sentience-centrists, and biocentrists. Anthropocentrists may be either contractualists, utilitarians, or deontologists and generally hold to a *conservationist* view of the environment: We ought to conserve natural resources for human good. Sentience-centrists, typically utilitarians like Peter Singer, extend conservationism to animals. But biocentrists are all deontologists, who adopt a *preservationist* view of the environment: We ought to preserve natural resources for their own sake. Nature has intrinsic value that ought to be respected in itself. Preservationists divide into two camps: individualist and holist. Biocentric individualists, such as Albert Schweitzer (reverence for life) and Paul Taylor (biocentric egalitarianism) hold that it is the individual life that is important—equal to every other individual life form. Biocentric holism, sometimes referred to as *ecocentric holism,* which includes the *Gaia hypothesis*, holds that it is ecosystems and species within those systems that have value and not just individuals themselves. We will examine this view in Chapter 9.

There is one further problem in environmental ethics that generates different strategies, depending on which ethical theory one embraces. To that we now turn.

OBLIGATIONS TO FUTURE GENERATIONS

Environmental ethics centers on our conserving resources and refraining from polluting. Environmentalists hold that one principal reason we ought not to degrade the Earth is because we have duties to future generations. We should leave the Earth in good shape for them. But is this really so? Do we have any obligations to future generations? Do future people have claims against us? Why should I care about posterity? What has posterity ever done for me?

The economist Robert Heilbroner points out that the question "Why should I care for posterity?" seems to have no rational answer. We seem to have no rational grounds for caring about future people, since they don't exist. Who is there to care about? What are their names, desires, weights, shapes, beliefs, values, and identities? It's hard to care about someone who doesn't exist—not even as a fetus.

So how can we be said to have moral obligations to future people? How can they have moral rights? Most philosophers believe that sentience is a minimal condition for rights, but future people aren't even sentient. In fact, unlike fetuses, they don't even have potential sentience, since there is nothing there for potentiality to predicate. Future people aren't people; they simply are not. So "they" don't have any rights, not even the right to exist, let alone the right to a clean environment. I violate no one's right if I choose not to have children. No one can reprimandingly cry to me from limbo, the storage house of unborn souls, "Have sex, you selfish egoist, so that I can be liberated from this shadowy existence." Heilbroner, himself committed to the prosperity of posterity, argues that there is no good argument for this thesis—it's just an existential choice. The voice of rationality, he says, is that of a distinguished political economist at London University:

> Suppose that, as a result of using up all the world's resources, human life did come to an end. So what? What is so desirable about an indefinite continuation of the human species, religious conviction apart? It may well be that nearly everybody who is already here on earth would be reluctant to die, and that everybody has an instinctive fear of death. But one must not confuse this with the notion that, in any meaningful sense, generations who are yet unborn can be said to be better off if they are born than if they are not.[12]

Another economist from Massachusetts Institute of Technology puts the matter this way:

> Geological time [has been] made comprehensible to our finite human minds by the statement that the 4.5 billion years of the earth's history [are] equivalent to once around the world in a Supersonic jet. . . . Man got on eight miles before the end, and industrial man got on six feet before the end. . . . Today we are having a debate about the extent to which man ought to maximize the length of time that he is on the airplane. According to what the scientists think, the sun is gradually expanding, and 12 billion years from now the earth will be swallowed up by the sun. This means that our airplane has time to go round three more times. Do we want man to be on it for all three times around the world? Are we interested in man being on for another eight miles? Are we interested in man being on for another six feet? Or are we only interested in man for a fraction of a millimeter—our lifetimes?
>
> That led me to think: Do I care what happens a thousand years from now? . . . Do I care when man gets off the airplane? I think I basically [have come] to the conclusion that I don't care whether man is on the airplane for another eight feet, or if man is on the airplane another three times around the world.[13]

Let us state the question even more starkly. Suppose that in 40 or 50 or 100 years from now people on Earth collectively do a cost-benefit assessment and unanimously decide that life is not worth its inherent suffering and boredom. Perhaps people become tired of their technological toys and fail to find anything worth living for, so they decide to commit collective suicide. Would this be immoral? If you don't like the idea of suicide, suppose that they all take a drug that will bring ecstatic happiness but renders them permanently sterile—and they knowingly take it. The result in either case would be the end of humanity. Again, I ask, have these people done anything immoral? Have they violated anyone's rights?

The philosopher Joel Feinberg, in a pioneer article, responded to this question this way, "My inclination . . . is to conclude that the suicide of our species would be deplorable, lamentable, and a deeply moving tragedy, but that it would violate no one's rights. Indeed if, contrary to fact, all human beings could ever agree to such a thing, that very agreement would be a symptom of our species' biological unsuitability for survival anyway."[14]

Notice that this problem is not like normal cases of suicide. We can sometimes argue that suicide is immoral because the people contemplating self-slaughter have responsibilities to others who will be harmed by the suicide. For example, the parent who decides to end his or her life may have an overriding obligation to care for the children who will be orphaned or who will suffer the shock of dealing with the suicide of a parent. But in this case there are no children who will be left behind. As soon as they come of age, they too agree to die or not to procreate. And there's no one else whose identity needs to be taken into account—or so it seems. Future people can't be consulted, of course, since they don't exist and therefore can't be identified.

So the first question to face is, Why would it be wrong, if it is, voluntarily to end the human race? Let us see how the four main types of ethical theories (the contractualist, the religious, the deontological, and the consequentialist-utilitarian) would respond to this question.

1. The contractualist would agree with Feinberg that the act of ending the human race is regrettable, but argue that it is morally permissible. Morality arises only in contractual relations, but since we have no one with whom to make the relevant contract, no contract is possible, and so no one's contract has been violated. We have no agreement with posterity to remain alive, let alone provide for its existence, since posterity doesn't exist. No promise to future generations has been broken. No responsibility to them has been violated, so no immoral act has been committed.

2. What about the religious ethicist? Note that the University of London economist qualifies his statement that there is nothing inherently bad about humanity's coming to an end with the phrase, "Religious convictions apart." That does make a difference. Religions and specifically the Judeo-Christian tradition specify a duty to be "fruitful and multiply and replenish the earth" (Gen. 1:29). God's command is sufficient to establish a duty to continue the human race until He decrees the coming of the kingdom of heaven. Humanity must not play God in usurping this divine prerogative. So the religious person has a duty issuing from God to perpetuate the species.

3. What would a deontologist say about this problem? There are two main varieties of secular deontological ethics: intuitionists and Kantian rationalists. The intuitionists could go either way; one intuitionist could have an intuition that bringing about the end of humanity is absolutely wrong, whereas another could have just the opposite intuition—that it is absolutely permissible. Of course, intuitionists would not put it so starkly. They are likely to say that on due reflection with full understanding . . . and then add their opposing judgments. But this abridgment of the reasoning process is a telling criticism against intuitionism. Moral philosophy should be able to do better than stop so quickly at the deliverances of intuition.

The Kantian, like Kant himself, might argue that suicide, whether it be collective or individual, is morally wrong because it contradicts the life principle; but most present-day Kantians would qualify the principle regarding suicide, arguing that so long as it is an autonomous act and one that can be universalized, it is morally permissible. So a Kantian might support the principle, "if any person P is in a situation where life holds no prospects for a high quality and P has no overriding obligations that would require that P continue to live, P may commit suicide." As far as I know, Kant never discusses future people, but seems to require that any object of my moral regard be an existing rational agent. It is well known that Kantianism can be appealed to in support of abortion, since fetuses are not rational agents. As we have no obligations to potential people in the form of fetuses, we have no obligations to those who aren't even conceived, who lack even biological identity. So a Kantian would likely allow for the end of humanity, as long as no moral principles were violated.

4. Consequentialists and utilitarians, who may have grave problems regarding justice and respect for individuals, provide the strongest philosophical justification for opposing the end of humanity. Utilitarians have one overriding duty—to maximize happiness—and consequentialists see the promotion of happiness as a prime duty (though it need not be maximized). As long as the quality of life of future people promises to be positive, we have an obligation to continue human existence, to produce human beings, and to take whatever actions are necessary to ensure that their quality of life is not only positive, but also high. It does not matter that we cannot identify these future people. We may look upon them as mere abstract placeholders for utility and aim at maximizing utility (or, as in the case of nonutilitarian consequentialists, simply providing a high level of utility).

If we want a theory that supports our conviction that we ought not allow the end of humanity, then we should choose either a religious perspective or consequentialism. If we think there are independent reasons for rejecting a religious answer, then some kind of consequentialism seems the best alternative. The core morality theory I outlined in Chapters 2 and 3, especially with principle 9, "Do good wherever feasible," can cover the problem of future generations. We should create conditions for well-being wherever possible, taking into account other rights and duties. Of course, you can reject this kind of reasoning altogether and simply assert that it is intuitively self-evident that we ought to continue the human race; but some of us would rather have a supporting argument, which religion or consequentialism provides.

These considerations of our obligations toward continuing the human race can help us address the question we began with: Why do we have obligations to posterity even though posterity has never done anything for us? We can see that contractualism, in the guise of Heilbroner and the economists, is unable to give us an answer to this question. The contractualist environmentalist Garrett Hardin tells the story of how he spent $1 to plant a redwood seedling that would take two thousand years to reach its full value of $14,000. He confesses that as an "economic man" he was being stupid in planting it, but he did so anyway. "It is most unlikely that any of my direct descendants will get [the value of the tree]. The most I can hope for is that an anonymous posterity will benefit by my act. . . . Why bother?" His answer is an admission of the failure of contractualist and economic reasoning—or of his own rationality. "I am beginning to suspect that rationality—as we now conceive it—may be insufficient to secure the end we desire, namely, taking care of the interests of posterity."[15] Of course, this recognition by itself is not sufficient to refute contractualism altogether, but it is a weighty objection, for we generally think we have duties to future people. Religion gives us two reasons to care for future generations: The first is that God commands us to continue the race. But religious believers have a second special reason to care for future people: God knows who will be born and loves these people as if they already existed. For God the whole temporal span of the world's existence is good. In serving God, as good stewards, we have a duty to Him to be good to the Earth, which includes leaving it healthy for future people who will be born and who are already loved by God. But all this supposes that God exists, which is a controversial question on its own.

Intuitionists, as is their wont, find it self-evident either that we do or that we don't have obligations to future generations, so they are not much help. Kantians, with their strong thesis of particular rational agency, have a particularly difficult time generating principles that would require duties to future agents. Kantians require identifiable people as the objects of our duties. Perhaps we could mount an argument on the basis of Kant's notion of imperfect duties to benevolence. Kant holds that we need not help any particular person or charity but that we should give to some charity or other. So it might seem that we have an imperfect duty to promote the well-being of people who will exist in the future—even if we don't know who they are. As long as we are reasonably sure that they will exist, we have a duty to see to it that their lives are reasonably good. Nevertheless, this principle seems weak insofar as it is only an imperfect duty, which is always overridden by perfect duties of justice to present people.

Some Kantians argue that we do have obligations to the future world in the persons of our children. We have an obligation to leave the environment in good shape for our children; and they, in turn, will have such an obligation for their children, and so on, so that the question of posterity is taken care of. The duty carries over to future generations. But this seems to commit a *fallacy of transitivity*.

If A has a duty X to B, and B has a duty X to C, A has a duty X to C.

This formula is invalid. I may have a duty to keep my promise to you, and you have a duty to keep your promise to your neighbor, but I don't have a duty to keep my promise to you to your neighbor. Similarly, our generation may have a duty to provide the next generation (our children) with an adequate living, but we don't have an obligation to provide the following generation (our grandchildren) with one. In fact, given scarce resources, it may be that simply sharing our resources with the next generation will use up a considerable amount, so that our children's generation will be unable to pass down a sufficient amount to their children.

John Rawls, writing within the Kantian framework, argues that "Each generation must not only preserve the gains of culture and civilization, and maintain intact those just institutions that have been established, but it must also put aside in each period of time a suitable amount of real capital accumulation."[16] But the criticism just described seems to apply to Rawls. Simply focusing on the next generation may not be adequate to deal with long-range problems. If we think that we have duties to enable people several generations hence to live well, we may have to make radical changes in our lifestyles, conserve our resources, and reduce our pollution in ways that go far beyond what would be necessary had we but the next generation in mind.[17]

As with the problem of continuing humanity, consequentialism has a compelling answer to the problem of why we should care about posterity. Morality is about promoting happiness or flourishing both in humans and other sentient beings, so it doesn't matter whether the beings who will exist are identifiable. Oxford University philosopher Derek Parfit has pointed out that in the standard, or deontological, view we hold that a crime or bad policy has to have a definite victim before we can call it a crime or bad policy or immoral. But the policy choices we make regarding energy, resources, and pollution will actually determine who is born, so that, in this view, as long as future people live worthwhile lives, they cannot blame us for our bad and wasteful policies. But this seems a mistaken view. Parfit offers the consequentialist principle (A): "It is bad if those who live are worse off than those who might have lived."

Parfit illustrates his principle this way: Suppose our generation has the choice between two energy policies—*safe energy policy* and *risky energy policy*. The risky policy promises to be safe for us but is likely to create serious problems for a future generation, say two hundred years from now. The safe policy won't be as beneficial to us, but promises to be stable and safe for posterity—those living two hundred years from now and beyond. We must choose, and we are responsible for the choice we make. If we choose the risky policy, we impose harms on our descendants, even if they don't now exist. In a sense we are responsible for the people who will live, since our policy decisions will generate different causal chains, resulting in different people being born. But more importantly, we are responsible for their quality of life, since we could have caused human lives to have been better off than they are.[18]

What are our obligations to future people? If consequentialists or religious theorists are correct, we have an obligation to leave posterity as good a world as we can. This would mean radically simplifying our lifestyles so that we use no more resources than are necessary, keeping as much topsoil intact as possible, protecting endangered species, reducing our carbon dioxide emissions, preserving the wilderness, and minimizing our overall deleterious impact on the environment in general—and, at the same time, using technology wisely. Exactly what all of this amounts to is the subject of the rest of this book. Perhaps it requires us to take a radically different view of nature and our relationship to nature. To that subject we turn in our next chapter.

Review Questions

1. Consider Jeremy Bentham's and John Stuart Mill's versions of utilitarianism. Evaluate the strengths and weaknesses of each. Then evaluate the main criticism of utilitarianism.
2. One criticism of utilitarianism is that it fails to protect people's rights. Consider five excitable sadists getting a total of 100 hedons while torturing an innocent victim who is suffering 10 dolors (units of pain). On a utilitarian calculus, this would result in a total of 90 hedons. If no other act would result in as many or more hedons, the utilitarian would have to endorse this act and argue that the victim has a duty to submit to the torture and that the sadists have a duty to torture the victim. What do you think of this sort of reasoning? How much does it count against utilitarianism?
3. Consider the case, discussed in the text, of the doctor needing organs for five needy patients, all of whom are in danger of dying unless they get suitable organs within the day. One needs a heart transplant, two need kidneys, one needs lungs, and another needs a liver. A homeless man who has no family walks into the hospital for minor emergency care (he cut his finger). By killing him and using his organs for the five, you could save five persons, restoring them to health. If you don't kill the man, are you negatively responsible for the death of the five patients? Explain your answer.
4. Rawls's false-analogy argument: John Rawls argues that utilitarianism errs in applying to society the principle of personal choice. That is, we all would agree that an individual has a right to forgo a present pleasure for a future good. I have a right to go without a new suit so that I can save the money for my college education or so that I can give it to my favorite charity. Utilitarianism, however, prescribes that we demand that you forgo a new suit for someone else's college education or for the overall good of the community—whether or not you like it or agree to it. That is, it takes the notion of an individual maximizing his or her utility and extends it to cover society in a way that violates the individual's rights. Is this a fair criticism?
5. Do you think that the Kantian argument that combines the categorical imperative with the principle of ends is successful? Is the notion of treating persons as ends clear enough to be a significant guiding action? Does it cover some intelligent animals but not severely retarded people? What about fetuses and infants? Are they included in it? Why, or why not?

6. Note the comments of the anti-Kantian Richard Taylor:

   If I were ever to find, as I luckily never have, a man who assured me that he really believed Kant's metaphysical morals, and that he modeled his own conduct and his relations with others after those principles, then my incredulity and distrust of him as a human being could not be greater than if he told me he regularly drowned children just to see them squirm.[19]

   He and others have criticized Kant for being too rigid. Many people use the idea of moral duty to keep themselves and others from enjoying life and showing mercy. Do you think that there is a basis for this criticism?

7. Kant has been criticized for stifling spontaneous moral feelings in favor of the deliberate will, so that the person who successfully exercises the will in overcoming a temptation is superior to the person who isn't tempted at all but acts rightly spontaneously. For example, the person who just barely resists the temptation to shoplift through a strenuous act of the will would be, on this criterion, morally superior to the person who isn't tempted to shoplift at all. Based on your analysis of Kant, do you think that this is a fair interpretation of Kant; and, if so, does it undermine his ethics?

8. Here is a question similar to the preceding. Kant holds that to be moral agents we must act with motive, act with the intention of doing the morally right thing simply because it is right and not because we are altruistic or benevolent or have good moral habits. He has been criticized for emphasizing the will too much and for rejecting the place of character and feelings in moral actions. Are these criticisms valid?

9. Many people besides Taylor have a negative reaction to Kant's moral theory. Evaluate this quotation from Oliver Wendell Holmes, Jr.:

   From this it is easy to proceed to the Kantian injunction to regard every human being as an end in himself and not as a means. I confess that I rebel at once. If we want conscripts, we march them up to the front with bayonets in their rear to die for a cause in which perhaps they do not believe. The enemy we treat not even as a means but as an obstacle to be abolished, if so it may be. I feel no pangs of conscience over either step, and naturally am slow to accept a theory that seems to be contradicted by practices that I approve.[20]

10. Compare the ethical theories discussed in the last two chapters on the problem of future generations. Which does the better job in setting forth a comprehensive and cogent answer to the question of which kinds of obligations, if any, do we have to future generations?

Suggested Readings

On Ethical Theory

Becker, Lawrence, and Charlotte Becker, eds. *A History of Western Ethics*. New York: Garland, 1992.

Brandt, Richard. *Ethical Theory*. Englewood Cliffs, N.J.: Prentice Hall, 1959.

Harris, C. E. *Applying Moral Theories*. 2nd ed. Belmont, Calif.: Wadsworth, 1992.

Pojman, Louis, ed. *Ethical Theory: Classical and Contemporary Readings.* 3rd ed. Belmont, Calif.: Wadsworth, 1997.

Sterba, James. *How to Make People Just.* Totowa, N.J.: Rowman & Littlefield, 1988.

Sterba, James, ed. *Contemporary Ethics: Selected Readings.* Englewood Cliffs, N.J.: Prentice Hall, 1989.

On Utilitarianism

Bentham, Jeremy. *Introduction to the Principles of Morals and Legislation.* Edited by W. Harrison. Oxford: Oxford University Press, 1948. Reprinted in *Ethical Theory: Classical and Contemporary Readings,* ed. L. Pojman. Belmont, Calif.: Wadsworth, 1998.

Brandt, Richard. "In Search of a Credible Form of Rule-Utilitarianism." In *Morality and the Language of Conduct.* Edited by H. N. Castaneda and George Nakhnikian. Detroit: Wayne State University Press, 1953.

Brock, Dan. "Recent Work in Utilitarianism." *American Philosophical Quarterly* 10 (1973).

Hare, R. M. *Moral Thinking: Its Levels, Method, and Point.* Oxford: Oxford University Press, 1981.

Mill, John Stuart. *Utilitarianism.* Indianapolis: Bobbs-Merrill, 1957.

Parfit, Derek. *Reasons and Persons.* Oxford: Oxford University Press, 1984.

Pojman, Louis, ed. *Ethical Theory: Classical and Contemporary Readings.* Belmont, Calif.: Wadsworth, 1998.

Quinton, Anthony. *Utilitarian Ethics.* London: Macmillan, 1973.

Smart, J. J. C., and Bernard Williams. *Utilitarianism For and Against.* Cambridge: Cambridge University Press, 1973.

Taylor, Paul. *Principles of Ethics.* Encino, Calif.: Dickenson, 1975.

On Kant and Deontological Ethics

Acton, Harry. *Kant's Moral Philosophy.* London: Macmillan, 1970.

Donagan, Alan. *The Theory of Morality.* Chicago: University of Chicago Press, 1977.

Feldman, Fred. *Introductory Ethics.* Englewood Cliffs, N.J.: Prentice Hall, 1978, chaps. 7 and 8.

Gewirth, Alan. *Reason and Morality.* Chicago: University of Chicago Press, 1978.

Harris, C. E. *Applying Moral Theories.* Belmont, Calif.: Wadsworth, 1992, chaps. 6 and 7.

Hill, Thomas E. *Dignity and Practical Reason in Kant's Moral Theory.* Ithaca, N.Y.: Cornell University Press, 1992.

Kant, Immanuel. *Foundations of the Metaphysics of Morals.* Translated by Lewis White Beck. Indianapolis: Bobbs-Merrill, 1959.

Wolff, Robert P. *The Autonomy of Reason: A Commentary on Kant's "Groundwork of the Metaphysics of Morals."* New York: Harper & Row, 1973.

On Duties to Future Generations

Partridge, Ernest, ed. *Responsibilities to Future Generations.* Buffalo, N.Y.: Prometheus Books, 1981.

Pojman, Louis, ed. *Environmental Ethics.* 2nd ed. Belmont, Calif.: Wadsworth, 1998.

Reichenbach, Bruce. "On Obligations to Future Generations." *Public Affairs Quarterly* 6:2 (April 1992).

Sikora, R. I., and Brian Barry, eds. *Obligations to Future Generations.* Philadelphia: Temple University Press, 1978.

Notes

1. Jeremy Bentham, *An Introduction to the Principles of Morals and Legislation* (1789), chap. 1.
2. Ibid., chap. 17.

3. John Stuart Mill, *Utilitarianism* (1863), chap. 2.
4. See Anthony Quinton, *Utilitarian Ethics* (London: Macmillan, 1973), p. 49f., for a good discussion of this and other similar points.
5. Arthur Koestler, *Darkness at Noon* (New York: Macmillan, 1941), p. 80.
6. This example and the trolley-car example are found in Judith Jarvis Thomson's "The Trolley Problem," in *Rights, Restitution, and Risk* (Cambridge: Harvard University Press, 1986), pp. 94–116.
7. Dietrich von Nieheim, Bishop of Verden, *De Schismate Librii,* iii, A.D. 1411, quoted in Koestler, *Darkness at Noon,* p. 76.
8. Thomson, "The Trolley Problem."
9. Søren Kierkegaard, *Either/Or,* vol. II, trans. Walter Lowrie (New York: Anchor Books, 1959), p. 271.
10. Immanuel Kant, *Fundamental Principles of the Metaphysics of Morals,* trans. T. K. Abbott (1873), sec. 1.
11. Immanuel Kant, *On a Supposed Right to Lie from Altruistic Motives* (1797), in *Immanuel Kant: Critique of Practical Reason and Other Writings in Moral Philosophy,* ed. Lewis White Beck (New York: Garland, 1976).
12. Quoted in Robert Heilbroner, "What Has Posterity Ever Done for Me?" *New York Times Magazine,* 19 January 1975.
13. Ibid.
14. Joel Feinberg, "The Rights of Animals and Unborn Generations," in *Philosophy and Environmental Crisis*, ed. W. Blackstone (Athens, Ga.: University of Georgia Press, 1974). Reprinted in *Responsibilities to Future Generations*, ed. Ernest Partridge (Buffalo, N.Y.: Prometheus Books, 1981).
15. Garrett Hardin, "Who Cares for Posterity," in *The Limits of Altruism* (Bloomington: Indiana University Press, 1977). Reprinted in *Environmental Ethics*, 2nd ed., ed. L. Pojman (Belmont, Calif.: Wadsworth, 1998), p. 281.
16. John Rawls, *A Theory of Justice* (Cambridge: Harvard University Press, 1971), p. 285.
17. In a thoughtful article, "Why Care About the Future?" in *Responsibilities to Future Generations*, ed. Ernest Partridge (Buffalo, N.Y.: Prometheus Books, 1981), Ernest Partridge argues that psychological health requires the kind of personal self-transcendence that is involved in caring about others, including future people. This is true, but it does not get us far enough. Such a condition of psychological health could be satisfied by my love for the next generation. In fact, I don't think that love is sufficient to get the job done. We need to recognize a deep moral duty to live the kind of lives that will enable people several generations down the line to flourish.
18. Derek Parfit, "Energy Policy and the Further Future: The Identity Problem," in *Energy and the Future*, ed. D. MacLean and P. G. Brown (Totowa, N.J.: Rowman & Littlefield, 1983). Reprinted in *Environmental Ethics,* ed. L. Pojman, pp. 289–96.
19. Richard Taylor, *Good and Evil* (Buffalo, N.Y.: Prometheus Books, 1984), p. xii.
20. Oliver Wendell Holmes, Jr., *Collected Legal Papers* (New York: Harcourt Brace Jovanovich, 1920), p. 340.

CHAPTER SIX

The Historical Roots of Our Ecological Crisis

So God created man in his own image, in the image of God he created them. . . . And God blessed them and said to them, "Be fruitful and multiply, and fill the earth and subdue it; and have dominion over the fish of the sea and over the birds of the air and over every other living thing that moves upon the earth."

—Gen. 1:27, 28

ANTHROPOCENTRISM AND THE CHRISTIAN WORLDVIEW

In 1967 Lynn White, professor of history at the University of California at Los Angeles, wrote an article which contended that our current ecological crisis was primarily due to "the orthodox Christian arrogance towards nature."[1] This arrogance, he argued, was rooted in a domineering, anthropocentric attitude which could be traced back to Genesis, especially 1:28, "God blessed [Adam and Eve], saying to them, 'Be fruitful and multiply, fill the earth and conquer it, and have *dominion* over the fish of the sea and over the birds of the air and over every other living thing that moves upon the earth.'" While it is generally appreciated that our belief systems shape our ecological practices, it is not always recognized how deep the influence of our belief systems is. White argues that the Judeo-Christian *dominion* thesis of man over nature has a profound effect on our treatment of nature, allowing us to exploit it with impunity: "Christianity is the most anthropocentric religion the world has seen. Man shares, in great measure, God's transcendence of nature." After all, in Christianity God becomes fully man in Christ: "Christianity, in absolute contrast to ancient paganism and Asia's religions, not only established a dualism of man and nature but also insisted that it is God's will that man exploit nature for his proper ends."

White contrasted the pagan *panpsychism* (the view that everything has spiritual or mental qualities, at least in primordial states) with the **anthropocentrism** (human-centeredness) of Christianity. In the pagan worldview animals, trees, hills, rivers, and streams are seen as being endowed with the sacred, so that it is evil to harm them without good cause and without going through proper rituals: "Before one cut a tree, mined a mountain, or dammed a brook, it was important to placate the spirit in charge of that particular situation, and to keep it placated." Sometimes the hunter asks the animal for forgiveness before killing it for food. In contrast to this, the Christian worldview promotes an indifference to nature: "To a Christian a tree can be no more than a physical fact. The whole concept of the sacred grove is alien to Christianity and to the ethos of the West. For nearly two millennia Christian missionaries have been chopping down sacred groves, which are idolatrous because they assume spirit in nature. . . . By destroying pagan animism, Christianity made it possible to exploit nature in a mood of indifference to the feelings of natural objects." A New Testament passage that seems deprecatory of animal worth is I Corinthians 9:9–10, where Paul quotes the Hebrew Bible (Old Testament): "For it is written in the law of Moses, 'You shall not muzzle an ox when it is treading out the grain,'" and then comments, "Is it for oxen that God is concerned? Does he not speak entirely for our sake?" Paul allegorizes the passage to mean that God's servants should be supported financially when plowing in the world of souls. White might have noted that the story of the flood (Noah's Ark) in Genesis 6–8 represents the greatest holocaust of animals and ecosystems in literature, if not history. White does give the example of the introduction in the seventh century of the plow with the vertical knife, which cut deep furrows into the Earth, something many Native American tribes would have deemed disrespectful of our Mother Earth. They abjured plowing, viewing cutting the earth as tantamount to striking one's mother. For Christians nature is a resource to be used for human purposes. Our science and technology have grown out of these Christian attitudes toward nature. One may ask how much of an exaggeration of this attitude is Francis Bacon's metaphor of *raping* nature for human purposes.

This radical dualism between humanity and nature is in sharp contrast with the views of many Native American tribes, who regarded every part of the land as sacred: "Every shining pine needle, every sandy shore, every mist in the dark woods, every meadow, every humming insect. All are holy in the memory and experience of my people. We know the sap which courses through the trees as we know the blood that courses through our veins. We are part of the earth and it is part of us. The perfumed flowers are our sisters. The bear, the deer, the great eagle, these are our brothers. The rocky crests, the juices in the meadow, the body heat of the pony, and man, all belong to the same family. . . . The rivers are our brothers. They quench our thirst. They carry our canoes and feed our children. So you must give to the rivers the kindness you would give any brother."[2]

For Native American tribes, according to White, the Earth is holy, our Great Mother, to be cherished. Before one kills a buffalo, one asks his forgiveness, acknowledging the symbiotic relationship between humans and other creatures. The sky is our Great Father, to be revered and protected from harmful pollutants. White has been criticized for romanticizing Native American attitudes toward nature. Subsequent scholarship has shown that their attitudes were varied and complex. Some tribes engaged in widespread slash-and-burn techniques, destroying the forests, and others slaughtered buffalo. William Cronon, however, has documented that the New England tribes had a very different attitude toward the land as property—sometimes called **usufruct**—which entails the right of enjoying all the advantages that come from using something that belongs to the community. For example, if one family used a given plot to grow vegetables, the other families would respect it and plant elsewhere. But if that family found a better plot somewhere, a new family would move in. The idea of an absolute right to property was unknown to them. Furthermore, New England Indians moved from place to place from season to season, having no fixed property. These Indians burned the forest undergrowth, but it was left to the colonists to burn whole forests. The Puritan Timothy Dwight wrote in 1638 of one Vermont town whose people had "cut down their forest with an improvident hand: an evil but too common in most parts of this country. Unhappily it is an increasing evil, and may hereafter put a final stop to the progress of population." Dwight also reported that in the 240-mile journey from Boston to New York a traveler passed through no more than 20 miles of wooded land, in 50 or 60 parcels.[3]

Cronon argues that it was the colonists' idea of treating the land as capital, as wealth, and using it to graze livestock, that changed the land. Forests were cleared for agriculture, and land became wealth. The seventeenth-century colonist Thomas Morton noted that the New England Indians with their simple lifestyles "lived richly" and had few wants or complaints.[4] Pierre Biard agreed and noted that the Indians went about their daily tasks with great leisure, "for their days are all nothing but pastime. They are never in a hurry. Quite different from us, who can never do anything without hurry and worry; worry, I say, because our desire tyrannizes over us and banishes peace from our actions."[5] Thomas Morton observed that the difference between the Indians and the colonists was not that one owned property and the other didn't, but that one loved property and the other had not yet learned to do so.

At the end of his article, White, himself a Christian, recommended that Christians follow the medieval monk St. Francis of Assisi (1181–1226), who preached to the birds and fellowshipped with the foxes, whose view of "nature and man rested on a unique sort of panpsychism of all things animate and inanimate, designed for the glorification of their transcendent Creator." By embracing this more holistic and symbiotic view of the relation of humanity to nature, we might be better armed to save our world.

Witness St. Francis's "Canticle of Brother Sun," which shows affinities with the Indian views described earlier:

> Praised be You, my Lord, with all your creatures,
> Especially Sir Brother Sun,
> Who is the day and through whom You give us light.

Praised be you, my Lord, through Sister Moon and the stars,
In heaven You formed them clear and precious and beautiful.
Praised be you, my Lord, through Brother Wind,
And through the air, cloudy and serene, and every kind of weather
Through which You give sustenance to Your creatures.
Praised be You, my Lord, through Sister Water,
Which is very useful and humble and precious and chaste.
Praised be You, my Lord, through our Sister Mother Earth,
Who sustains us and governs us,
And who produces varied fruits with colored flowers and herbs.[6]

White's article immediately created an uproar. White was denounced and even called "a junior Anti-Christ." On the other hand, radical environmentalists and some not so radical wholeheartedly embraced his position. Some Christians were awakened by his essay and resolved to take more notice of a theology of nature. Others, like Patrick Dobel, agreed that there was much insight in White's analysis, but that the biblical picture was not as crudely domineering as White claimed. The biblical picture of humanity's relation to the Earth is not anthropocentric, but theocentric, based not on a commercial contract, but on a *covenant*, wherein the parties pledge themselves to each other forever. We human beings are the highest form of God's creation, but as such we have special responsibilities to the rest of creation. We are to be good stewards of the Earth, for the Earth is not ours but God's and created for the good of all people and animals, including those who will live in the future. Natural resources are God's gifts, which we may use but not abuse. According to Dobel, "the preservation of what is given 'in trust' demands a recognition of the owner's dictates for the resources. We must know the limits and laws of the world in order to use them wisely. Our actions must be guided, in part, by concerns for future generations. Above all, we must never knowingly exhaust or ruin what has been given to us. If doing so is absolutely necessary to sustain life, then equity demands that we must leave some equally accessible and beneficial legacy to replace what has been exhausted."[7]

Others, such as the sociologist Lewis Moncrief, argued that White's analysis was too simple and that a complex web of forces, including democracy, technology, urbanization, and an aggressive attitude toward nature, accounted for the ecological mess in which we now find ourselves. Moncrief argues that White's analysis misses the essential point that human beings have been altering the environment from their beginning. He asks, "if our environmental crisis is a 'religious problem,' why are other parts of the world experiencing the same environmental problems that we are so well acquainted with in the Western world?"[8] A more plausible account of the causes of our crisis is complex and has to do with the nature of capitalism, technology, democratization, urbanization, and individualism. Christianity may have provided a suitable soil for the growth and flourishing of capitalism and the industrial revolution which developed from capitalism, but the link is contingent, not necessary or even very strong. Capitalism, with its emphasis on individualism and self-interested entrepreneurship, gives rise to resource consumption and pollution. Technology exacerbates the exploitation of resources and production of pollution.

Where does the truth lie in these matters? First, White seems correct in claiming that Christianity is an anthropocentric religion. While only God has infinite value, we read in the Bible that "God made man in His image and likeness" (Gen. 1:26) as a "little less than God, and crowned him with glory and honor," giving him "dominion over the works of [God's] hands." "Thou has put all things under his feet, all sheep and oxen, and also the beasts of the field, the birds of the air, and the fish of the sea, whatever passes along the part of the sea" (Ps. 8). God became a human being in Christ, highlighting the intimate relationship and high objective value in humanity. Indeed, how else can one attribute positive objective value to human beings except by identifying an intrinsically valuable property, such as the image of God? Second, in the Judeo-Christian tradition the primary goal of life is salvation, eternal life in heaven, not ecological harmony. Third, there seems to be a strong dualism in our culture between humanity and nature. The paradigm relationship in literature, for instance, is not "Man in Nature" but "Man against Nature." Nature is an enemy, which drags us down, fighting against our upward quest for heaven. It taints the soul. Nature, at best, is a recalcitrant resource, separate from humanity and to be used by us in our attempt to climb upward toward God, our Father. After Adam sins, God pronounces his task in relationship to the Earth: "Cursed is the ground because of you; in toil you shall eat of it all the days of your life; thorns and thistles it shall bring forth to you and you shall eat the plants of the field. In the sweat of your brow you shall eat bread" (Gen. 3:17–19). Yet a recognition of our linkage to the Earth remains. After all, we were created from the Earth. God finishes the last sentence: "In the sweat of your brow you shall eat bread, till you return to the ground, for out of it you were taken. You are dust, and to dust shall you return" (Gen. 3:19).

Summing up these points, I think we can identify four basic theses in the dominion theory:

1. Human beings are the only creatures on Earth with objective positive value. We are the apex of God's creation, made in His image and likeness. Though created from the Earth, we can transcend it. Animals lack intrinsic value altogether, having only instrumental value. The classic Augustinian hierarchy is shown in Table 6-1.

For the Augustinian Christian, the more reality, the more value; and so God, who is ultimate reality, has infinite value, and evil, which is a complete absence of God and so of reality, is a nothing that affects us by its

Table 6-1 Classic Augustinian Hierarchy

Infinite value/reality	*God*
Increase in value/reality	Angels Good humans Bad humans Devils (fallen angels) Bodies/physical objects/nature
No value/no reality	Evil

absence of the good. Angels have more reality/value than humans, good humans more than bad ones, but both more than devils or mere material nature.

2. Nature is humanity's resource. It is given by God to serve humanity's needs. Humans are lords of nature as God is lord of humanity. Here a chain of command appears. God's method of subduing us is His word and power of grace and judgment. Our method of subduing nature is our deeds, tools, and technology.

3. Nature is recalcitrant and must be subdued by humanity (especially after the fall). Nature is not passive, but active and evil. The Earth was created good, but after the fall, it is under the power of the devil and is now waiting for the redemption from God.

4. God will bring His redemption of nature in His good time. Our job is to prepare ourselves for heaven, rejecting sinful nature and living according to the spirit.

Accordingly, it was not altogether an aberration when former Secretary of the Interior, James Watt, an evangelical Christian, declared in 1980 that there was nothing wrong with strip-mining in the western states, even though it would destroy the environment, because, after all, Christ would be coming soon and would transform the Earth to the glorious kingdom of God. He testified before Congress that "I do not know how many future generations we can count on before the Lord's return," implying that it could be soon, so that we should not worry too much about using up the natural resources in the West.[9] As you can imagine, environmentalists were stunned, likening Watt's appointment to the act of putting the wolf in charge of the chicken coop. Furthermore, the fact that Christianity is a religion that permits meat eating tends to undermine any notion of protecting animals, let alone seeing them as having rights.

And yet Dobel is correct. There is another side to this issue. The model of the good steward has always had a place in Judaism and Christianity. In

Genesis Adam and Eve are placed in God's garden as gardeners to tend and nurture it (Gen. 2:15). Jews and Christians are not to cause wanton destruction of God's gift of nature but are to use it wisely, leaving its rich resources for future generations. Furthermore, no one knows when Christ will return, so that it is presumptuous to squander and denude our inheritance in the way James Watts advocated. Nevertheless, though this stewardship interpretation is embedded within the Judeo-Christian tradition, it may not have been adequately emphasized. Although I attended American Christian churches as I grew up, I can recall hearing only a single sermon on environmental responsibility before the 1980s. The single exception concerning anything close to an environmental challenge was a sermon on Albert Schweitzer's *reverence for life*. The typical attitude was *otherworldly*, concentration on heaven, not this corrupt Earth. "Kill everything in you that belongs only to the earthly life. . . . Let your thoughts be on heavenly things, not on the things that are of the earth" (Col. 3:2–5). James Watt's disregard of the environment in the name of the second coming of Christ, who would make things whole, was replicated over and over again. At the very least, the Judeo-Christian stewardship model was not emphasized. What I learned about environmental integrity I learned on Boy Scout camping trips.

APPROACHING A BALANCED REGARD FOR ALL SPECIES

We have been examining the historical roots of our ecological crisis. The conclusion seems to be that Judaism and Christianity, while holding a stewardship model of our relationship to the environment, did not emphasize our covenantal responsibility as they should have. But there is a deeper issue here. That is, is the very anthropocentrism embodied in these religions, as well as in the secular Enlightenment which challenged them, sufficient to motivate us to be environmentally responsible? If our belief systems have implications for our behavior, does even a religious anthropocentrism, as just described, put too much weight on humanity over against the Earth and its biotic systems? Do we need a deeper worldview, one closer to that of the Native Americans or Eastern philosophy, where nature and animals take on greater importance, where nature is seen as having not mere instrumental value, but rather intrinsic and objective value?

These are the questions we shall be addressing in subsequent chapters. The next two chapters will debate which is the correct theory of nature and our relationship with it. Here are some of the questions we shall be discussing:

1. What is the correct attitude toward nature? Should we regard nature as spiritually empowered? as inherently valuable or only instrumentally valuable? as god-indwelt or as simply a wild force to be subdued?

2. How can we *know* which attitude is the correct one? Suppose we conclude that it were better for us all if almost everyone believed in animism (the view that spirits dwell in natural objects), but concluded that animism was in fact false? Should we try to get ourselves to believe something we have no evidence for simply in order to save ourselves and nature? What if we discover that the best way to save the environment is to draw up a new myth or invent a new religion (or change the one we believe in in order to get the same result). Should we do it?

3. Can an anthropocentric philosophy, which values only humans, possibly be sufficient to save the environment? Is enlightened self-interest adequate to resolve our environmental needs?

4. What is our obligation to animals, especially higher animals? Should we be as concerned about their welfare as about human welfare? Do animals have rights—not to be unnecessarily harmed, experimented on, hunted, or eaten?

5. What is our obligation to preserve other species? Is there a sacred holism that suffers as species are forever destroyed? Is there a hierarchy of being that is preserved in the variety of species? How do we measure the short-term economic benefits to humans of cutting down the rain forests versus eliminating rare species?

6. What is our obligation to future and distant people with regard to the environment? Do future people have rights, so that we must be frugal (and even sacrifice) with regard to natural resources and prevent large-scale pollution? Or is the notion of rights an unwarranted moral view? Or do we at most have duties to the next generation—to leave it enough resources to live on?

These are some of the central questions that we will examine in the first part of this book. We will endeavor to represent various points of view with the best arguments available. Your own position cannot be secure until you have considered all sides of an issue; and, furthermore, unless you have undermined the best case your opponent can make, you can never be sure it has been defeated. Philosophy is the quest for truth and the best position that reason can arrive at.

Review Questions

1. Do you agree with White's assessment that it is our Judeo-Christian "dominance" model that has led to our ecological crisis? Read Genesis 1–3 and compare it with White's analysis. Is his account a correct interpretation of the text?
2. What is the appeal of St. Francis? Reflect on his "Canticle of Brother Sun." What sort of environmental attitude does St. Francis's poem em-

body? Compare it with Native American views on our relationship to nature. Discuss the implications of such views. Are they helpful? Are they true? Explain your judgment.

3. If one does not accept a theistic version of creation, does the stewardship (or gardener) model make any sense? A steward is one who manages the household affairs of another person. If there is no God, the Earth is not God's household. But then whose is it? To whom are we stewards?

Suggested Readings

Cobb, John B., and L. Charles Birch. *The Liberation of Life: From Cell to Community*. Cambridge: Cambridge University Press, 1981.

Cronon, William. *Changes in the Land*. New York: Hill & Wang, 1986.

Dobel, Patrick. "The Judeo-Christian Stewardship Attitude Toward Nature." *Christian Century*, 12 October 1977.

Elsdom, Ron. *A Bent World: A Christian Response to the Environmental Crisis*. Downers Grove, Ill.: InterVarsity Press, 1981.

Fowler, Robert Booth. *The Greening of Protestant Thought*. Chapel Hill, N.C.: University of North Carolina Press, 1995.

Passmore, John. *Man's Responsibility for Nature*. New York: Scribner's, 1971.

Pojman, L. ed. *Environmental Ethics: Readings in Theory and Application*. 2nd ed. Belmont, Calif.: Wadsworth, 1998.

Pyne, Stephen J. *Fire in America: A Cultural History of Wildland and Rural Fire*. Princeton: Princeton University Press, 1982.

Notes

1. Lynn White, "The Historical Roots of Our Ecological Crisis," *Science* 155 (10 March 1967), reprinted in *Environmental Ethics: Readings in Theory and Application,* 2nd ed., ed. L. Pojman (Belmont, Calif.: Wadsworth, 1998), pp. 15–21.
2. Chief Seattle's alleged letter to the president of the United States in 1852. We know now that Chief Seattle did not write this letter. Nevertheless, I think it is an accurate representation of Native American attitudes. See Stephen J. Pyne, *Fire in America: A Cultural History of Wildland and Rural Fire* (Princeton: Princeton University Press, 1982) for a balanced discussion of Native American burning practices.
3. William Cronon, *Changes in the Land* (New York: Hill & Wang, 1986), p. 121.
4. Ibid., p. 80.
5. Ibid.
6. St. Francis of Assisi, *The Canticle of Brother Sun,* trans R. J. Armstrong and I. Brady (New York: Paulist Press, 1982).
7. Patrick Dobel, "The Judeo-Christian Stewardship Attitude Toward Nature," *Christian Century* (12 October 1977), reprinted in *Environmental Ethics,* ed. L. Pojman.
8. Lewis Moncrief, "The Cultural Basis of Our Environmental Crisis," *Science* (30 October 1970), reprinted in *Environmental Ethics,* ed. L. Pojman.
9. Quoted in Robert Booth Fowler, *The Greening of Protestant Thought* (Chapel Hill, N.C.: University of North Carolina Press, 1995), p. 47.

CHAPTER SEVEN

Animal Rights: Sentience as Significant

Brute beasts, not having understanding and therefore not being persons, cannot have any rights. . . . The conclusion is clear. We have no duties to them,—not of justice, . . . not of religion. . . .
—JOSEPH RICKABY, S.J., *Moral Philosophy of Ethics and Natural Law,* 1889

In their behavior toward creatures, all men are Nazis.
—ISAAC BASHEVIS SINGER, *Enemies: A Love Story*

ALL IS NOT WELL IN THE ANIMAL KINGDOM

Every minute of the day, 24 hours a day, 100 animals are killed in laboratories in the United States. Fifty million experimental animals are put to death each year.[1] Some die in the testing of industrial and cosmetic products, some are disposed of because they are female, some are killed after being force-fed or after being tested for pharmaceutical drugs. Insecticides, pesticides, anti-freeze chemicals, brake fluids, bleaches, Christmas tree sprays, silver cleaners, oven cleaners, deodorants, skin fresheners, baby preparations, bubble baths, bath salts, freckle creams, eye makeup, crayons, fire extinguishers, inks, suntan oils, nail polish, mascara, hair sprays, zipper lubricants, and paints are all tested on animals before humans are allowed to use them. Although not required in all cases, many companies have traditionally performed animal tests to gain approval of their products by the government. A legal requirement that animals be anesthetized can be circumvented in many experiments, and in others the nature of the tests preclude the use of anesthesia. In the Draize eye test, concentrated solutions of commercial products are instilled into rabbits' eyes and the damage is then recorded according to the size of the area injured. Monkeys have also been used for these experiments. After the tests, the animals are destroyed.[2]

Civet cats are confined in small cages in dark rooms where the temperature is 110°F until they die. The musk, which is scraped from their genitals once a day for as long as they can survive, makes the scent of perfume last a bit longer after each application. In recent years, several cosmetic companies have begun to abandon the use of animals for testing in cosmetic products—a sign of progress.

In military primate-equilibrium studies, monkeys are set in simulated flying platforms and tested for their ability to keep their balance under duress. They are subjected to high doses of radiation and chemical warfare agents to see how these would affect their ability to fly. When they become nauseated and begin to vomit from the doses of radiation, they are forced to keep their platform horizontal through the inducement of electric shocks.

In July 1973 Representative Les Aspin of Wisconsin discovered that the United States Air Force was planning to purchase 200 beagle puppies whose vocal cords were tied to prevent normal barking; they would be used for testing poisonous gases. At the same time the army was preparing to use 400 beagles for similar tests. More recently army laboratories fed 60 beagles doses of TNT to determine the effects of the explosive on animals.[3]

At several universities dogs, monkeys, and rats have been confined to small rooms—unable to escape from the electric shocks that emanate from the steel grid floors—in order to determine how they will react to unavoidable pain. In toxicity tests animals are placed in sealed chambers and forced to inhale sprays, gases, and vapors. In dermal toxicity tests rabbits have their fur removed so that a test substance can be placed on their skin. The skin may bleed, blister, and peel. In immersion studies animals are placed in vats of diluted substances, sometimes causing them to drown before the results can be obtained.

Rhesus monkeys were given massive doses of cocaine until they began to mutilate themselves; eventually they died of cocaine abuse. In 1984 at a major university, experiments were performed on baboons in order to determine brain damage from head injuries. The subjects were strapped down in boxlike vises and had specially designed helmets glued to their skulls. Then a pneumatic device delivered calibrated blows to the helmet, causing brain injuries to the baboons. Videotapes showed that while the head injuries were being inflicted the researchers stood by joking about their subjects.[4]

Neither is all well down at MacDonald's Farm. Factory farming, with its high-tech machinery, has replaced the bucolic pleasantries of free-range agriculture. Farmer MacDonald doesn't visit his hens in barns in order to pick an egg from the comfortable nest. Now as soon as the chicks are hatched, they are placed in small cages. Between five and eight (up to nine) chickens are pressed close together in cages about 18 inches by 10 inches over thin wired floors that hurt their feet and don't allow them to move around. They are painfully debeaked so that they cannot attack each other in these unnatural quarters. In other chicken factories the chickens are hung by their feet from a conveyer belt that escorts them through an automatic throat-slicing machine. Three billion chickens are killed in the United States each year. Likewise, pigs and veal calves are kept in pens so small that they cannot move or turn around and develop muscles. They are separated from their mothers so they cannot be suckled and are fed a diet low in iron so that we can eat very tender meat.[5] James Rachels describes the process this way:

> Veal calves spend their lives in pens too small to allow them to turn around or even to lie down comfortably—exercise toughens the muscles, which reduces the "quality" of the meat, and besides, allowing the animals adequate living space would be prohibitively expensive. In these pens the calves cannot perform such basic actions as grooming themselves, which they naturally desire to do, because there is not room for them to twist their heads around. It is clear that the calves miss their mothers, and like human infants they want something to suck: they can be seen trying vainly to suck the sides of their stalls. In order to keep their meat pale and tasty, they are fed a liquid diet deficient in both iron and roughage. Naturally they develop cravings for these things, because they need them. The calf's craving for iron is so strong that, if it is allowed to turn around, it will lick at its own urine, although calves normally find this repugnant. The tiny stall, which prevents the animal from turning, solves this "problem." The craving for roughage is especially strong for bedding, since the animal would be driven to eat it, and that would spoil the meat. For these animals the slaughterhouse is not an unpleasant end to an otherwise contented life. As terrifying as the process of slaughter is, for them it may actually be regarded as a merciful release.[6]

Are animal experiments justified? If so, should there be constraints on them in order to minimize suffering? Is it morally permissible to eat animals? Why or why not? In sum, do animals have rights? These are the questions we will discuss in this chapter. We will first examine five major theories regarding our obligations to animals and then examine specific issues, such as vegetarianism, animal experimentation, and hunting for sport.

FIVE THEORIES OF OBLIGATION TO ANIMALS

Five theories on the moral status of animals appear in the history of Western philosophy and religion, from assigning animals no status on the one extreme to assigning them equal status with humans on the other extreme.

The No-Status Theory

The *no-status theory* was set forth by the French philosopher René Descartes (1596–1650), who held that animals had no rights or moral status because they had no souls. Since, according to Descartes, the soul is necessary to consciousness, animals cannot feel pain or pleasure. They are mere machines: "From this aspect the body is regarded as a machine which, having been made by the hands of God, is incomparably better arranged, and possesses in itself movements which are much more admirable than any of those which can be invented by man."[7]

In Descartes's view, animals are automata who move and bark and utter sounds like well-wound clocks. Because they lack a soul, which is the locus of consciousness and value, they have no moral status whatsoever. It is no more morally wrong to pull the ears off a dog or eat a cow than it is to kick a stone or eat a carrot.

In Descartes's time animals were subjected to excruciating tortures in physiological experiments. Dogs were restrained by nailing their paws to boards so that they could be more easily observed while being cut open. Sometimes their vocal chords were cut so that their shrieks would not disturb the anatomists. In the next century Nicholas Fontaine observed the following:

> [The anatomists] administered beatings to dogs with perfect indifference, and made fun of those who pitied the creatures as if they felt pain. They said the animals were clocks; that the cries they emitted when struck were only the noise of a little spring that had been touched, but that the whole body was without feeling. They nailed poor animals up on boards by their four paws to vivisect them and see the circulation of the blood which was a great subject of conversation.[8]

We now know that these philosophers were wrong about the mental life of animals. Animals do feel pain and pleasure. They have consciousness and engage in purposeful behavior. Dogs and cats manifest intelligence; gorillas and chimpanzees exhibit complex abstracting and reasoning abilities and have the capacity to communicate through language. The differences between humans and other mammals are more a matter of degree than of kind.

The nineteenth-century British philosopher William Whewell seems to have held a slightly different version of the no-status view. "The pleasures of animals are elements of a very different order from the pleasures of man. We are bound to endeavor to augment the pleasures of men, not only because they are pleasures, but because they are human pleasures. We are bound to men by the universal tie of humanity, of human brotherhood. We have no such tie to animals."[9]

This modified Cartesian view is reflected in the claim of Nobel Prize–winning microbiologist David Baltimore, that no moral issue is involved in animal research, and in the writings of psychologists G. Gallup and S. D. Suarez, who say that "the evolution of moral and ethical behavior in man may be such that it is not applicable to other species." Similarly, the veterinarian F. S. Jacobs writes that "domestic animals exist in this world because they fulfill man's needs. . . . Therefore it is meaningless to speak of their rights to existence, because they would not exist if man did not exist."[10]

According to the no-status theory, animals have no rights, and we have no positive obligations to them. They are merely resources for us to dispose of as we will.

The Indirect-Obligation Theory

The dominant position in Western philosophy and religion has been the view that while animals have no inherent rights, we ought to treat them kindly. We have duties to them because we have obligations to rational beings, God, and people who own animals. Furthermore, if we treat animals cruelly, we may adopt bad habits and so treat humans cruelly.

The creation story of Genesis 1 supports a stewardship model of creation. All of nature has been given to human beings for their use, but humanity must use it properly for God's sake: "And God blessed [man and woman] and God said to them, 'Be fruitful and multiply, and fill the earth and subdue it; and have dominion over the fish of the sea and over the birds of the air and over every living thing that moves upon the earth.' And God said, 'Behold, I have given you every plant yielding seed which is upon the face of all the earth, and every tree with seed in its fruit; you shall have them for food.'"[11]

The philosophers Thomas Aquinas (1225–74) and Immanuel Kant (1724–1804) hold that cruelty to animals is wrong because it forms bad character and it leads to cruelty to human beings. Aquinas holds a hierarchical view that permits humans to use animals for human purposes:

> There is no sin in using a thing for the purpose for which it is. Now the order of things is such that the imperfect are for the perfect . . . thing, like plants which merely have life, are all alike for animals, and all animals are for man. Wherefore it is not unlawful if men use plants for the good of animals, and animals for the good of man, as the Philosopher [Aristotle] states.
>
> Now the most necessary use would seem to consist in the fact that animals use plants, and men use animals, for food, and this cannot be done unless these be deprived of life, wherefore it is lawful both to take life from plants for the use of animals for the use of men. In fact this is in keeping with the commandment of God himself. (Genesis 1:29, 30 and Genesis 9:3)

Aquinas continues: "If any passages of Holy Writ seem to forbid us to be cruel to dumb animals, for instance to kill a bird with its young: this is either to remove man's thoughts from being cruel to other men, and lest through being cruel to animals one become cruel to human beings: or because injury to an animal leads to the temporal hurt of man, either of the doer of the deed, or of another."[12]

Kant's conclusion is similar to Aquinas's view. We have "no direct duties" to animals, for they "are not self-conscious and are there merely as a means to an end."

> The end is man. . . . Our duties toward animals are merely indirect duties toward humanity. Animal nature has analogies to human nature, and by doing our duties to animals in respect to manifestations of human nature, we indirectly do our duty to humanity. . . . If a man shoots his dog because the animal is no longer capable of service, he does not fail in his duty to the dog, for the dog cannot judge, but his act is inhuman and damages in itself that humanity which it is his duty to show toward mankind. If he is not to stifle his human feelings, he must practice kindness toward animals, for he who is cruel to animals becomes hard also in his dealing with men.[13]

The difference between Aquinas and Kant on this issue is that whereas for Aquinas humans, but not animals, are endowed with immortal souls, for Kant humans, but not animals, are endowed with reason and free will.

The weakness of the indirect-obligation view is that it makes rational self-consciousness the sole criterion for being morally considerable. While such self-consciousness may be the criterion for having full-blooded rights and for being a morally responsible agent, it is not the only thing that is of moral importance. Causing pain and suffering are bad in themselves, and we have duties not to do such things but to ameliorate and eliminate pain and suffering.

The Equal-Status View

K. McCabe, the codirector of the animal rights group People for the Ethical Treatment of Animals (PETA) has said, "There is no rational basis for separating out the human animal. A rat is a pig is a dog is a boy. They're all mammals. . . . In time, we'll look on those who work in [animal laboratories] with the horror now reserved for the men and women who experimented on Jews in Auschwitz. . . . That, too, the Nazis said, was 'for the greater benefit of the master race.'"[14]

We call the view that equates human beings with animals the equal-status thesis. Its foremost proponent is the philosopher Tom Regan, who holds three goals on the treatment of animals: (1) the total abolition of the use of animals in science, (2) the total dissolution of commercial animal agriculture, and (3) the total elimination of commercial and sport hunting and trapping. Even though Regan concedes that some individual uses of animals for biomedical experimentation might be justified and free-range grazing is better than factory farming, all of these uses constitute infringements on animal rights, and the exceptional cases are so isolated as to serve only to confuse the issue.

What is wrong is not the pain caused, the suffering, or the deprivation, though these compound the wrong. What's fundamentally wrong is "the system that allows us to view animals as *our resources*, here for us—to be eaten, or surgically manipulated, or put in our cross hairs for sport or money."[15]

Why is it wrong to treat animals as our resources? Because, according to Regan, they have *inherent value* and are ends in themselves just as we are. They are of *equal worth* to human beings.

> To say we have such value is to say that we are something more than, something different from, mere receptacles. Moreover, to insure that we do not pave the way for such injustices as slavery or sexual discrimination, we must believe that all who have inherent value have it equally, regardless of their sex, race, religion, birthplace, and so on. Similarly to be discarded as irrelevant are one's talents or skills, intelligence and wealth, personality or pathology, whether one is loved or admired—or despised and loathed. The genius and the retarded child, the prince and the pauper, the brain surgeon and the fruit vendor, Mother Teresa and the most unscrupulous used car salesman—all have inherent value, all possess it equally, and all have an equal right to be treated with respect, to be treated in ways that do not reduce them to the status of things, as if they exist as resources for others.[16]

What is the basis of the equal inherent value? Just this: "we are each of us the experiencing subject of a life, each of us a conscious creature having an individual welfare that has importance to us whatever our usefulness to others. We want and prefer things; believe and feel things; recall and expect things."

Regan's deontological egalitarianism (rights and respect) argument can be stated thus:

1. All humans (or all subject-of-life humans) have equal positive value.
2. There is no morally relevant difference between humans and (some) animals (such as mammals).
3. Therefore all (some) animals have equal positive worth with humans.
4. Moral rights derive from the possession of value.
5. Since humans have rights (to life, not to be harmed, and so forth), animals have those same rights.

Several problems arise in Regan's theory of equal inherent value. First, he hasn't explained why being an experiencing subject entails possessing inherent value. The reply that has been given by his defenders is "The subject experiences the value—hence the subject has inherent value." But this seems confused. One can't validly argue from valuing something *V* to being *V* without additional premises. The subject may be unhealthy but still value health, or I may value reason without being rational. Further, how do we know that being "subject-to-a-life" grants one such inalienable positive value? Is it supposed to be intuitively self-evident? If so, then it would also seem self-evident to some that merely being conscious entails less value than being *self-conscious*, especially *rationally* self-conscious. Someone in a daze or dream may be minimally conscious, but that is a state less valuable than being fully self-conscious with plans and projects. It is desirable to have more reason or intelligence rather than less reason or intelligence. Intelligence, knowledge, and freedom are inherent values, but animals have less of them than humans. It's true that humans have varying degrees of them, but as a species (or on average) we have more of what makes for worth than other species, so that there would seem to be degrees of value interspecies and intraspecies. If so, we must feel dissatisfied with Regan's assessment. He simply has not given any evidence for the thesis that all animals have equal worth and are to be treated with equal respect.

Regan rejects the notion of differing degrees of inherent value based on differing degrees of self-awareness or some other mental capability, affirming that this leads to the view that mentally superior people have stronger moral rights than mentally inferior people.

There are at least two ways to respond to Regan here. First, following deontologists like Kant and Rawls, some may appeal to the threshold view of self-consciousness and argue that all and only those who are capable of rational deliberation and life plans are to be accorded a serious right to life. While there may be differences among humans with regard to the ability to reason, almost all (excepting infants, the mentally ill, the senile, the seriously retarded and brain damaged) have sufficient ability to be counted within the circle of full moral citizenry. Some higher animals, such as dolphins, gorillas, and chimpanzees, may also belong to this group. But the threshold argument seems arbitrary, ad hoc, and, hence, unsound. It invites the question, Why draw a line where none exists in reality? If self-consciousness and rationality are decisive qualities for moral appraisal, then isn't more better than less, so that those individuals with more of the qualities should be respected more highly than those with less?[17] Alternatively, we might argue that what gives any being special value is its capacity for moral action, so that the more ability and actualization of this capacity, the better that being is. Either way, egalitarianism is to be rejected in favor of a hierarchy—whatever has more of the requisite quality, the better that thing is.

The second way to respond to Regan's antihierarchical notion of value is to take a contractarian approach to ethics, arguing that there are no inherent values and that animals are not normally part of the social contract. Since rights derive from contract, animals do not have any rights.[18] Of course, the contractualists may still recommend kindness to creatures outside of the contract; but where human interests are compelling, animals may be sacrificed to those interests in ways that humans may not.

What seems to underly Regan's egalitarianism is the sanctity of life principle, or rather the sanctity of a *subject-of-a-life* principle, since he views being "subject-of-a-life" (that is, having beliefs and desires) as intrinsically valuable. But just as the sanctity of life principle fails to distinguish between quality of life (the life of a bacterium is equal to that of Mother Teresa), the subject-of-a-life does the same within a narrower scope. Whereas the former presupposes an outmoded vitalism,[19] the latter presupposes a deeper metaphysic than Regan offers. Not only does he fail to provide a foundation for his egalitarianism, but he also fails to provide reasons to accept any of his absolutist goals: the total abolition of the use of animals in scientific experimentation, the total dissolution of commercial animal agriculture, and the total elimination of commercial and sport hunting and trapping. Until Regan gives us reason to accept his sanctity of a subject-of-a-life principle, his arguments will fail to support any of these aims.

There is one further problem with Regan's approach that should be mentioned. It fails to explain why we shouldn't intervene in the animal world, eliminating animal cruelty. Shouldn't radical zoophiles go into the wild and protect helpless rabbits, deer, and birds from marauding predators, members of the cat family (lions, tigers, leopards), wolves, and other carnivores? Perhaps carnivores could be confined to separate quarters and fed the carcasses of other animals, including human beings. How can these animal egalitarians rest until all of nature is turned into the peaceable kingdom, where the lion is made to lie down with the lamb? The suffering that wolves cause rabbits or leopards and tigers cause antelope and deer is far more devastating than what the clean shot of an expert hunter's rifle inflicts.

Regan, as well as biocentric egalitarians (like Paul Taylor; see Chapter 10), have difficulty with the violent behavior of wild animals. It doesn't fit their vision of a peaceable kingdom, but they can't say that we have duties to eliminate all predatory behavior, let alone all carnivores, to make the world safe for pacifist herbivores and plants, for carnivores can't help their need for meat. Their "respect for nature" doesn't allow humans to intervene as environmental imperialists. So what are we to say about the rights of sheep and rabbits not to be torn asunder and harmed by wolves and other predatory animals? Here is Regan's reply. Even though the sheep and rabbits have a right to life and a right not to be harmed, the wolf has no duty to respect those rights, since *ought* implies *can* and the wolf is only doing what is natural and cannot do otherwise. If we intervene we are violating the wolf's right to dinner. Here is Regan's argument:

> [I have] defended [the thesis] that moral patients have *no* duties and thus do not have the particular duty to respect the rights of others.* *Only moral agents can have duties,* and this because only these individuals have the cognitive and other abilities necessary for being held morally accountable for what they do or

---

* A *moral patient* is an object of moral consideration. It can be a person, animal, or ecosystem.

> fail to do. Wolves are not moral agents. They cannot bring impartial reason to bear on their decision making—cannot, that is, apply the formal principle of justice or any of its normative interpretations. That being so, wolves in particular and moral patients generally cannot *themselves* meaningfully be said to have duties to anyone, nor, therefore, the particular duty to respect the rights possessed by other animals. In claiming that we have a prima facie duty to assist those animals *whose rights are violated,* therefore, we are not claiming that we have a duty to assist the sheep against the attack of the wolf, since the wolf neither can nor does violate anyone's rights. The absurd results leveled against the attribution of rights to animals simply do not materialize.[20]

I think that Regan is wrong here. If the sheep has a positive right to live in peace, then we have a duty to try to help it. If a boulder came hurling down and was about to crush the sheep, wouldn't we be remiss in our duty if we didn't take reasonable steps to get the sheep out of harm's way? Likewise, even if the wolf's actions are natural, we have a duty to the sheep to save it from the wolf.

Furthermore, can't Regan's argument be used to prohibit us from saving humans from wild animals? from bacteria? from insane humans who are only following their nature in raping and killing? Consider the police officer coming to the distraught parents of a child who has been brutally raped and killed by a homicidal maniac. "Well, I would have intervened when I saw Mr. Smith brutalizing your daughter, Mrs. Brown, but then I realized that he was only following his nature as a violent animal, so I was obliged to leave him alone." Or to paraphrase Regan, "You see, Mrs. Brown, in claiming that we have a prima facie duty to assist those children *whose rights are violated,* we are not claiming that we have a duty to assist the child against the attack of the madman, since the madman neither can nor does violate anyone's rights."

One should note that Regan is inconsistent, for (only eleven pages after saying that we may not intervene in favor of the sheep) he argues that we may kill a rabid dog when it attacks a human. How is this different from a rabid wolf or dog's attacking a sheep?

In the end Regan hasn't justified either his position or the implications of that position. He holds a deep intuition that all people are of equal worth and appeals to a philosophical consensus on that assumption, which provides a basis for his entire moral theory. From there he argues that since there is no relevant difference between humans and mammals (over the age of one), we should treat all such mammals equally, as possessing equal worth. But if there is no reason to believe that all people, let alone all mammals over the age of one, possess equal positive value, what is left of Regan's moral system? Are we left with any morality at all? Does his system lead to moral nihilism? I must leave you to decide the issue.[21]

The Equal-Consideration Theory

The equal-consideration theory was first set forth by Jeremy Bentham (1748–1832), the father of classical utilitarianism, and developed by Peter Singer in his epoch-making book *Animal Liberation* (1975). The theory says that animals are just like us in basic morally relevant ways and so merit our moral regard. As a hedonistic utilitarian, Bentham believed that the essence of morality was to maximize pleasure or happiness and minimize pain and suffering. Furthermore, each sentient being was to count for one, and no one was to count as more than one in the utilitarian calculus. In a classic passage written in 1789, Bentham compares the irrationality of our views toward animals with the irrationality of our views toward other races:

> The day *may come,* when the rest of the animal creation may acquire those rights which never could have been withholden from them but by the hand of tyranny. The French have already discovered that the blackness of the skin is no reason why a human being should be abandoned without redress to the caprice of a tormentor. It may come one day to be recognized, that the number of the legs [or] the [color] of the skin are reasons equally insufficient for abandoning a sensitive being to the same fate. What else is it that should trace the insuperable line? Is it the faculty of reason, or, perhaps, the faculty of discourse? But a full-grown horse or dog is beyond comparison a more rational, as well as a more conversable animal, than an infant of a day, or a week, or even a month, old. But suppose the case were otherwise, what would it avail? The question is not, Can they *reason*? nor Can they *talk*? but, Can they *suffer*?[22]

Animals are *sentient beings,* capable of experiencing pleasure and pain, capable of being conscious of their suffering. Since sentience is the sole criterion for moral consideration, they are morally considerable in the same way that human beings are.

Bentham's theory was criticized in his own day as crudely *hedonistic,* for reducing all morality to considerations of pleasure and pain. If this were so, then it might be morally right for five sadists who receive a total of 100 hedons in the process to torture a child who only suffers 50 dolors (antihedons). Or if a burglar will get more pleasure from my artwork than I do, he has a right to steal it (if in doing so he doesn't produce additional harm).

Sometimes utilitarians like Bentham write as though the single concern of morality was to eliminate suffering in the world. If this were morality's only concern, we would have an obligation to eliminate all sentient life as painlessly as possible, since the only way to eliminate suffering (even gratuitous suffering) is to kill every living organism.

At other times Bentham sounds as if promoting positive pleasure (and in the process eliminating pain) were the essence of morality. For this reason his version of utilitarianism was labeled "the pig philosophy" by his contemporary critics. Better a pig satisfied than Socrates dissatisfied! In an experiment done on rats, electrodes are wired to the limbic areas of their cerebral cortex, and the rats are shown a button, which, when pressed, will stimulate a reward or pleasure center in the brain. I'm told that the rats in the experiment become addicted to the need for the stimulation so that they lose interest in food and sex and spend most of their time pressing the stimulation button until they die. If the maximization of pleasure is what morality is all about, we ought to plug everyone, animals and humans, up to these pleasure machines.

Peter Singer, aware of the dangers of classical utilitarianism, has modified Bentham in order to present a more plausible version of utilitarianism. But the one thing in Bentham that Singer does accept is the idea that all sentient beings are linked together by their capacity to suffer. From this Singer develops his notion of animal liberation, a global theory of duties to animals. He has four essential points:

1. *The principle of equality:* Every sentient being deserves to have his or her needs and interests given equal consideration. Following Bentham, each sentient being is to count for one and only for one. Status and privilege should play no part in doling out benefits. Rather, we should distribute goods on the basis of need and interest.

> Equality is a moral idea, not an assertion of fact. There is no logically compelling reason for assuming that a factual difference in ability between two people justifies any difference in the amount of consideration we give their needs and interests. *The principle of equality of human beings is not a description of an alleged actual equality among humans: it is a prescription of how we should treat them.*[23]

2. *The principle of utility:* The right act is the one that maximizes utility or happiness (or minimizes suffering or unhappiness). Singer holds a version of *preference utility*. We ought to satisfy the actual preferences of sentient beings.

3. *Rejection of speciesism,* which constitutes a violation of the principle of equality. Singer compares speciesism, the arbitrary favoring of one's species, with racism: "The racist . . . [gives] greater weight to the interests of members of his own race, when there is a clash between their interests and the interests of another race. Similarly the speciesist allows the interests of his own species to override the greater interests of members of other species."[24]

4. *The relevance of self-consciousness:* There is a difference between our equal ability to suffer and our equal worth as rational, self-conscious agents. Singer values rational self-consciousness over mere sentience, but he agrees that sentience, or the ability to suffer, gives us a baseline equality for some considerations.

Singer makes a distinction between activities that cause suffering and those that cause death. It may be worse to cause animals to suffer than to kill them, since killing may be done with minimum pain and since the animal generally does not have an understanding of life and death. With humans the situation is the other way around. Since humans generally prize life with pain (up to a limit) to death, it is worse to kill humans than to cause them suffering.

Not all animals are of equal worth. As Singer says,

> A rejection of speciesism does not imply that all lives are of equal worth. While self-awareness, intelligence, the capacity for meaningful relations with others, and so on are not relevant to the question of inflicting pain—since pain is pain, whatever other capacities, beyond the capacity to feel pain, the being may have—these capacities may be relevant to the question of taking life. It is not arbitrary to hold that the life of a self-aware being, capable of abstract thought, of planning for the future, of complex acts of communication, and so on, is more valuable than the life of a being without these capacities. To see the difference between the issues of inflicting pain and taking life, consider how we would choose within our own species. If we had to choose to save the life of a normal human or a mentally defective human, we would probably choose to save the life of the normal human; but if we had to choose between preventing pain in the normal human or the mental defective—imagine that both have received painful but superficial injuries, and we only have enough painkiller for one of them—it is not nearly so clear how we ought to choose. The same is true when we consider other species. The evil of pain is, in itself, unaffected by the other characteristics of the being that feels the pain; the value of life is affected by these other characteristics.[25]

So it is irrelevant whether animals and humans are of equal worth *when it comes to suffering*. In suffering we are to be given equal consideration. Since sentience lies at the core of our moral thinking, and language and intelligence lie nearer the periphery, a large part of our morality will have to do with liberating people and animals from suffering.

Regarding animal experimentation, Singer points out that our current practices are speciesist. He admits that some animal experimentation is justified on utilitarian grounds, but points out that this argument also justifies experimentation on small and retarded children. When the advocate of animal experimentation asks, "Would you be prepared to let thousands of humans die if they could be saved by experimenting on a single animal?" Singer responds, "Would the experimenter be prepared to perform his experiment on an orphaned human infant, if that were the only way to save many lives?" If the experimenter is not prepared to use an orphaned human infant, then his readiness to use nonhumans is simple discrimination, since adult apes, cats, mice, and other mammals are more aware of what is happening to them, more self-directed and, so far as we can tell, at least as sensitive to pain, as any human infant.[26]

Although Singer's views on the prejudice of speciesism compel us to rethink the moral status of animals, they still have some of the weaknesses attributed to Bentham—weaknesses inherent to utilitarianism—that if jealous John will suffer 100 dolors if he is denied my $100, but I will only suffer 80 dolors if he steals my $100, John has a moral right to steal my money.

But there is a deeper problem with Singer's preference act utilitarianism that resides in the very notion of *preference* itself. For Singer the morally right act is the one that yields the highest preference satisfaction. No doubt that an animal's desire not to suffer outweighs my desire for the taste of meat, but this doesn't get us to the conclusion that preferences should be considered equally or that preference maximization is the only relevant criterion to be considered.

Even if a dog or mouse is suffering more pain than a child, it is, nevertheless, deeply counterintuitive to say that we have a duty to give the single pain reliever to the dog or mouse. If the child is my child, then I think I have a strong duty to give it to him or her—even if the dog is my dog. The utilitarian might respond that this may be true because the child will live longer than the dog or mouse and have memories of the pain for a longer time. Well, then imagine that it is an elderly man who is not likely to live longer than the puppy. I still sense that it would be right (or at least not wrong) to give the pill to the elderly man.

In fact, it seems to me that I should prefer my grandfather's vital interests more than the preferences of all the dogs and mice in the world. If my grandfather were in danger of losing a leg and forty dogs were each in danger of losing a leg, and I could either save my grandfather's leg or all of the legs of the forty dogs, I wouldn't hesitate to operate on grandfather. No matter how many dogs' legs I could save, I'd feel it my duty to save grandfather's.

Singer, of course, might argue that this only proves that I am still an immoral speciesist. Isolated intuitions prove nothing. And so they don't. So let's proceed further.

Suppose that ten people will inherit my grandmother's vast wealth. Each has a strong preference that my grandmother die so they can pay their bills. The total of their preference units is 1,000. Suppose my grandmother's desire to continue to live is only 900 units. Wouldn't I—supposing that I, as a utilitarian, can pull it off without too much guilt—have a duty to kill grandmother as painlessly as possible?

"No," the preference utilitarian exclaims. "For you have forgotten the unintended side effects of creating fear in the minds of others that this might be done to them and so magnifying the negative preference points."

Well, then, all I have to do is be pretty sure that the poison I administer in grandmother's tea, which induces a heart attack, is the kind that leaves no trace.

In fact, can't the preference utilitarian do anything he or she wants, including killing people, taking their property, and breaking promises to them, just so long as his or her preference (or the aggregate preferences of the utilitarian's group) outweigh the preference of the victim? If you suddenly become despondent and no longer value life, why can't I utilitarianly kill you if it satisfies a preference to do so? If I can get you to care less than you do about your Mercedes Benz and can ensure that only minimally bad side effects will follow, can I not steal your car with complete utilitarian approval?

Perhaps you would reply that this would be exploitation. Well, why is exploitation wrong? If I desire to exploit you more than you dislike being exploited, doesn't my preference win out over yours?

Likewise, couldn't we justify slavery just by brainwashing the slaves to prefer slavery to freedom or at least not to value freedom more than we value having slaves?

In the end Singer doesn't gain much, if any, advantage over Bentham's crude hedonic calculus. Instead of people being *pleasure/pain* receptacles, now they simply become *preference* receptacles.

One final criticism that deeply affects Singer's argument turns on the fact that his principle of equality—the principle of equal consideration—is not really about equality at all. Equal consideration according to what? It is the *what* that is doing all the work. It is the *what* of being rational and living in a specifically complex social structure that justifies us in sending humans, but not dogs or even monkeys, to college. Self-consciousness gives higher animals more consideration than merely sentient ones. The principle of equality is merely the rule of impartiality: Apply your principles in a disinterested manner, according to the relevant criteria, not according to irrelevant ones. In the words of Aristotle, "Treat equals equally, and unequals unequally." It is purely a formal principle without any substantive force, one that could be used (as it was by Aristotle) to justify slavery and to justify the secret killing of innocents. It is compatible with Nietzsche's Superman, with his magnificent will (read "preference") for power.

Let's sum up our discussion on Singer's utilitarianism version of animal liberation. He has made a good case for rejecting speciesism and for regarding animals as morally considerable. Probably some higher animals are persons with strong claims to better treatment than we are affording them. What isn't clear is that his or any brand of utilitarianism, which compares hedonic states of humans and animals, is a plausible moral theory. Most of us will find it counterintuitive to prefer dogs and mice to our friends and relatives or any normal human at all. Whether we are just speciesists is a question we must investigate further. We turn next to a theory that tries to deal with this issue within the framework of broad anthropocentrism.

The Split-Level Theory

Consider the following true story. In January 1974 a boat sank off the eastern coast of the United States. The captain, his wife, their 80-pound Labrador retriever, and an injured crewman occupied a lifeboat to which two youths, aged 19 and 20, and a 47-year-old Navy veteran, tied themselves with ropes while floating in the freezing waters. The captain refused repeated requests by the swimmers that he throw the dog overboard to make room for some of those tied to the lifeboat. After nine hours, the youths perished and the veteran struggled aboard. All the occupants, including the dog, were rescued. After an initial investigation, the Coast Guard recommended that no criminal action be brought against the captain, who explained that he simply couldn't bring himself to sacrifice his dog. He also stated that he feared the lifeboat would capsize if he took the youths on board. Sixteen months later, in May 1975, the captain was indicted in a federal court for manslaughter, for refusing to eject his dog in favor of some of the swimmers who died.[27]

Most of us find something morally offensive in this incident. We think that the captain should have preferred the lives of the youths and veteran to that of his dog. But the two egalitarian theories we have examined (Regan's equal-rights view and Singer's utilitarian view) would support the captain's decision. Are they right contrary to our common intuitions? The split-level theory of morality attempts to explain why our intuitions are correct in favoring humans over animals, while still granting animals moral status.

The **split-level theory** is the name of a position, first set forth by Martin Benjamin and Donald VanDeVeer, which aims at correcting the preceding four positions.[28] The Cartesian no-status view and the indirect-obligation view contain the insight that rational self-consciousness endows human beings with special worth, but they err on two counts: (1) in holding that animals don't have this quality at all (we know now that some do to some extent) and (2) in holding that only rational self-consciousness gives one any rights or makes one morally considerable. At the other extreme, the equal-status view and the equal-consideration view recognize the importance of sentience and the ability to suffer as morally considerable, but these views tend to underemphasize the aspect of rational self-consciousness as setting the majority of humans apart from the majority of animals.

The split-level view combines the insights of both types of theories. It is non-speciesist in that it recognizes some animals, like chimpanzees and dolphins, may have an element of rational self-consciousness and some humans may lack it (say, fetuses and severely retarded people). The split-level view recognizes that both sentience and rational self-consciousness are important in working out a global interspecies morality. This view not only rejects Regan's sanctity-of-life egalitarianism, but it also rejects Singer's equal-consideration-of-interests principle. Rational self-consciousness does make a difference. A higher sort of being does emerge with humanity (and perhaps some higher primates and dolphins), so that we ought to treat humans with special respect.

This theory distinguishes between *trivial needs* and *important needs*. It says that with regard to important needs, human needs override animal needs, but animal important needs override human trivial needs. For example, the need for sustenance and the need not to be harmed are important needs, whereas the need for having our tastes satisfied is a trivial need. So while humans have the right to kill animals if animals are necessary for health or life, we do not have the right to kill higher animals simply to satisfy our tastes. If there are equally good ways of finding nourishment, then humans have an obligation to seek those ways and permit animals to live unmolested.

The above applies to higher animals who have a developed nervous system, enabling them to suffer and develop a sense of consciousness. Since no evidence shows that termites or mosquitoes have a sense of self, it is permissible to exterminate the termites and kill the mosquitoes when they threaten our interests. Of course, if we someday discover that termites and mosquitoes are highly self-conscious, we will be obliged to act differently, but until we have evidence to that effect, we may continue our present practices.

The split-level theory has been criticized as being unabashedly anthropocentric. Why should human criteria, such as self-consciousness and rationality, be the litmus tests for moral consideration?

Most of us who think humans should have greater consideration than animals and plants have a difficult time explaining why. We can appeal to sentience and say that sentient beings can suffer and enjoy pleasure, so that all sentient beings, animals and humans, are more morally considerable than plants, species, and ecosystems that cannot suffer or enjoy pleasure. We can say that humans typically are more sensitive, future-oriented, and so forth, so that their interests should receive some added weight. But there is a more important difference between humans and most animals. Humans are capable of moral deliberation and moral action. They have moral capacities. They merit higher status by virtue of their moral capacity—this ability to live morally. Since morality is the most important quality in the world, humans have by nature the most important quality in the world. Having moral capacity *merits* special consideration. Although this basic quality is not deserved—we did nothing to earn it—it is the basis of desert.[29] If humans actualize their moral capacity, they are deserving of special status in proportion to how moral they become. They not only merit but also deserve special consideration over animals. If they fail to actualize their moral capacity in becoming virtuous persons, they forfeit their merit and merit turns to a demerit. They deserve reprobation in that they have squandered their birthright and become immoral. So a hierarchy exists. Only rational beings, humans and other beings capable of moral choice and action, deserve special consideration, happiness, or well-being in the highest sense.[30] Animals, as innocent, deserve to be left alone in pursuing their good—even if it means killing each other for survival—though it might be supererogatory of humans to protect and help them flourish—say in safari parks or as domesticated animals. But immoral people, insofar as they are immoral, deserve punishment, because they have forfeited their value. They are lower than the animals in the value hierarchy.

If this reasoning is sound, preferring human interests to nonhuman is not chauvinistic or speciesist, but is based on the nature of morality itself as the highest good in the world. This merit theory is not anthropocentric, though it may well be **anthropogenic**—discovered and invented by humans, even as humans discovered the use for and invented the wheel.

It follows that humans have special responsibilities to care for nature, for distant people, for animals, and for future people, since these duties follow from the nature of morality—as the promotion of well-being and amelioration of suffering.

The merit theory provides justification for the split-level theory of morality: Humans by nature merit higher regard, but they must earn that high status by actualizing it via inculcating the moral law, by becoming virtuous beings who do their duty to all of nature.

Which of these five theories is correct? The no-status theory and the indirect-obligation theory seem to underrate the worth of animals, but the equal-status theories seem to overrate it, leaving some form of the split-level theory a victor by virtue of tying moral standing to actual moral goodness.

VEGETARIANISM

In the beginning of this chapter we noted some facts regarding animal mistreatment. Free-range grazing and farming has been largely replaced by the complicated technology of agribusiness, which promotes an impersonal process that tortures chickens, turkeys, cows, calves, pigs, and other animals. Since we have a moral obligation to eliminate gratuitous suffering whenever possible, we should cease to participate in the factory-farming food production of our society. Rather, we should oppose it. Since more than 95% of the (non-fish) meat that reaches the market is factory-bred, we have a strong duty to refrain from eating most of the meat that is available today. But there is good reason to refrain from eating all meat. Why kill any

THE FAR SIDE By GARY LARSON

"Lord, we thank thee."

animal when it's not necessary to do so? With very little effort we can have a healthy diet using vegetables, fruits, grains, and nuts. If our health depended on meat, as is the case in arctic climates and some less developed cultures, the injunction against causing unnecessary pain and suffering would be overridden. But this is not our culture. In fact, we can live healthier lives without meat. Given the hormones and antibiotics routinely added to cattle and poultry feed and the chemicals that may be added later as the meat is processed and readied for sale, meat is becoming a health hazard. Since most dairy-product production also involves cruel treatment of animals in factory farms, we should try to use only milk products produced from free-range farm animals.

Why do otherwise good people eat meat? When I ask my friends and students why they eat meat, they say they've just never thought about it—so deeply is it ingrained in our culture. Yet they would not think of eating dogs—who are typically not as intelligent as the pigs they eat. Or they say that they have a passion for steak or veal. Or they have developed the habit of meat eating. But habit and taste preference are not good moral reasons. Habit is simply a culturally derived practice, and taste is an aesthetic consideration. Morality trumps both cultural practices and aesthetics—what would you think if I said I love the look of red blood that bursts out of the veins of my victims as I cut their arteries? You'd rightly think that I was either crazy or morally depraved. But with the backing of culture and in the name of big business (which the meat industry is), we kill helpless animals for aesthetic reasons. Looked at objectively, from the point of view of a common core morality that would require us to avoid causing unnecessary pain and suffering and that would require us to do good whenever we can do so with minimal effort, meat eating is a vicious practice. Reflective moral people in a society like ours, where other kinds of food are readily available, will become vegetarians.

Vegetarianism makes good sense from an energy-conservation point of view. We would save 90% of nutritional energy if we ate the grain and beans now fed to the factory animals instead of the animals themselves. The argument, called the trophic-levels argument,* goes like this:

No transfer of energy from one **trophic level** to another is 100%. In fact, only about 10% of the chemical energy available at one trophic level gets transferred and stored in usable form in the bodies of the organisms at the next trophic level. In other words, about 90% of the chemical energy is degraded and lost as heat to the environment. This is sometimes called the ten percent law. The percentage transfer of useful energy between trophic levels is called *ecological efficiency,* or *food-chain efficiency.* For example, suppose a man eats some grain or rice containing 10,000 units of energy (calories). Ninety percent of the energy is lost in the transfer, so only 1,000 units reach him. Now suppose that a steer eats that same grain, containing 10,000 energy units and the steer is slaughtered and eventually fed to the man. The steer gets the 1,000 units, of which the man gets only 100 units, 90% of the energy being lost (as heat energy) in the transference process. Figure 7-1 illustrates this pyramid of energy transfer in aquatic grazing systems. Original producers of food are the phytoplankton, which are eaten by zooplankton, which in turn are eaten by fish, which in turn are eaten by humans. There is a 90% usable energy loss (which is turned into heat) at each trophic level. In this case the fishermen get only about 10 units of energy for every 10,000 original units. It would be 1,000 times more energy efficient if we could acquire a taste for phytoplankton salad!

Decreased efficiency occurs as we increase the distance between the primary producers of food and the ultimate consumers. We would get a lot more food value if we ate the beans and grains that are fed to animals,

---

* *Trophic,* from the Greek *trophe,* "to feed," means pertaining to nutrition, having to do with the process of nutrition.

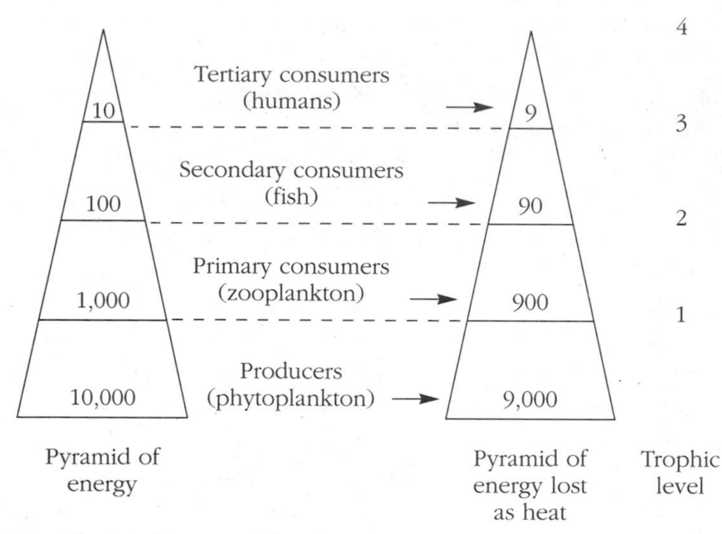

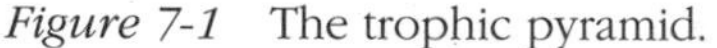

Figure 7-1 The trophic pyramid.

rather than the animals themselves. More than 80% or 90% of food energy could be saved in the process, and thus more food could be made available to feed the global population.

So vegetarianism is both rational and moral. There are no compelling reasons for prohibiting eating dead farm animals, of course, but I doubt that this will hold much interest for meat-lovers. Land presently used to graze animals can often be better used to grow food crops. Mary Ann Warren has pointed out that there may be environmental reasons for refraining from grazing free-range animals, for they greatly damage the ecology of fragile arid lands. She writes, "It might be better to do less grazing of domestic animals on marginal land, and go on feeding some animals with grain."[31] A compromise could be made, letting animals graze freely but feeding them mostly cultivated crops; those animals could provide our eggs and dairy products.

Some philosophers argue that if we didn't eat large quantities of meat, many domestic animals, such as cows and pigs, would never live, since they are raised for food. Given that some life is better than none, we are justified in killing them for food. But this is a bad argument. We wouldn't breed extra people simply because "more life is better than none." It were better that we didn't bring domestic animals into the world than to do so and kill them for food.

There are exceptions to strict vegetarianism. Sometimes animal populations need to be culled—for example, many of our forests have too many deer. So hunting deer for food is sometimes ecologically warranted. I suppose eating insects, phytoplankton, and small sea creatures who lack a developed nervous system that produces a minimal sense of self-consciousness is permissible. Anyone care for termites for lunch?

ANIMAL EXPERIMENTS

We have noted that millions of animals die in laboratory experiments each year. Richard Ryder puts the figure between 100 and 300 million worldwide.[32] Many of these experiments have been shown to be unnecessary and unjustified. More needs to be done to avoid unnecessary duplication and to enforce the laws requiring the use of anesthetics. Andrew Rowan and others have argued that alternative, nonanimal testing procedures, such as CAT and PET scans, *in vitro* cell and tissue cultures, and autopsy studies, can replace animal experiments in many instances. For example, CAT and PET scans have been valuable in the study of Parkinson's disease, visual physiology, and musculoskeletal tumors.[33]

Nevertheless, Regan's principle of rejecting all animal experimentation is to be rejected. Most proponents of alternative nonanimal testing procedures concede that the cessation of animal experimentation at the present time or foreseeable future would have serious adverse consequences for humans. If we accept the split-level theory, discussed earlier, then experimentation, with safeguards, should continue.

It is vitally important that we eliminate or control serious disease (cancer, heart disease, and infectious diseases like malaria and AIDS), and to do this animal experiments are often necessary. Medical research on animals has helped bring about treatment of diabetes, cancer, stroke, and heart ailments. "Dogs were used in the discovery of insulin and monkeys were used in the development of a polio vaccine."[34] We have no reason to believe that we can discover new improved treatments for present diseases without experimenting on animals. Nonetheless, we should use every safeguard in preventing abuse as we carry out our experiments. Most, if not all, university research centers now have institutional animal care and use committees to review and approve experimental protocols. But we can eliminate animal experiments in testing cosmetics, football helmets, and other items not vital to human flourishing. Either do without these luxuries (cosmetics and football helmets—we might have to resort to touch football) or find other ways of testing the products.

Singer, more moderate than Regan, would accept that some experimentation is justified. "Experiments serving no direct and urgent purpose should stop immediately, and in the remaining areas of research, methods involving animals should be replaced as soon as possible by alternative methods not involving animals."[35] But this may be too restrictive. As Henry Sackin has pointed out,

> Very often we simply do not know whether a path of scientific inquiry will ultimately lead to something important or useful. The discovery of the structure of DNA in 1953 would have been regarded by animal advocates as inappropriate and irrelevant. Yet it forms the basis of molecular biology that now makes it possible to genetically engineer pest-resistant crops and provide adequate and inexpensive insulin for diabetics. Only those of us who possess divine omniscience can argue that certain animal experiments are less important than others. We simply do not know what is or is not important until after we understand the implications of that knowledge. This may explain why Nobel Prizes are often awarded years after the original discoveries were actually made.[36]

HUNTING FOR SPORT, FURS, AND ZOOS

Unless we subscribe to an egalitarian theory of animal rights, hunting for food seems morally permissible, and in some cases hunting may be warranted in keeping the animal population down; but hunting primarily for sport, trophies, or the fun of killing are activities much harder to justify from a moral perspective.[37] Perhaps they are impossible to justify. Here is a hunter's description of his killing of an elephant:

> The elephant stood broadside to me, at upwards of one hundred yards, and his attention at the moment was occupied with the dogs. I fired at his shoulder, and secured him with a single shot. The ball caught him high on the shoulder-blade, rendering him instantly lame; and before the echo on the bullet could reach my ear, I plainly saw the elephant was mine. I resolved to devote a short time to the contemplation of this noble elephant before laying him low; . . . I quickly kindled a fire and put on the kettle, and in a few minutes my coffee was prepared. There I sat in my forest home, cooly sipping my coffee, with one of the finest elephants in Africa awaiting my pleasure beside a neighboring tree. . . .
>
> Having admired the elephant for a considerable time, I resolved to make experiments for vulnerable points. [The hunter misses the vulnerable points, further wounding the elephant, but finally succeeds in delivering a fatal shot.] Large tears now trickled from [the elephant's] eyes, which he slowly shut and opened, his colossal frame quivered convulsively, and falling on his side, he expired.[38]

Reflecting on this passage should be sufficient reason to condemn hunting for trophies.

If furs are necessary for people to keep warm, we may legitimately kill animals for their furs; but if less violent ways of producing clothes are available, we should take them. I confess that, as a hiker, my feet prefer leather shoes, but we probably can produce suitable nonleather alternatives. Of course, we could use the carcasses of dead animals for our leather. Aside from that, if we can produce good synthetic shoes, we should begin to do so. Leather sofas and coats probably are not morally acceptable since we can produce good-quality sofas and coats in other ways.

Zoos present a special problem. On the one hand, they provide a splendid source of amusement and education and sometimes even opportunities for scientific research. On the other hand, keeping wild animals in captivity involves considerable suffering. Animals must be captured, removed from their native habitats, transported in confining vehicles, and maintained in an artificial environment. Do the benefits of zoos outweigh the suffering caused in removing animals from the wild and maintaining them in alien environments? Dale Jamieson has argued that they do not, concluding that zoos are not morally justified.[39] He argues that our amusement is not a good reason to perpetrate suffering, that we might be better educated about animals by viewing films and listening to lectures, and that very little research really goes on in most zoos. Moreover, zoos encourage dangerous speciesism:

> [Zoos] teach us a false sense of our place in the natural order. The means of confinement mark a difference between humans and animals. They are there at our pleasure, to be used for our purposes. Morality and perhaps our very survival require that we learn to live as one species among many rather than as one species over many. To do this, we must forget what we learn at zoos. Because what zoos teach us is false and dangerous, both humans and animals will be better off when they are abolished.[40]

Is Jamieson correct in advocating the abolition of zoos? While his considerations have merit, the counterarguments need to be heard. Granted, the amusement factor is not itself a good reason for causing suffering; nonetheless, zoos can be made enjoyable habitats for at least some animals. Forestlike habitats for chimpanzees and monkeys, large prairie areas for lions and tigers, and so forth can approximate the wild, providing spacious prairies in which to roam, while at the same time protecting them from predators of the wild. Granted, they may become partially domesticated in zoos. Why is that wrong? While there may be a presumption against taking an animal out of its native habitat, surely this can be overridden if it will benefit the animal itself. Their life expectancy is typically greater in the zoo than in the wild. They are assured food and shelter and may very well flourish. At least, we must do all we can to assure that they do. I grant that there's a deep issue here about what is "natural" and how we are to understand "benefiting animals." Perhaps it will turn out that some species of animals are benefited and some are not. A lot more research on animal psychology needs to be done, but I confess I see no argument enjoining an absolute prohibition on domesticating wild animals.

Furthermore, Jamieson may underestimate the educational value of zoos. Here I have no statistics, but can only attest to the joy and wonder I felt as a child—and still feel—when I visited the Brookfield Zoo in Illinois. Watching monkeys at play, noting the resemblance between chimpanzees and humans, hearing the lion roar, gazing at the long neck of the giraffe, studying sharks and dolphins, and learning to identify snakes constituted one of the most profound learning experiences of my childhood, and I imagine that others can testify to the same experience.

Granted, animals in zoos should be treated kindly and conditions must be such that they benefit from these conditions; but if we can do this, then probably zoos can be justified. Perhaps safari parks should replace zoos for larger animals.

Let us sum up our discussion.

All conscious beings are morally considerable, since they can suffer and be aided in flourishing; but, unless we want to stop killing harmful (to us) bacteria, we can agree that some animals are more valuable than others. Those who have a sense of personal identity, who are rationally self-conscious and have plans and projects, deserve special consideration. So the split-level view seems, on balance, to be the correct position. Regarding important needs, human needs in general override the needs of lower animals; but regarding trivial needs, the important needs of animals override these nonvital needs of humans. Speciesism is a vice, like racism, if it gives an irrational sense of privilege to humans simply because they are human; but because humanity as a species is superior to other species (based on its moral sense), no evil is involved in developing general policies that reflect that fact.

The principle that should be minimally at work in our relations with animals is the principle of nonmaleficence. We ought not to cause unnecessary harm or suffering. Is there a better way to live and eat? If so, then we should take it. Do we need to do all of the research and animal experimentation we are presently engaged in? If not, we should cut out the unnecessary experimentation. Are hunting and trapping necessary for human flourishing? If not, we should work toward the elimination of these activities. For the rest, we should be kind to all of the creatures of the earth, trying to make this cruel jungle into a closer approximation of the Garden of Eden.

Look again at Regan's three goals: (1) the total abolition of the use of animals in science, (2) the total dissolution of commercial animal agriculture, and (3) the total elimination of commercial and sport hunting and trapping. If the split-level analysis is correct, all of these goals may be rejected as absolutist, but each has a point. Except in situations of special need, we should cease eating meat, but free-range-produced dairy products are permissible. Abuses must be eliminated in research and animal factories abolished. Hunting and trapping for sport should in general be prohibited, but exceptions might be permitted—for example, when a deer population is reaching a dangerous surplus and a dieback is likely. Animals might be trapped as humanely as possible for fur where severe climates exist.

Schematically put, here are the results of our analysis:

JUSTIFIED ACTIVITIES

1. Using milk products from free-range animals, especially if they are necessary for a community's well-being or survival. Note that since cows and goats eat grass, which are not digestible by humans, no energy loss occurs in eating cows and goats.
2. Important biomedical experiments to increase scientific knowledge and find cures for diseases. But safeguards must be built in to these experiments, minimizing the harm done to the animals. Unnecessary duplication should be prevented.
3. Zoos in order to enable city dwellers to learn about animals. They must meet minimum standards of decency for animals, including spacious areas for cats like lions and tigers to roam.
4. Hunting when necessary for food or when an animal population gets out of control, but not for sport.
5. Trapping for fur where severe climatic conditions exist and perhaps where this is the traditional clothing of indigenous people.

ACTIVITIES NOT JUSTIFIED

1. Animal factories—where the suffering of animals is terrible and unnecessary.
2. Eating meat that involves killing animals, except when hunting in order to keep a population under control.
3. Nonvital experimentation—cosmetic experiments, football helmets.
4. Other luxury activities where animals are unnecessarily exploited—hunting for sport, trapping for fur for fashion wear, bullfights.

At the end of *Animal Liberation* Singer issues a challenge for us all:

> Human beings have the power to continue to oppress other species forever, or until we make this planet unsuitable for living beings. Will our tyranny continue, proving that we really are the selfish tyrants that the most cynical of the poets and philosophers have always said we are? Or will we rise to the challenge and prove our capacity for genuine altruism by ending our ruthless exploitation of the species in our power, not because we are forced to do so by rebels or terrorists, but because we recognize that our position is morally indefensible?[41]

While we have warrant for using some animals for human good, we ought to modify many of our practices, realizing that animals have interests and are morally significant.[42]

Review Questions

1. Discuss the two quotations at the opening of this chapter. Are either of them true or close to the truth? Explain.
2. What are the implications of the moral principle "Do no unnecessary harm"? Would it lead us to abolish factory farming? eating meat? some animal experimentation? all animal experimentation? hunting? Explain your position.
3. Do animals have rights? Are they morally considerable? Discuss the five theories on the moral status of animals outlined in this chapter. Which position seems the best to you? Would that position call for changes in your actions?
4. Is it morally permissible to eat meat? Under what circumstances would it be moral or immoral to do so?
5. Discuss the trophic pyramid. What are the implications for our diets?
6. Is hunting for sport morally permissible? Consider the description of a hunter's killing an elephant in the last section of this chapter.
7. Robert White and Henry Sackin, among others, have argued that without experimentation on animals the cures for many diseases would not have been discovered. Does this justify animal experimentation? Explain your answer.
8. Discuss the case of the captain who chose to save his Labrador dog rather than two youths in light of the theories concerning the moral status of animals discussed in this chapter. What do you think was the morally right thing to do. Why?

Suggested Readings

Abbey, Edward. *Desert Solitaire*. New York: Ballantine, 1968.

Benjamin, Martin. "Ethics and Animal Consciousness," in *Social Ethics,* ed. Thomas Mappes and Jane Zembaty. Englewood Cliffs, N.J.: Prentice Hall, 1982.

Frey, R. G. *Rights, Killing, and Suffering*. Oxford: Blackwell, 1983.

Kleinig, John. *Valuing Life*. Princeton: Princeton University Press, 1991.

Regan, Tom. *The Case for Animal Rights*. Berkeley: University of California Press, 1983.

Rohr, Janelle, ed. *Animal Rights: Opposing Viewpoints*. San Diego: Greenhaven Press, 1989.

Singer, Peter. *Animal Liberation,* 2nd ed. New York: *New York Review of Books,* 1990.

———. "Animals and the Value of Life," in *Matters of Life and Death*. Edited by Tom Regan. New York: Random House, 1980.

Taylor, Paul. *Respect for Nature*. Princeton: Princeton University Press, 1986.

Notes

1. The Office of Technology Assessment (OTA) puts the number of animals bred for laboratory experiments each year between 35 and 50 million. Andrew Rowan, who has made

one of the most thorough studies of the matter, told me that he had downscaled his estimate from between 60 and 70 million animals to a figure close to that of the OTA. For helpful discussions of the scientific literature on the ethics of animal research see M. T. Phillips and J. A. Sechzer, *Animal Research and Ethical Conflict* (New York: Springer-Verlag, 1989) and Andrew Rowan, *Of Mice, Models, and Men: A Critical Evaluation of Animal Research* (Albany, N.Y.: State University of New York Press, 1984).

2. The material in this section on animal experimentation is taken from Richard Ryder, *Victims of Science: The Use of Animals in Research* (London: National Anti-Vivisection Society, 1983); Andrew Rowan, *Of Mice, Models, and Men;* Peter Singer, *Animal Liberation,* 2nd ed. (New York: *New York Review of Books,* 1990); Muriel the Lady Dowding, "Furs and Cosmetics: Too High a Price?" in *Animals, Men, and Morals,* ed. Stanley Godlovitch, Rosline Godlovitch, and John Harris (New York: Taplinger, 1972); and Philip Zwerling, "Animal Rights, Human Wrongs," *Animal Agenda,* December 1985.
3. Singer, *Animal Liberation,* p. 29f. Other reports tell of dogs that are driven insane with electric shocks so that scientists can study the effects of insanity. Cats are deprived of sleep until they die. Primates have been restrained for months in steel chairs allowing no movement, and elephants have been given LSD to study aggression. Legs have been cut off mice to study how they walk on the stumps, and polar bears have been drowned in vats of crude oil to study the effect of oil spills in polar regions.
4. Singer, *Animal Liberation,* p. 80.
5. The material in this paragraph is based on John Robbins's *Diet for a New America* (Walpole, N.H.: Stillpoint, 1987).
6. James Rachels, "Vegetarianism," in *World Hunger and Moral Obligation,* ed. William Aiken and Hugh LaFollette (Englewood Cliffs, N.J.: Prentice Hall, 1977), pp. 180–93.
7. René Descartes, *Discourse on Method* in *The Philosophical Works of Descartes,* trans. E. S. Haldane and G. R. T. Ross (Cambridge: Cambridge University Press, 1911), vol. 1.
8. From Nicholas Fontaine's *Memoirs,* quoted in Peter Singer, *Animal Liberation* (New York: Avon, 1976), p. 220. The nineteenth-century British philosopher P. Austin modifies this Cartesian view (named after Descartes) by stating that while we have no positive duties to animals, we ought not to be cruel to them: "Animals should be treated with personal indifference; they should not be petted, they should not be ill-treated. It should always be remembered that they are *our slaves*, not our equals, and for this reason it is well to keep up such practices as hunting and fishing, driving and riding, merely to demonstrate in a practical way man's dominion."
9. William Whewell, *Lectures,* cited in *Animal Rights and Human Obligations,* ed. Peter Singer and Tom Regan (Englewood Cliffs, N.J.: Prentice Hall, 1976), p. 131.
10. Cited in M. T. Phillips and J. A. Sechzer, *Animal Research and Ethical Conflict,* p. 75.
11. Some Christian vegetarians point to this passage, which instructs humans to eat only plants, as a justification for vegetarianism. Meat eating was permitted only after the flood in Genesis 9.
12. These passages from Thomas Aquinas are from *Contra Gentiles* (London: Benziger Brothers, 1928), Book III, Part II, ch. 112.
13. Immanuel Kant, *Lectures on Ethics*, trans. Louis Infield (New York: Harper & Row, 1963), p. 239.
14. K. McCabe, "Who Will Live, Who Will Die," *The Washingtonian* (August 1986).
15. Tom Regan, "The Case for Animal Rights," in *In Defense of Animals*, ed. Peter Singer (Oxford: Blackwell, 1985), pp. 13–26.
16. Ibid., p. 13.
17. For a fuller discussion of this point, see my article "On Equal Human Worth: A Critique of Contemporary Egalitarianism," in *Equality*, ed. L. Pojman and R. Westmoreland (Oxford: Oxford University Press, 1997).

18. There are exceptions to this point. We do establish social contracts with our pets. If I give my dog or cat reasonable expectations and he or she does the same for me, so that we rely on one another, this may well constitute a moral contract.

19. Vitalism is the doctrine that processes of life are not explicable by the laws of physics and chemistry alone and that life is in some sense self-determining.

20. Tom Regan, *The Case for Animal Rights* (Berkeley, Calif.: University of California Press, 1983), p. 284f.

21. Donald VanDeVeer challenges Regan's all-or-nothing notion of inherent value. He argues that inherent value can come in degrees. He begins the following quotation with Regan's thesis:

    "The question of whether anything *has* moral standing *remains,* in my estimation, an all or nothing matter. . . . Either a thing has inherent value or it does not." Regan takes this fact to be a reason to claim that "Inherent value is thus a categorical concept. . . . It does not come in degrees." This view . . . seems to arise from a logical confusion between two distinct questions. The first is the question of whether an entity *possesses* a certain property, given a certain precision in the specification of the property. Either the entity will or it will not. The other question is *whether the property when possessed can vary in magnitude or degree.* . . . Either there is something of monetary value in my pocket or there is not. If there is, we may ask how valuable it is. Such a question is not precluded by the nature of the *concept* of monetary value. I maintain that we also can sensibly inquire how intrinsically valuable is a living individual, and we may maintain that a life has inherent value even if its inherent value is not great. I know of no plausible analysis of "inherent value" or "inherent worth" such that these very *concepts* preclude judgments of varying levels of value or worth. (D. VanDeVeer, "Interspecific Justice and Intrinsic Value," *Electronic Journal of Analytic Philosophy* 3, 1995)

22. Jeremy Bentham, *The Principles of Morals and Legislation* (1789), ch. 17, sec. 1. The main weakness of Bentham's account is that it equates humans and animals, so that if a chicken and a child are suffering equal pain, and we only have one pain reliever, we have a genuine moral dilemma on our hands. We might consider giving it to the chicken.

23. Peter Singer, *Animal Liberation,* p. 5.

24. Peter Singer, "Animals and the Value of Life," in *Matters of Life and Death*, ed. Tom Regan (New York: Random House, 1980), p. 236.

25. Peter Singer, *Animal Liberation,* p. 22.

26. Peter Singer, *Animal Rights and Human Obligation* (Englewood Cliffs, N.J.: Prentice Hall, 1976).

27. "Captain Indicted in Deaths at Sea," *New York Times,* 21 May 1975, p. 93; and "Captain Says He Feared Lifeboat Would Capsize," *New York Times,* 13 May 1976, p. 39.

28. Martin Benjamin, "Ethics and Animal Consciousness," in *Social Ethics,* ed. Thomas Mappes and Jane Zembaty (Englewood Cliffs, N.J.: Prentice Hall, 1982); Mary Anne Warren, "A Critique of Regan's Animal Rights Theory," *Between the Species* 2 (1987); Donald VanDeVeer, "Interspecific Justice," in *People, Penguins, and Plastic Trees,* ed. Donald VanDeVeer and Christine Pierce (Belmont, Calif.: Wadsworth, 1986); Louis Lombardi, "Inherent Worth, Respect, and Rights," *Environmental Ethics* 5 (Fall 1983). See also James Sterba, "Environmental Justice: Reconciling Anthropocentric and Nonanthropocentric Ethics" in *Environmental Ethics,* ed. L. Pojman (Belmont, Calif.: Wadsworth, 1998). Warren's position is that all sentient animals, capable of having experiences, have moral rights, but they are not of equal strength as those of persons.

29. I distinguish *merit* from *desert.* Merit consists of any property that is valuable to something else. It is a consequentialist value. Desert consists in what one intends, whether one has a good will, whether one makes an effort to do a certain thing. It is deontological.

30. Frans de Waal, in his remarkable work *Peacemaking Among Primates* (Cambridge: Harvard University Press, 1989), makes a good case for a moral consciousness in chimpanzees. If this is so, we should afford chimpanzees the same moral status as humans.
31. Mary Anne Warren in correspondence, 25 January 1991.
32. Ryder, *Victims of Science.*
33. Rowan, *Of Mice, Models, and Men.*
34. Henry Sackin, "An Ethical Basis for Animal Experimentation" (unpublished paper).
35. Peter Singer, *Animal Liberation,* p. 32.
36. Sackin, "An Ethical Basis for Animal Experimentation."
37. See Robert Loftin, "The Morality of Hunting," *Environmental Ethics* 6 (Fall 1984).
38. Quoted in Richard Carrington, *Elephants* (New York: Chatto & Windus, 1958), p. 158.
39. Dale Jamieson, "Against Zoos," in *Reflecting on Nature*, ed. Lori Gruen and Dale Jamieson (New York: Oxford University Press, 1994), pp. 291–98.
40. Ibid., p. 298.
41. Peter Singer, *Animal Liberation,* p. 258.
42. John Kleinig, Michael Levin, Trudy Pojman, and Sterling Harwood made helpful criticisms on a draft of this chapter, for which I am grateful.

CHAPTER EIGHT

Does Nature Have Objective Value?

THE CONCEPT OF NATURE

Environmental ethics has to do with the value of nature and our relationship with nature. But what is nature? The Latin *natura* or *natus* derives from the verb "to grow" and "to be born," which in turn is related to the Greek *gignomai* "to be born," roots that survive in *pregnant, genesis,* and *native.* Nature is whatever has been generated and come to be. Sometimes it contrasts with *supernatural,* that which transcends nature, and at other times it contrasts with *artificial* or *cultural.* One of the main dictionary definitions states that it is the material or physical world, especially that unaffected by man, including plants, animals, geographical features, and so on or the places where these exist largely free of human influence. But *nature* also refers to the laws and principles believed to govern the universe, including the laws that *ought* to be followed by human beings in their conduct. The Cynics and Stoics advocated living "according to Nature," as though there were an invincible **natural law** that governed human behavior, the violation of which spelled disaster. The Greek Cynic Diogenes advocated living according to nature, simple and animal-like. He lived in a large wine cask, wrapped a strip of cloth around his body in place of fur, and lapped water like a dog. He withdrew from society, rejecting its pleasures and claims. Christian monks were to follow his example, though for religious reasons. Cicero spoke of a *lex naturale* (a universal law of nature) that was equivalent to the *jus naturale,* the universal justice, the law that should rule all society. The natural light of reason (*lumen naturale*) was able to discover it. Thomas Reed would later call this natural light "common sense," and Bolingbroke said, "The tables of natural law are so obvious that no man who is able to read the plainest character can mistake them, and therefore no political society ever framed a system of law in direct and avowed contradiction to them."[1] For Rousseau, the essence of this pristine goodness was found in the "noble savage" unspoilt by civilization. Alexander Pope wrote

> Lo, the poor Indian! whose untutored mind
> Sees God in clouds, or hears him in the wind;
> his soul proud Science never taught to stray
> Far as the solar walk or milky way;

yet simple nature to his hope has giv'n,
Behind the cloud-topped hill, an
humbler Heav'n.

(*Essay on Man,* Epistle I: 99–104)

For the Greeks and Romans, especially the Stoics, the natural is the good, the original, the unspoilt. It is superior to the artificial or merely conventional. Hence the universal moral law, discoverable by reason or intuition, is better than humanmade law, convention. But, good as the natural was, the supernatural was even better. In the Judeo-Christian tradition, nature is good—"God made the heavens and the earth . . . and saw that it was good" (Gen. 1:1, 4)—but sin in the guise of human disobedience corrupts it. Accordingly, homosexual behavior was bad because it did not fit the natural purposes of sexual intercourse. Nature itself is regular, reliable, lawlike, obedient to the rules laid down by God; but a wild irregularity has been produced by sin, so that nature is out of joint, filled with restless violence, manifest in earthquake, tornado, hurricane, volcano, and predation. Nature is seen "groaning in travail . . . waiting with eager longing" for redemption. It will be set "free from bondage to decay" in a glorious redemption (Rom. 8:19–22). The idea is that nature is alienated from God, wild and restless, but will be restored as a peaceable kingdom, where the lion shall lie down with the calf, "where the wolf and the lamb shall feed together, the lion shall eat straw like the ox; and the dust shall be the serpent's food. They shall not hurt or destroy in all my holy mountain" (Isa. 65:25).

As far back as the twelfth century and reflected in the work of Giordano Bruno and Benedict Spinoza, nature is seen as having two aspects: an active and a passive, what is called *natura naturans* and *natura naturata*. Active nature is an omnipotent, infinite energy, or force, that acts, molding its will upon reality, which is itself nature as object. Passive nature, *natura naturata,* "nature natured," refers to nature as product, what is produced by itself. In Spinoza the two natures are fused as two dimensions of God, a pantheistic reality. "God or Nature," writes Spinoza, signifying they are the same. It follows that all is good, since it proceeds from the perfect nature of God, who does nothing in vain. He who understands this truth cannot but love God as he loves himself. "This love of God cannot be tainted with emotions of envy or jealousy, but is the more fostered as we think more men to be joined to God by this same bond of love (*The Ethics,* V: 20). This is pure intellectual love of God, wherein we see that our mind "is in God and is conceived through God." This is the way of highest possible pleasure, true blessedness and contentment (V: 30, 32). It follows, since God and nature are one, that true blessedness is attained by becoming one with nature.

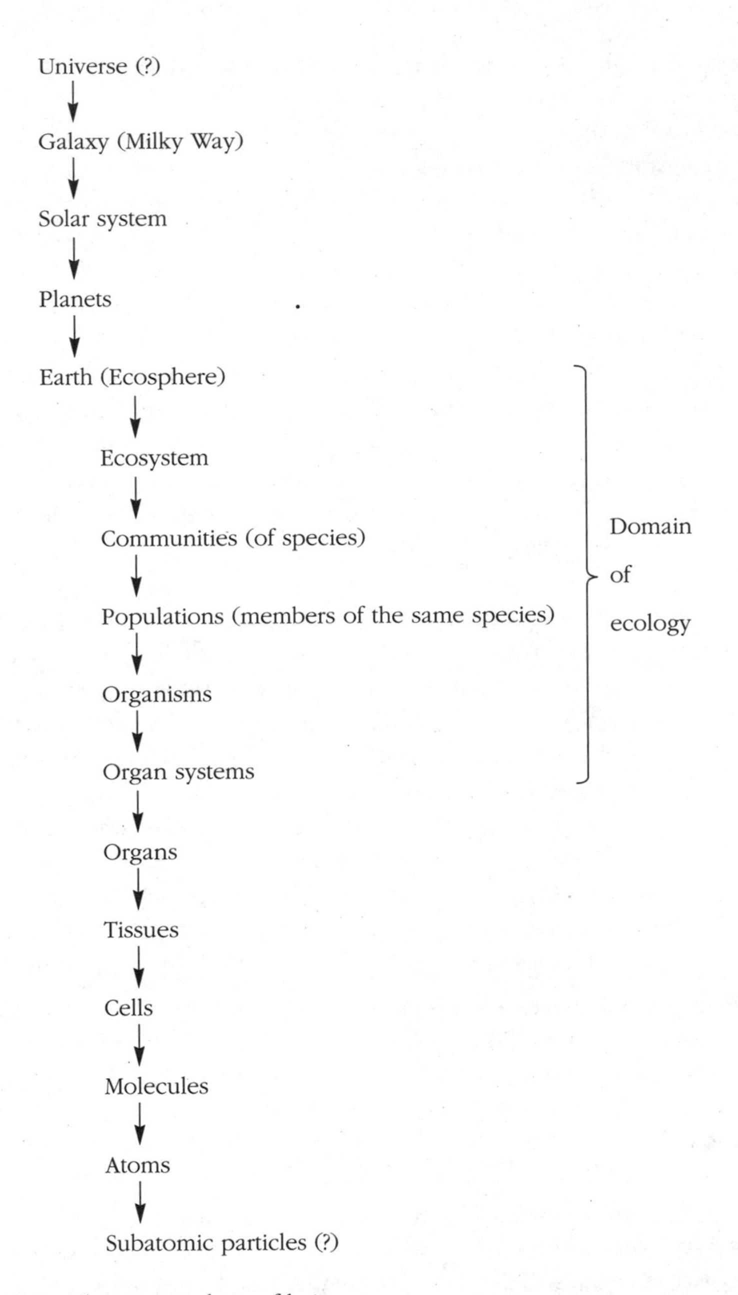

Figure 8-1 The cosmic chain of being.

We will come back to the status of nature later in this chapter and the next. For now we may identify nature minimally and roughly with the physical and biological environment, with what the ecologist is interested in. For the ecological philosopher, nature refers to the chain of being—from the universe, the largest unit, to the subatomic particle, whatever is the smallest unit of reality and in the interaction of all things. Figure 8-1 is a representation of this cosmic chain of being.

We should define our terms. *Ecology* is derived from the Greek words *oikos* ("home") and *logos* ("study of"). It's the study of where and how organisms live. The place where they live is called their *habitat*. It is concerned with the relationship between organisms and their environment, how they interact with one another and with their physical and chemical surroundings. An *organism* is any form of life. Organisms can be classified into *species*. They are members of a common *species* if under normal conditions they can breed with one another and produce live, fertile offspring. Members of a common species in the same area are called a *population*. Populations of different species living in the same area are called a biological *community*. An *ecosystem* is a community of different species interacting with one another and with the abiotic (nonliving) environment. The *ecosphere* consists of all the Earth's ecosystems. The scope of nature that concerns environmental philosophers is, for the most part, that zone between the ecosphere and organ systems that we refer to as the *domain of ecology*. Environmental ethics has to do with human responsibility to the subjects within that domain.

DOES NATURE HAVE OBJECTIVE VALUE?

In the second book of the *Republic,* Plato wrote that there were three kinds of value—those things that were good in themselves and for no other reason, such as a simple pleasure; those not good in themselves but good for what they bring, instrumentally good, such as a key or money; and those good both in themselves and for what they bring, both instrumentally and intrinsically good, such as happiness or knowledge. Plato also thought there were eternal forms, immutable, and transcendent, beyond the *World of Appearance;* they constituted true *Reality.* The greatest of the forms was the form of the Good, the highest form, in which all the other forms participated. It was both the supreme good in itself and instrumentally good, since it brought about all other goods. This notion of value existing in a world of its own, objectively, whether or not anyone recognizes it, is the paradigm of Value.

This way of putting the problem of value is not complete, for its example of an intrinsic value is an experience, namely, pleasure, rather than a thing, an object. Both subjectivists and objectivists can agree that some *experiences* are good in themselves. What is at issue, however, is whether values exist *objectively,* like table and chairs, stars and atoms, penguins and people, or whether they exist only *subjectively,* in the minds of appraisers (self-conscious evaluators like human beings) or experiencers (sentient beings like dogs, mice, and frogs). So it is really the question of the objectivity of values that we are interested in. We could, however, still use the term *intrinsic* to signify the objectivity of values if we applied it to *objects,* as when we ask, Do plants or landscapes have intrinsic value? I think it is better to distinguish the terms, however, so I will use the four terms as follows:

1. *Intrinsic value:* An object or experience valued for its own sake, such as pleasure, a moment of insight, a religious experience, or a sense of well-being. It need not refer to its contributions to other values.
2. *Extrinsic value:* Something valued for the sake of some other value, such as medicine, taken in order to promote health or well-being. We may also refer to this as instrumental value.[2]
3. *Subjective value:* Something, say a landscape, that becomes valuable by being appraised by an evaluator. My pleasure at seeing a landscape endows it with value. Value is in the eyes (or mind) of the beholder (or evaluator).
4. *Objective value:* Something that is valuable whether or not anyone appraises it as such. The landscape is good because of the features it possesses, whether or not anyone is there to see it. Life is good whether or not it is sentient, let alone whether anyone is able to appreciate it.

The dominant categories are subjective and objective value. An intrinsic or instrumental value may be either subjective or objective. For example, a person may subjectively value an abstract painting for its own sake (intrinsically), but the painting may or may not have objective value. Conversely, a person may only value knowledge instrumentally, whereas some knowledge may well have objective value. Objective values are like Plato's forms. They exist in their own right whether or not anyone notices them.

Recall the problem, first set forth in Plato's dialogue *Euthyphro,* as to whether the Good must exist independently of God, so that even God must measure up to this ideal. Is the Good good because God loves it, or does God love the Good because it is good? The objectivist answers that the Good is an ideal, independent of God, so that even God can't make what is evil good or vice versa (for example, God can't make it the case that torturing children is good). The subjectivist might be a divine-command theorist, who holds that God's valuing the Good is what makes it good (if God doesn't exist, neither does the Good). As far as values are concerned, we are all gods, creating good and evil. Our valuing some object or experience is what makes it good. So we can apply the *Euthyphro* problem to value in general: Is the Good good because a conscious being values it, or do conscious beings value the Good or good things because they are objectively good?

Now the question before us is whether nonsentient nature, the Earth with its plants, trees, geological formations, majestic mountains, grasslands, and rivers, has objective value? Or is value in the eye of the beholder; that is, is value something that we give nature by appreciating it? William James gives expression to the subjective view:

> Conceive yourself, if possible, suddenly stripped of all the emotions with which your world now inspires you, and try to imagine it *as it exists,* purely by itself, without your favorable or unfavorable, hopeful or apprehensive comment. It will be almost impossible for you to realize such a condition of negativity and deadness. No one portion of the universe would then have importance beyond another; and the whole collection of its things and series of its events would be without significance, character, expression, or perspective. Whatever of value, interest, or meaning our respective worlds may appear imbued with are thus pure gifts of the spectator's mind.[3]

Value inheres not in nature itself but in the appreciating mind. As R. B. Perry argued, objects acquire value if and only if a conscious being takes *interest* in it.[4] We may call this the *subjective theory of value.* It is diametrically opposed to both the Platonic and the Judeo-Christian concept of value, the *objective theory of value,* which holds that value exists objectively, whether or not anyone recognizes it.

This subjective view has been popular in contemporary philosophy, perhaps because of the lack of confidence in a transcendent world, let alone a God. *There is no value without an evaluator,* without at least a sentient being "desiring" or consciously appraising something as desirable. For *anthropocentrists,* human beings are the center of the world and, as such, are the sole valid appraisers of the world. As the Greek Sophist Protagoras said, "Man is the measure of all things, both those which are, that they are and those which are not, that they are not." Only humans, as self-conscious evaluators, can endow the world (or objects in it) with instrumental value by the act of valuing the world (or objects in it).

But several environmental philosophers have challenged this anthropocentric subjectivity. One group of philosophers rejects the narrowness of perspective in confining the evaluators to human beings. They would include animals as legitimate evaluators. We may call this the *sentience-centric view of values.* They include those who hold that animals have rights (see Chapter 7). A second group goes even further and rejects both the anthropocentrism and the criterion of subjectivity as the necessary locus of value. We call these *ecocentrists* or *biocentrists,* because they hold that the ecosystem, or all living things, has value, objective value.

The German philosopher-theologian-medical doctor Albert Schweitzer (1875–1965) is the most prominent thinker holding this type of view, one that he named *reverence for life,* the view that all of life is sacred and we must treat each living being as an inherently valuable "will-to-live." Schweitzer, an extraordinary versatile genius—a concert organist, a musicologist, a renowned New Testament scholar, a foremost Christian theologian, a philosopher and physician—at the age of 40 renounced fame as a scholar and dedicated his life to the amelioration of suffering in a hospital he founded in Lambarene in Gabon, West Africa. One day in 1915, while on a river journey to assist a patient, he had the following experience:

> At sunset of the third day, near the village of Igendja, we moved along an island in the middle of the wide river. On a sandbank to our left, four hippopotamuses and their young plodded along in our same direction. Just then, in my great tiredness and discouragement, the phrase "Reverence for Life" struck me like a flash. As far as I knew, it was a phrase I had never heard nor ever read. I realized at once that it carried within itself the solution to the problem that had been torturing me. Now I knew a system of values which concerns itself only with our relationship to other people is incomplete and therefore lacking in power for good. Only by means of reverence for life can we establish a spiritual and humane relationship with both people and all living creatures within our reach. Only in this fashion can we avoid harming others, and, within the limits of our capacity, go to their aid whenever they need us.

Schweitzer contrasts this view with that of typical rationalist philosophers with their anthropocentric notion of the self as an abstract, isolated entity, sufficient unto itself. Descartes is a prime example:

> Descartes tells us that philosophizing is based on the judgment "I think therefore I am." From this meagre and arbitrarily selected beginning it is inevitable that it should wander into the path of the abstract. It does not find the entrance to the ethical realm, and remains held fast in a dead view of the world and of life. True philosophy must commence with the most immediate and comprehensive facts of consciousness. And this may be formulated as "I am life which wills to live and I exist in the midst of life which wills to live." This is no mere excogitated subtlety. Day after day and hour after hour I proceed on my way invested in it. In every moment of reflection it forces itself on me anew. A living, world- and life-view, informing all the facts of life, gushes forth from it continually, as from an eternal spring. A mystically ethical oneness with existence grows forth from it unceasingly.[5]

Just as you and I seek happiness, Schweitzer sees each will-to-live yearning for a primordial version of that state of being, if only in the form of pleasure. Just as you and I flee pain and experience "terror in the face of annihilation and injury," so does every will-to-live do the same.

Ethics consists in practicing the same "reverence for life toward all will-to-live, as toward my own." Schweitzer states his fundamental principle of morality: "It is *good* to maintain and cherish life; it is *evil* to destroy and to check life." A person is truly ethical when he obeys the obligation to do what is in his power to aid all life in need and when he avoids doing unnecessary harm.

> He does not ask how far this or that life deserves sympathy as valuable in itself, nor how far it is capable of feeling. To him life as such is sacred. He shatters no ice crystal that sparkles in the sun, tears no leaf from its tree, breaks off no flower, and is careful not to crush any insect as he walks.* If he works by lamplight on a summer evening, he prefers to keep the window shut and to breathe stifling air, rather than to see insect after insect fall on his table with singed and sinking wings. . . . Should he pass by an insect which has fallen into a pool, he spares the time to reach it a leaf or stalk on which it may clamber and save itself. Ethics is in its unqualified form extended responsibility with regard to everything that has life.

Schweitzer ordered his life according to this principle. Instead of choosing the comfort of a cool breeze in his rooms, including his operating room, in hot equatorial Africa, he kept the windows shut so as not to attract insects and flies who would otherwise have immolated themselves against the kerosene lamps or hot lightbulbs.

Schweitzer's reverence for life has been heralded as the essence of morality, a recognition of our symbiotic relationship with nature, a benevolence to all living things. His theory is a version of *vitalism,* the idea that all living things are valuable. But is this true?

Note that in the preceding passage Schweitzer goes beyond the idea that all life is sacred to include inanimate objects such as ice crystals as part of the moral domain. But ignoring this problem and concentrating on his vitalism, one wonders if he ever weeded his garden or exterminated termites? Did he ever use bacteria-killing antibiotics to help cure his sick patients? According to Schweitzer's principle, all life is equally sacred, so that we may not choose between the life of a phytoplankton or a malaria-carrying mosquito on the one hand, and saving the life of a human being on the other. If life is the only relevant consideration and questions of quality do not enter in, two cockroaches are worth more than one human being. The more living things, the better, regardless of levels of sentience or consciousness.

Schweitzer's own answer to this problem was to admit that life is tragic. We must kill in order to survive. But we should recognize the deep tragedy while killing and kill no more than we absolutely have to in order to survive. A sense of remorse should accompany our daily endeavors. For our lives are of no more (or less) worth than the bacteria, weeds, trees, and nonhuman animals we kill. All of life is sacred—yet we cannot help but destroy these holy oracles.

Most of us will hesitate before accepting this mystic vision of the sacredness of all life. But perhaps we need to pause and ask ourselves what

---

* One may note that ice crystals are not alive, let alone sentient. Schweitzer seems to be going beyond reverence for life to reverence for all of nature.

really is valuable? Life surely is. It is wonderful! The sprouting of a plant, the blooming of a flower, the emergence of a chick from an egg, the birth of a baby—who would not stand in wonder at these miracles? The mystery of life is the ultimate miracle, one that must have amazed cave dwellers millennia ago as much as it does us today. Even those with little religious feeling stand in awe of the birthing process, and most people have a sense of wonder, viewing the birth of a baby as a gift of God, and so feeling that only God should be able to end human life. Murder, the unwarranted killing of human life, is universally condemned as the worst crime of all.

There is a primordial tendency to treat life as sacred, to surround life with a mystic aura and set high value on it. Edward Shils suggests that it is a universal, self-evident intuition that even predates organized religion. It is the "proto-religious" feeling. Thus it is not surprising that virtually every major religion has espoused a version of the idea of the sanctity of life. Indeed, this idea is essentially a religious one. Its roots are in a religious worldview. Animists believe that all things have souls and are to be revered. Vitalists believe that all biological life contains a dynamic force that cannot be reduced to the elements of chemistry or physics. Hindus and Jains believe that all animals have sacred souls. Each person is endowed with an immortal soul that is of infinite value. It is our true self and that which gives us whatever worth we have.

However, Schweitzer's reverence for life theory rests on a mystical insight. Not much argument is provided for it. For that we need to turn to another philosopher.

THE ENVIRONMENTAL PHILOSOPHY OF HOLMES ROLSTON III

The philosopher who has done most to develop Schweitzer's ideas in a philosophically rigorous way is Holmes Rolston III of Colorado State University. Rolston argues that values exist in the world objectively, apart from human choice or consciousness or even animal consciousness. He is concerned to reject *anthropocentrism,* the idea that human beings are the center of the world and that without us the world has no value. It is anthropocentrism that has brought on our present environmental crisis. He calls this type of subjectivism the "light-in-the-refrigerator theory. Nothing inside is of value until I open the door and the light comes on."[6] But Rolston thinks that values are also independent of animal consciousness or sentience. All life has objective value.

Rolston starts his defense of *objectivism* with an analogy to the Lockean distinction between primary and secondary qualities. The British empiricist John Locke (1632–1704) distinguished between those qualities that were inseparable from the bodies themselves and those that were caused in us by the bodies:

> Qualities thus considered in bodies are, First, such as are utterly inseparable from bodies, in what estate soever it be; such as, in all the alterations and changes it suffers, all the force can be used upon it, it constantly keeps; and such as sense constantly finds in every particle of matter which has bulk enough to be perceived, and the mind finds inseparable from every particle of matter, though less than to make itself singly be perceived by our senses; e.g., take a grain of wheat, divide it into two parts, each part has still *solidity, extension, figure, and mobility;* divide it again, and it retains still the same qualities: and so divide it on till the parts become insensible. They must retain still each of them all those qualities. For, division . . . can never take away either solidity, extension, figure, or mobility from any body, but only makes two or more distinct separate masses of matter of that which was one before; all which distinct masses reckoned as so many distinct bodies, after division, make a certain number. These I call *original* or *primary* qualities of body, . . . solidity, extension, figure, motion, or rest, and number.[7]

Primary qualities are really in their objects, and inseparable from them, and so our ideas of them truly represent the objects. Such qualities are solidity (or bulk), extension, figure (form), movement (and rest), and number. These are the true building blocks of knowledge, because they accurately represent features in the world. Ultimately, the world is made up of indivisible, minute atoms, which underlie physical objects. Secondary qualities are not in the things themselves but are powers to cause various sensations in us. The powers themselves are in the object. What is not in the object is the representation which it causes. It is in our mind.

For example, an object's color is the power of the primary qualities to cause color ideas in us. They do this by reflecting certain kinds of light off surfaces, due to the structure of those surfaces and the structure of our minds. When under normal circumstances, we look at an object, it looks a certain color—say, red. The redness that we are acquainted with is not in the object itself but in the way the light reflects off the object into our eye and is communicated to our brain. Colors, sounds, tastes, touch, and smell are all secondary qualities. Strictly speaking, the secondary qualities don't exist in the objective world apart from consciousness; but given consciousness, they are discovered as projections of the real world in conjunction with our noetic structure (having to do with our beliefs and worldviews). That is, they are relational, depending on the relationship of world to perceiver. They are close approximations of what the world really is, but not exactly so.

Rolston suggests a similar analogy with value. The beauty of a landscape may result from the relationship between the objective features of the landscape and our minds. The luscious taste of an apple, which is an almost universal experience, is caused by the way the apple impinges on our taste

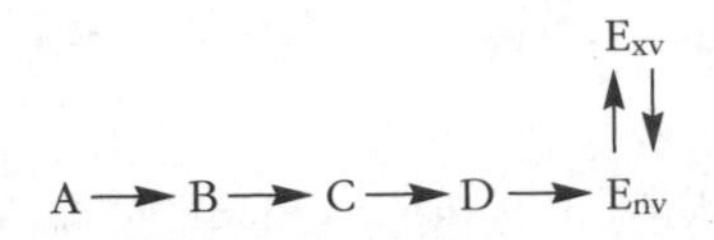

Figure 8-2 The evolutionary model.

buds. We have evolved in such a way so as to recognize beauty and goodness—that is, value—in the world. The value we recognize may be only a close approximation of what is really (objectively) valuable, but it is nonetheless genuine, pointing to something truly valuable in and of itself.

Rolston's thesis is that value should be seen as holistic, relational, emergent, and evolutionary. The evolutionary process yields creatures like us who exist in relationship with environments. Figure 8-2 illustrates the evolutionary process, where causal sequences (A, B, C, D) lead to the production of an event associated with natural value (E_{nv}), which produces an event of experienced value (E_{xv}), such as our appreciation of a beautiful landscape or a harmonious ecosystem (such as a rain forest).

The upward arrow indicates that the natural value causes the experience of value, and the downward arrow indicates our active response to the natural value. We do appraise it accurately and freely.

More accurately, Rolston says, would be Figure 8-3, which depicts the holistic relationship between our value and the holistic field of our value-laden world. Rolston writes:

> The setting is a given fact, a datum of nature, even though the subject must respond in imaginatively resourceful ways. I see things out there in the "field" which I choose to value or disvalue. But on deeper examination I find myself, a valuing agent, located within that circumscribing field. I do not have the valued object in "my field," but find myself emplaced in a concentric field for valuing. The whole possibility is among natural events, including the openness in my appraising. John Dewey remarked that "experience is *of* as well as *in* nature." What seems a dialectical relationship is an ecological one.[8]

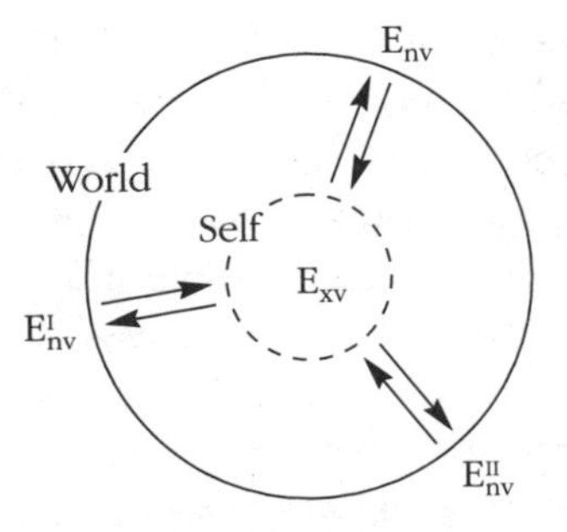

Figure 8-3 The holistic model.

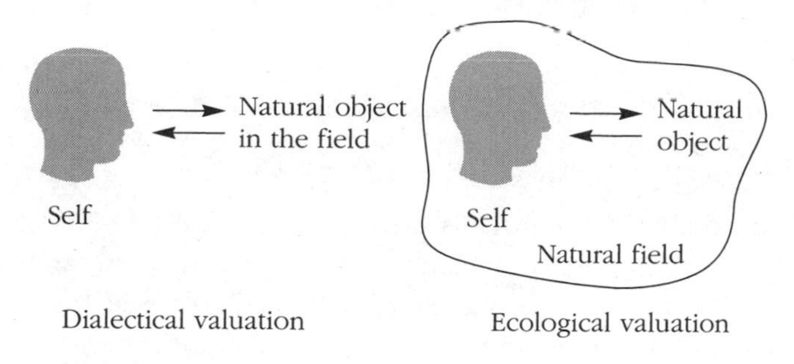

Figure 8-4 The ecological model.

The ecological model is depicted in Figure 8-4.

Rolston asks, "How do we humans come to be charged up with values, if there was and is nothing in nature charging us up so?" We need not believe in a Platonic or even theistic special creation of values; an evolutionary account will suffice. There are, as it were, "clots" in nature:

> Every natural affair does not have value, but there are "clots" in nature, sets of affinities with projective power, systems of thrust, counterthrusts, and structure to which we can attach "natures" in the plural. There are achievements with beginning, ending, and cycles, more or less. Some do not have wills or interests, but rather headings, trajectories, traits, and successions which give them a tectonic integrity. They are projective systems, if not selective systems. This inorganic fertility produces complexes of value—a meandering river, a string of paternoster lakes—which are reworked over time. Intrinsic value need not be immutable. Anything is of value here which has a good story to it. Anything is of value which has intense harmony, or is a project of quality. There is a negentropic [negative entropy] constructiveness in dialectic with an entropic tear down, a mode of working for which we hardly yet have an adequate scientific, much less a valuational, theory. . . . These performances are worth noticing—remarkable, memorable—and they are not worth noticing just because of their tendencies to produce something else, certainly not merely because of their tendency to produce this noticing in our subjective human selves. All this gets at the root meaning of nature, its power to "generate" (Latin: *nasci, natus*).[9]

Subjectivists err in inferring from the silence of science about natural value that nature has no value, but this is a non sequitur: A nonanswer is not the same as a negative answer. "The logic by which one reads values out of nature is no less troublesome than the logic by which one finds values there and lets them stay."[10] Furthermore, our intuitions corroborate the view that there is some good about nature itself or natural systems, such as rain forests and harmonious landscapes. Rolston asks us to consider the following thought experiment: Suppose that after a nuclear war, the radioactive fallout sterilizes the genes of humans and mammals, but is harmless to the flora, invertebrates, reptiles, and birds. Would the last group of valuers be morally justified in destroying the remaining biosphere? Would it not be objectively better for the remaining ecosystem to survive, even minus conscious evaluators like us? But suppose, further, that all sentient beings were also eliminated? Would it not still be wrong to destroy the nonsentient ecological systems?

Rolston thinks the answer is obvious to our reflective intuitions. He concludes that "Never have humans known so much about, and valued so little, in the great chain of being. As a result, the ecological crisis is not surprising. To devalue nature and inflate the human worth is to do business in a false currency. This yields a dysfunctional, monopolistic world view. We are misfits because we have misread our life-support system. We rationalize that the place we inhabit has no normative structures, and that we can do what we please. Afterwards, this view sinks down into the hinterlands of our minds, an invisible persuader which silently shapes an ethic."[11]

Ernest Partridge has attacked Rolston's *objectivity theory,* arguing that a properly structured *subjective theory of values* is sufficient to ward off the anthropocentrism that environmental philosophers are so concerned to repudiate. Subjectivity can lead to biocentrism. Partridge calls his position *subject/object-dyadism* (S/O-D) to be distinguished from various *objectivist theories.*[12] He holds that values are relational—existing between sentient beings (evaluators) and objects—but he distinguishes his position from the purely subjectivist theories that are represented by our previous quotation from William James or by David Hume, who wrote, "Vice and virtue . . . may be compared to sounds, heat and cold, which, according to modern philosophy, are not qualities in objects, but perceptions in the mind."[13] He calls this type of S/O-D theory "soft S/O-D" and says it is more likely to be "associated with relativistic, subjectivist, anthropocentric and noncognitive theories of value." It holds that values are *entirely* mind-dependent. He labels his view "hard S/O-D," for, though it is relational, it emphasizes the objective, categorical, and cognitive aspects of value. Values are not entirely mind-dependent, not simply objects of emotions or desire. They are discoverable, not invented or simply the objects of our choice (we could choose wrongly). He quotes John Laird approvingly:

> There is beauty . . . in sky and cloud and sea, in lilies and in sunsets, in the glow of bracken in autumn and in the enticing greenness of a leafy spring. Nature, indeed, is infinitely beautiful, and she seems to wear her beauty as she wears color or sound. Why then should her beauty belong to us rather than to her?

Rolston would also endorse this statement. The difference is that Partridge would say that Nature's beauty is only *potential,* until evaluators appreciate it. This beauty or value is thrust upon us (under normal conditions—so we may call values rational and universal, given our common humanity), but the value is not actualized until and unless an evaluator comes on the scene.

Partridge also appeals to thought experiments to bring home his point. Imagine a planet like Jupiter, inhospitable to life, with several moons and with the kind of atmosphere sufficient to create landscapes, sunrises, and lunar phases that would be of incredible beauty were there anyone to observe and enjoy them. But since this planet is inhospitable to life, are these conditions really beautiful? Have they any value at all? If an "asteroid collides with the planet, altering its physical and chemical conditions and shrouding it in an opaque atmosphere, thus obliterating all sunsets and landscapes forever," is this a calamity? To whom? He answers—"To no one! But if to no one, why regard it as a calamity?"

He offers other thought experiments. Suppose we destroy all sentient life in a nuclear holocaust, leaving only what Jonathan Schell calls a "republic of insects and grasses," so that sentient life will never again evolve. Does it matter whether the Taj Mahal crumbles to ruin? Would the world really be more valuable if it did not? Or ask yourself, "Was the Grand Canyon 'magnificent in itself' before it was seen by any human being? Has the surface of Mars been 'littered' and 'despoiled' (made less valuable) by the spacecrafts that have struck its surface? Does sugar sit sweetly on the cupboard shelf?"[14]

What is the answer to these questions? Partridge thinks it is obvious upon reflection that we will answer "no" to all of them. Since Rolston confines his analysis to the biotic community, not inorganic entities, he would agree with Partridge regarding most of the examples given. So the Jupiter-like planet, inhospitable to life as it is, does not have a good; but if it did have something living on it, even a plant or bacterium, its destruction would be bad. The universe is poorer for its loss. He writes in response, "[T]hough things do not matter *to* plants, a great deal matters *for* them. We ask, of a failing plant, what's the matter *with* that plant? If it is lacking sunshine and soil nutrients and we arrange for these, we say, 'The tree is benefiting from them'; and *benefit* is—everywhere else we encounter it—a value word. Objectively, biologists regularly speak of the "survival value" of plant activities."[15] He certainly believes that the Grand Canyon, as a self-organizing, self-sustaining ecosystem, was magnificent as such before any human being came along. And, similarly, Venus's integrity, supposing it has an ecosystem with biotic forms, has been damaged by the garbage of crashed spacecraft. He responds that Partridge's mistake is to insist on the necessity of a subjective evaluator. "I too insist on a subjective evaluator for some kinds of natural values, such as a person enjoying a sunset. But not for all kinds. 'The sugar is not *sweet* unless tasted,' but the sugar is valuable to the plant whether or not we ever capture it for *our* tastes. . . . The plant values its own life intrinsically . . . and makes instrumental use of the nutrients and the sunshine."

How do you respond to these thought experiments? Is this a puzzle wherein people will simply have different fundamental intuitions? Let's hope that more can be said than this. First of all, the comparison of values with secondary properties, which both sides appeal to, is a useful analogy and may shed light on the matter. According to the Lockean view, secondary qualities don't exist in their own right, but only in relationship with an observer. The color red doesn't exist in the world, but is a phenomenon created in the mind as light is reflected off surfaces and received by the optic nerve and sent to the brain for interpretation. If values are like this, they are caused by the world but dependent on our minds for their actualization. To this extent the subjectivist seems right. But given standard human nature and given normal lighting conditions, we cannot help but see red. It's not a subjective choice. To this extent the objectivist seems right—we don't invent values any more than we invent our perception of colored objects (or sounds or tastes). They are given to us by the way the world relates to us and we to the world. This comparison with secondary qualities has been used by John McDowell to provide the basis for an objectivist ethics.[16] Moral principles are just those principles which social reality causes human beings like ourselves to discover. If minds were constructed differently, we would not see the color red. Likewise, if our nature were different—say, we had exometalic skeletons so that we didn't feel pain—some parts of morality would be different; for example, physical beatings wouldn't be morally wrong.

This relational view is neither purely objective nor subjective. It is a combination; and, depending on which feature we emphasize, our theory will look more objectivist or subjectivist, as the case may be. But as long as we recognize the truth of the relationship, we are not really disagreeing with each other—merely pointing to a different aspect of the relationship.

If this analysis is correct, Partridge's soft subject/object dyadism would seem to be closer to the Lockean secondary-qualities analogy. Rolston really wants to go beyond the merely perceptual metaphor and state that we are enmeshed in a value field from which our value judgments arise. The problem I find with Rolston's challenging account is that it sloughs over an important distinction between what causes or contributes to value and what is objectively valuable. The dark shadings of a Rembrandt painting may contribute to the holistic beauty of the painting without themselves being beautiful. The atoms that make up a living body are necessary for life, but they are not living. Likewise, our ecological field may be necessary for value to arise, but it may have only contributory value (a type of instrumental value) and lack objective value.

The question—what kind of value does nature have?—is vitally important to environmental ethics. Indeed, some environmental philosophers hold that it is the primary question, and some believe that environmental ethics sets itself apart by holding that nature has objective value.[17] We will consider other theories about the value of nature, especially biocentric theories, in the next two chapters. Let's hope that the issues and arguments will become clearer.

But first let us conclude and summarize this chapter. We saw two reactions to *anthropocentrism,* the view that all valuing is centered in human evaluation, so that if there were no human beings, there would be no value. (We will come back to this view in the next chapter.) The first reaction is Rolston's *objectivity theory of value.* Nature has value due to its own self-creating, self-sustaining activity. We may call this *an ecocentric theory of value,* because it places value in natural systems. Actually, Rolston divides nature up into three levels:

1. Sentient beings: higher zoology
2. Insentient beings: lower zoology and botany
3. Inanimate nature: geomorphology, astronomy, physics, and chemistry

Rolston affirms value at all three levels. Value at level 2 denies that value must be felt experience. Rolston's critics, like Partridge, generally accept level 1, but deny 2 and 3. Partridge also rejects anthropocentrism, preferring a *subject/object-dyadism,* which holds that there must be an evaluator to discover *potential* values in the world, though these appraisers need not be human—but may be animals. We may call this a *sentience-centric theory of value,* since it holds that sentience, rather than simply life, is the locus of values, that which is necessary to make them actual.[18]

Review Questions

1. What is nature? What does it contrast with? How do various thinkers use the term?
2. What is value? What kinds of value are there? What is intrinsic value? What is objective value? Do you believe that anything has objective value? What does William James think? Do you agree with him?
3. Explain Albert Schweitzer's view of reverence for life. Does it have some attraction for you? Explain.
4. Analyze the view of Holmes Rolston III on intrinsic (objective) value in nature. You might consult Note 11 for his own latest words on the subject.
5. Analyze Ernest Partridge's critique of Rolston's views and his own nonanthropomorphic sentience-based system. Which view is the more plausible and why?

Suggested Readings

Bond, E. J. *Reason and Value.* Cambridge: Cambridge University Press, 1983.
"The Intrinsic Value of Nature," *The Monist* 75, 2 (April 1992).

Lovejoy, Arthur. *The Great Chain of Being*. New York: Harper & Row, 1960.
Perry, R. B. *A General Theory of Value*. Cambridge: Harvard University Press, 1926.
Rescher, Nicholas. *Introduction to Value Theory*. Englewood Cliffs, N.J.: Prentice Hall, 1982.
Rolston, Holmes III. "Yes, Value Is Intrinsic in Nature." In *Environmental Ethics*. Edited by L. Pojman. Belmont, Calif.: Wadsworth, 1998.
Schweitzer, Albert. *Civilization and Ethics*. Translated by A. Naish. London: Black, 1923.
Taylor, Paul. *Respect for Nature*. Princeton: Princeton University Press, 1986.
Taylor, Richard. *Good and Evil*. New York: Macmillan, 1970.

Notes

1. Bolingbroke, *Fragments, or Minutes of Essays* xvi in *Works,* 1809, quoted in Arthur O. Lovejoy, *The Great Chain of Being* (New York: Harper & Row, 1960). In his *Essays in the History of Ideas* (Baltimore: Johns Hopkins University Press, 1948) Lovejoy identifies sixty-six different meanings of "nature," attesting to the systematic ambiguity of the idea.
2. We could have a fifth type of value—*contributory*. An object contributes to the worth of a whole of which it is a part by enhancing its value. For example, a good shortstop contributes to the total worth of a baseball team and a dark texture may contribute to the beauty of a painting. I think that contributory value can be accommodated within the scheme I set forth.
3. William James, *Varieties of Religious Experience* (New York: Longman Green, 1925), p. 150, quoted in Holmes Rolston, "Yes, Value Is Intrinsic in Nature," *Environmental Ethics,* ed. L. Pojman (Belmont, Calif.: Wadsworth, 1998), p. 71.
4. R. B. Perry, *A General Theory of Value* (Cambridge: Harvard University Press, 1926).
5. Passages from Albert Schweitzer are from *Civilization and Ethics,* trans. A. Naish (London: Black, 1923); a selection of this work is reprinted in *Environmental Ethics,* ed. L. Pojman, pp. 93–98.
6. Holmes Rolston, "Values at Stake: Does Anything Matter: A Response to Ernest Partridge," in *Environmental Ethics,* ed. L. Pojman, p. 90.
7. John Locke, *An Essay Concerning Human Understanding* (1689), II.viii.9.
8. Holmes Rolston, "Yes, Value Is Intrinsic in Nature," p. 76.
9. Ibid., p. 78.
10. Ibid., p. 79.
11. Ibid., p. 80. Rolston writes, "My most frequent line of argument [for intrinsic value in nature] is that value language is used all the time about plants and their activities. The plant needs sunshine, nutrients, photosynthesizes, makes sugar, secretes toxins to prevent insect damage, makes seeds. Most of biology is at the non-sentient level. What genes do is conserve value. I try to get these folk to see that any account of an organism defending its life is value-laden. Insects as well as plants. If they concede, well, yes plants do value sugar, then the whole question of value has to be opened wide up. En route will come species, ecosystems, still in the living world. Later will come canyons, mountains, rivers, crystals, leaving the living world. At this point I move to thinking of the creative process in, with, and under the formation of the planet with its powers of achievement as valuable." (Letter to author, 15 February 1998).
12. Ernest Partridge, "Values in Nature: Is Anybody There?" in *Environmental Ethics,* ed. L. Pojman, p. 82.
13. David Hume, *A Treatise of Human Nature,* Book III, 1.
14. Ibid.
15. Holmes Rolston, "Values at Stake: Does Anything Matter," p. 89.
16. John McDowell, "Values and Secondary Qualities," in *Morality and Objectivity,* ed. Ted Honderich (London: Routledge & Kegan Paul, 1985).

17. John O'Neill begins his essay "The Varieties of Intrinsic Value" by stating, "To hold an environmental ethic is to hold that non-human beings and states of affairs in the natural world have intrinsic value" (*Monist* 75: 2; April 1992). In the same issue Robert Elliot in "Intrinsic Value, Environmental Obligation, and Naturalness" argues for a similar thesis. I prefer a broader understanding of environmental ethics, as having to do simply with our ethical duties regarding the environment, leaving these metaethical issues outside the definition of the subject.

18. Holmes Rolston III, William Throop, and two anonymous reviewers made helpful comments on a draft of this chapter, for which I am grateful.

CHAPTER NINE

Ecocentric Holism: The Land Ethic

In wildness lies the preservation of the world.

—HENRY DAVID THOREAU

HETCH HETCHY AND JOHN MUIR

> *As to my attitude regarding the proposed use of Hetch Hetchy by the city of San Francisco . . . I am fully persuaded that . . . the injury . . . by substituting a lake for the present swampy floor of the valley . . . is altogether unimportant compared with the benefits to be derived from its use as a reservoir.*
>
> —GIFFORD PINCHOT, 1913

> *These temple destroyers, devotees of ravaging commercialism, seem to have a perfect contempt for Nature, and instead of lifting their eyes to the God of the Mountain, lift them to the Almighty Dollar.*
>
> —JOHN MUIR, 1912

In the early years of this century (about 1901–13), a battle was waged that was to divide the environmental movement in the United States for the rest of the century. San Francisco was growing and needed water. The most convenient watershed was 150 miles east in the newly designated Yosemite National Park, where the Tuolumne River formed a small lake surrounded by high-walled rock cliffs. The area was known as Hetch Hetchy. The first attempts by the leaders of San Francisco to persuade the federal government to dam up the lower end of the lake to form a reservoir were rebuffed; but after the earthquake of 1906 devastated the city, sympathy abounded. In 1908 Secretary of the Interior James R. Garfield approved the application, declaring, "Domestic use is the highest use to which water and available storage basins can be put."[1] The war was on; and educators, politicians, environmentalists, and citizens lined up behind opposing banners: those of the *conservationists* and *preservationists* (Table 9-1). The leaders of these two movements, Gifford Pinchot (1865–1946), head of the National Forest Service, and John Muir (1838–1914), founder of the Sierra Club, had been friends up to this point. No longer. They waged what may be the most tenacious political war over the environment that our nation has ever witnessed. The conservationists argued that while nature and its accompanying wilderness had significant aesthetic value, their greatest use

Table 9-1 Attitudes Toward Nature

Consumption	*Conservation*	*Preservation*
The attitude of using up a resource without regard to the foreseeable future. Consumption of non-renewable resources such as coal, oil, and minerals (iron and copper, for instance) diminshes their total supply. Species can be eliminated by over-consumption. Thc land can be premanently ruined by overgrazing, erosion, removal of top-soil, or nutrient loss.	This attitude recognizes that resources are limited and that we have obligations to pass these resources on to posterity. It seeks frugality with regard to nonrenewable resources and wise use with regard to scarce renewable resources, such as watcr, air, and animal species. It represents an "enlightened anthropocentrism" that seeks to manage the wilderness for the good of humanity. Nature has no objective value, only subjective, extrinsic value.	The attitude of saving the wilderness or species for their own sake. Nature has objective value, so deserves to be protected for its own sake. The wilderness and members of other plant and animal species have a right to exist, so that we violate their rights by exploit ing them for our use or for the sake of future generations

Our discussion takes seriously only two attitudes or theories, conservationism and preservationism, but actually there is a third theory that is most prominent of all, consumptionism. We will discuss the practical aspects of these throries in Chapters 16 and 17.

was as a *resource* for the human good. Pinchot wrote, "Forestry is the knowledge of the forest. In particular, it is the art of handling the forest so that it will render whatever service is required of it without being impoverished or destroyed. Forestry is the art of producing from the forest whatever it can yield for the service of man."[2] This was a utilitarian argument. The Hetch Hetchy would serve human need far more as a reservoir than as a mere aesthetic attraction. The conservationists' motto was "Make wise use of nature." Don't be wasteful, don't unnecessarily destroy the wilderness, but integrate it with service to the nation. A reservoir would not destroy Yosemite, they argued, but maximize its potential.

The preservationists, led by Muir, disagreed. Nature had objective value, value in her own right. Humans were part and parcel of nature and to harm her was not only evil in itself, but also would have long-term repercussions for posterity. We owed future generations a pristine wilderness. Let San Francisco get its water elsewhere or, failing that, limit its population to what the area's carrying capacity could sustain. Economics should not dictate environmental policy—at least not when it came to something as sacred as the wilderness.

President Theodore Roosevelt, himself a lover of nature, was caught up in the midst of this battle and felt himself pulled by both sides. Roosevelt consulted expert engineers and, taking their advice, reluctantly sided with the conservationists against his friend Muir, assuring him that everything possible would be done to maintain the park. Roosevelt confided that after he made his decision, he doubted it was right, but he stuck to it. Meanwhile, the preservationists made the damming of the Hetch Hetchy an aesthetic and moral cause: nature versus mammon, the beauty of the unspoilt wilderness versus economic greed. It was wrong to turn every tree, river, and mountain into dollars and cents. Muir preached the "Tuolumne gospel" against San Francisco, referred to as "the prince of the powers of Darkness and Satan and Co. who scorned the God of the Mountains." He wrote, "We may lose this particular fight but truth and right must prevail at last. Anyhow we must be true to ourselves and the Lord."[3] And again, "Dam Hetch Hetchy! As well dam for water-tanks the people's cathedrals and churches, for no holier temple has ever been consecrated by the heart of man." The protest was effective and in December 1908 even President Roosevelt retreated from his decision to dam Hetch Hetchy.

The people of San Francisco felt bewildered and betrayed and renewed their fight for the Hetch Hetchy, arguing that a reservoir would not spoil the beauty of Yosemite but rather enhance it. Woodrow Wilson became president in 1913 and had the House Committee on Public Lands open hearings on the issue. Gifford Pinchot was the star witness. He began by testifying to his love of nature and his commitment to the conservationist cause: Value nature; don't squander her resources. He continued, "As to my attitude regarding the proposed use of Hetch Hetchy by the city of San Francisco . . . I am fully persuaded that . . . the injury . . . by substituting a lake for the present swampy floor of the valley . . . is altogether unimportant compared with the benefits to be derived from its use as a reservoir. The fundamental principle of the whole conservation policy is that of use, to take every part of the land and its resources and put it to that use in which it will serve the most people." In other words, be a utilitarian with regard to nature. On September 3, 1913, members of the House of Representatives passed the Hetch Hetchy bill 183 to 43, with 203 members abstaining. No repersentative from a Western state voted against it. The battle moved to the Senate, where the bill also passed on December 6. Not long after, Muir died, a casualty of the lost battle.

Who was John Muir? Born in Scotland in 1838, Muir immigrated with his family to Wisconsin in the 1850s. Brought up a strict Scottish Presbyterian, he was compelled to memorize the New Testament. Partly under the influence of the transcendental philosophers Henry David Thoreau and Ralph Waldo Emerson and partly because of his mystical experiences with nature, he rejected his Calvinist roots, deeming them too anthropocentric, for a more ecocentric religious perspective. While in his twenties, after leaving the University of Wisconsin, he went on long hiking trips along the Canadian shores of Lakes Huron and Ontario, then a wilderness devoid of human habitation. On one such trek through swampland, he came unexpectedly upon an exquisite flower. He broke down and wept. Suddenly, he felt transformed, one with nature, illuminated with the truth of the universe. The flower epiphany was the turning point in his life. Henceforth he became an advocate for a natural religion—something between *pantheism* and *panentheism*.[4]

Pantheism is the doctrine that everything there is constitutes a unity and that this unity is divine. Pantheists thus deny the radical distinction between God and creatures that is drawn by monotheistic religions. The philosopher Benedict Spinoza held that there is only one substance and that it is divine. In Hinduism God is seen as a world soul, a oneness of all things. Multiplicity is an illusion. Individuals are like waves of the great ocean, mere temporal forms of a tiny part of the whole.

Panentheism is the doctrine that God is *in* all things, but not to be identified with all things. He still enjoys independent existence as a person. As the *Cambridge Dictionary of Philosophy* puts it, "Just as water might saturate a sponge and in that way be in the entire sponge, but not be identical with the sponge, God might be in everything without being identical with everything." Panentheism treats God as the soul of the world that accompanies the world even as the soul or mind accompanies the human body. Panentheism sees the world as God's body. Hence, we should be reverent toward nature because it is God incarnate. We will discuss a version of panentheism at the end of this chapter, when considering the Gaia hypothesis.

Muir illustrates the pantheist's faith in seeing all of nature as expressing a divine reality. He rejects as anthropocentric questions like, How could a good God allow so much evil and waste in the world, like poison ivy, rattlesnakes, and weeds? He writes of poison ivy: "It is somewhat troublesome to most travelers, inflaming the skin and eyes, but blends harmoniously with its companion plants, and many a charming flower leans confidingly upon it for protection and shade. . . . Like most other things not apparently useful to man, it has few friends, and the blind question, 'Why was it made?' goes on and on with never a guess that first of all it might have been made for itself."[5] Poison ivy, rattlesnakes, and "weeds" were all part of an organic whole. It is only the narrow horizon of egoistic and arrogant humans ("Lord Man," he deridingly labels us) that fails to see the glory and validity of these things. True, the universe would be "incomplete without man; but it would also be incomplete without the smallest transmicroscopic creature that dwells beyond our conceitful eyes and knowledge."[6]

Nature is God's temple. "The clearest way into the Universe," he wrote, "is through a forest wilderness."[7] And again, "Plants are credited with but dim and uncertain sensation, and minerals with positively none at all. But why may not even a mineral arrangement of matter be endowed with sensation of a kind that we in our blind exclusive perfection can have no manner of communication with?"[8]

Again, he sees nature in religious terms: "When I entered this sublime wilderness the day was nearly done, the trees with rosy, glowing countenances seemed to be hushed and thoughtful, as if waiting in conscious religious dependence on the sun, and one naturally walked softly and awe-stricken among them. I wandered on, meeting nobler trees where all are noble, subdued in the general calm, as if in some vast hall pervaded by the deepest sanctities and solemnities that sway human souls."[9]

Muir's life incarnated Thoreau's dictum, "In wildness lies the preservation of the world."

ALDO LEOPOLD AND LAND ETHICS

Muir's successor as the spiritual leader of environmentalism was Aldo Leopold (1887–1947). Leopold worked for the U.S. Forest Service before becoming the first professor of wildlife management at the University of Wisconsin. He is considered the father of "the land ethic," set forth in his main work, *Sand County Almanac* (1947), one of the classics of environmentalism.[10] Like Muir, Leopold was distressed at the degradation of and callous disregard for the environment and sought to articulate our symbiotic relationship with "the land" or the biotic community. He set forth a theory of environmental ethics that included a dialectical tension between prudential reasons (we harm ourselves when we harm the land) and deontological reasons (the biotic community has inherent value and should be respected and preserved for its own sake). We humans must come to see ourselves not as atomistic individuals or conquerors of the land but rather as plain members and citizens of the biotic community. What is distinctive about Leopold's environmental ethic is its thoroughly holistic nature.

In reading Muir, one sometimes gets the impression that an adequate environmental ethic would be so radical as to border on misanthropy. His ethic sets nature in opposition to the human world: "The God of the Mountain" versus "mammon" and "Lord Man." Domesticated animals, especially compared to their relatives in the wild, are only "half alive." Sheep are characterized as "hooved locusts." At first glance, Leopold seems less radical, arguing that the land ethic is just an extension of ordinary ethics. He shows how humankind has gradually extended the scope of moral considerability from one's own tribe, to one's nation, to one's race, to all human beings, regardless of race and gender. Now we are challenged to extend consideration to animals and to the whole biotic community. "The land ethic simply enlarges the boundaries of the community to include soils, waters, plants, and animals, or collectively: the land."

He extended rights to animals and plants in the biotic community: "Of the 22,000 higher plants and animals native to Wisconsin, it is doubtful whether more than 5% can be sold, fed, eaten, or otherwise put to economic use. Yet these creatures are members of the biotic community, and its (as I believe) stability depends on its integrity; they are entitled to continuance." Yet these rights are not absolute. Leopold recognized the legitimate role of the predator. Predators are part of the natural cycle of life, of the biotic pyramid. We all need to eat, and some are carnivores by nature. We all die and refertilize the land. Hunting deer or other game is morally acceptable behavior for the human predator. Accordingly, his views have not been welcome to the animal rights philosophers (see Chapter 7 for a discussion of animal rights). A deep division presently exists between land ethicists and zoocentric sentientists, who tend to espouse a more traditional ethics based on the good of the individual rather than the whole community. It is Leopold's *holism* that both sets his philosophy apart from others and makes it controversial.

In his holism, Leopold was influenced by the British ecologist Arthur Tansley, who in 1935 first set forth the concept of an ecosystem rather than the individual organism as the fundamental ecological idea. Tansley wrote:

> But the more fundamental conception is, as it seems to me, the whole *system* (in the sense of physics), including not only the organisms-complex, but also the whole complex of physical factors forming what we call the environment of the biome—the habitat factors in the widest sense. It is the systems so formed which, from the point of view of the ecologist, are the basic units of nature on the face on the earth. The *ecosystems*, as we may call them, are of the most various kinds and sizes. They form one category of the multitudinous physical systems of the universe, which range from the universe as a whole down to the atom.[11]

Tansley's point is that while individuals have life, their life cannot be sustained alone apart from a network of interacting living and nonliving objects. Together these constitute a flow of energy and the cycling of chemicals that support individual life.

Consider Figure 9-1, a simplified diagram of the structure and energy flow of an **ecosystem.** An ecosystem consists of living, or *biotic*, components (plants, animals, and microorganisms) and nonliving, or *nonbiotic*, components (nutrients, air, water and solar energy). Living organisms are divided into *producers* and *consumers*. On the right of the diagram we see producers (plants) producing food. They combine inorganic (abiotic) nutrients in the

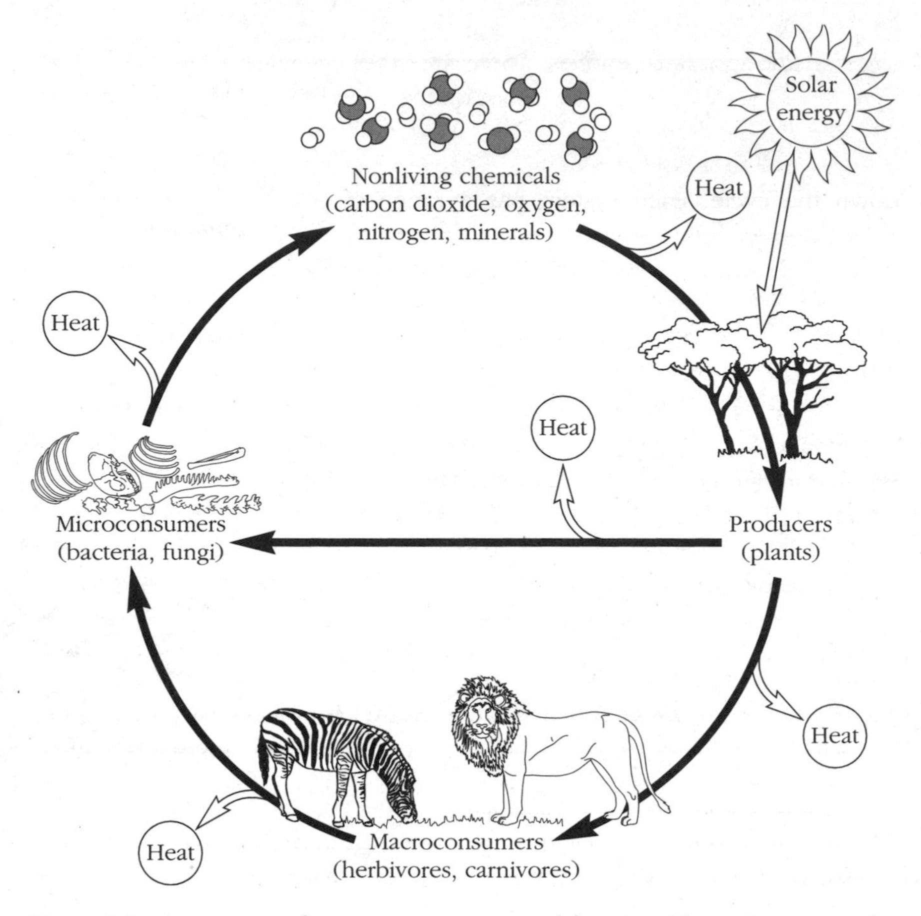

Figure 9-1 A summary of ecosystem structure and function. The major structural components (energy, chemicals, and organisms) are connected through the functions of energy flow (unshaded lines) and chemical cycling (shaded lines).

environment with solar energy in a process called *photosynthesis*, converting the nutrients to chemical energy. All other members of the ecosystem are consumers. *Macroconsumers* are either carnivores or herbivores or both (omnivores). Pigs, rats, foxes, squirrels, bears, cockroaches, and humans are omnivores. Other consumers (such as vultures, flies, crows, hyenas, and ants) are scavengers, which feed on dead organisms (*detritus*). On the left of Figure 9-1 we see *microconsumers,* or *decomposers,* bacteria and fungi, which also live off the parts of dead organisms, breaking down detritus and thus recycling nutrients, such as carbon dioxide, oxygen, nitrogen, and other minerals (see top of the diagram), which in turn are used in photosynthesis to produce chemical energy or food in plants. The decomposers are themselves a food source for worms and insects.

Ecosystems are *superorganisms*, groups of organisms that involve complex and often highly specialized relationships between various organic and inorganic entities. There are other chemical cycles within this grand ecological energy cycle. Under optimal conditions, the cycle functions well, and a balance with nature is maintained. But if there should be too little or too much sunlight, water, oxygen, or nutrients, disharmony breaks down the cycle. Each system has a range of tolerance—some systems flourish under bright sunlight, others in the shade. Some like it hot, others cool. Some survive desert conditions, and other require marshlands. There is room for variation. But when the necessary conditions vary too radically from the ideal, the system breaks down. Pollution by humans can cause such a breakdown.

A study of energy flow within ecosystems gives us a sense of the complexity and delicate balance of something resembling a superorganism. We see that all life is connected and that we are interdependent within a web of life—not self-sufficient individuals who can survive without an enormous support system. The case for holism is compelling. "No man is an island entire of itself, but each a piece of the whole," to quote John Donne.

Leopold developed Tansley's idea of an interdependent holistic ecosystem and made it the key principle in his land ethic:

> The key-log which must be moved to release the evolutionary process for an ethic is simply this: quit thinking about decent land-use [the Conservationist concern] as solely an economic problem. Examine each question in terms of what is ethically and esthetically right, as well as what is economically expedient. A thing is right when it tends to preserve the integrity, stability, and beauty of the biotic community. It is wrong when it tends otherwise.[12]

In the final analysis, it seems that Leopold's holism is not just a corrective to simpleminded individualism, but also a radical rejection of the longstanding principle of the primacy of the individual, whether human or animal. This passage seems to undermine human concerns, subordinating them to those of biosystems. The new morality requires that we give up not only individualism but also the idea of the primary importance of human beings, and that we make the ecosystem the fundamental unit of moral concern. Following this logic, just as it is permissible—even laudable—to cull excess sheep or deer when they interfere with the integrity, stability, or beauty of the biotic community, it would be permissible—even laudable—to cull humans who interfere with the integrity, stability and beauty of the biotic community. Suppose a biosystem (say, a swampland situated in a forest) would remain stable, united (maintaining its integrity), and beautiful for 100 years with humans but could maintain these virtues for 150 years if we killed all the humans who inhabit it. What would the land ethic require us to do? This seems anti-anthropocentric with a vengeance—hardly an extension of our morality, but a reductio ad absurdum of it.

Baird Callicott, the leading contemporary exponent of the land ethic, develops the implications of the theory. Putting the ecosystem's good first, we must revise our views of death, which is to be seen as a good thing, part of the natural cycle of life, growth, old age, and death. It's part of the natural process—as is killing animals in order to eat. So hunting animals is good. Similarly, it would seem to follow that killing humans who interfere with the "integrity, stability, and beauty" of the ecosystem would also be a moral obligation.

Callicott notes that "The extent of misanthropy in modern environmentalism thus may be taken as a measure of the degree to which it is biocentric. Edward Abbey [an environmentalist writer] in his enormously popular *Desert Solitaire* bluntly states that he would sooner shoot a man than a snake. . . . [T]his is perhaps only his way of dramatically making the point that the human population has become so disproportionate from the biological point of view that if one had to choose between a specimen of *Homo sapiens* and of a rare even if unattractive species, the choice would be moot."[13]

Callicott goes on to reject the idea of equality of persons, as set forth in deontological ethical systems like Kant's and utilitarian systems like Bentham's and Mill's: "The land ethic does not accord equal moral worth to each and every member of the biotic community; the moral worth of individuals (including . . . human individuals) is relative, to be assessed in accordance with the particular relation of each to the collective entity which Leopold called the 'land.' . . . [T]he preciousness of individual deer, *as of any other specimen*, is inversely proportional to the population of the species."[14] In other words, the good of the ecosystem is the only value.

In emphasizing holism, the land ethic subordinates or even rejects the validity of individual rights. Like utilitarianism, holism is an absolutist, consequentialist ethic, replacing utilitarianism's principle "do whatever will maximize happiness to sentient beings" with the principle "do whatever promotes the greatest integrity, stability, and beauty of the ecosystem." The founder of utilitarianism, Jeremy Bentham, rejected the notion of individual rights as "nonsense on stilts." The land ethicist would very likely concur. Individualism is a deontological concept—respect individuals (whether they be plants, animals, or human beings is left open at this point), even if it means less overall utility. Individuals have rights that ought never to be overridden—or, at least, only rarely, when survival or enormous good is at stake. The same kind of criticism that is leveled against utilitarianism can be leveled at holism: It sacrifices the individual for the sake of the whole.

Tom Regan sets forth this criticism this way:

> The difficulties and implications of developing a rights-based environmental ethics . . . include reconciling the *individualistic* nature of moral rights with the more *holistic* view of nature. . . . Aldo Leopold is illustrative of this latter

> tendency. . . . The implications of this view include the clear prospect that the individual may be sacrificed for the greater biotic good, in the name of "the integrity, stability and beauty of the biotic community." It is difficult to see how the notion of the rights of the individual could find a home within a view that . . . might be fairly dubbed "environmental fascism."[15]

In the face of such criticisms, land ethicists generally reply that they are only advocating that humans recognize their connection to holistic systems without which life cannot be sustained. In the *Republic*, Plato required citizens to sacrifice themselves for the good of the whole community; the land ethic expands the community for which we must sacrifice ourselves to include the biosystem.[16] The question is, What exactly is our responsibility to nature or to a particular ecosystem? What is the relative value of humanity vis à vis natural processes? Is the human being of special value in nature, or is he or she simply a "plain member and citizen of it," to be accorded no more rights or value than any other functioning part of the whole?

Callicott supports his theory of ecological holism with an argument based on quantum physics. He notes that at the subatomic level of quanta we are continuous with all of nature. As a physicalist, Callicott rejects the idea of a soul in favor of a materialist notion of selfhood as part of our body (located in the brain). So it follows that you and I are continuous with all of nature. Quoting Alan Watt, Callicott infers that "the world is our body."[17] Assume next, as most of us will, that the self has intrinsic value (what I have called *objective value* in Chapter 8). "Nature is intrinsically valuable to the extent that the self is intrinsically valuable." Since I and the environment are one, "the injury *to me* of environmental destruction is primarily and directly to my extended self, and to the larger body and soul with which I am continuous."[18]

What should we make of this argument? Not much, I'm afraid. It is an example of the fallacy of composition. Its form is:

> If A has Value V and if B is part of A, then B must also have Value V.

But there are counterexamples to this. Consider one exemplification of this formula: If Jones is good and if Jones has cancer, cancer must be good. Jones may be good without its being the case that everything about him is good—he may be a bad actor or a poor shortstop or have cancer, migraine headaches, HIV, or a diseased heart. Simply because we all share quanta doesn't mean that everything made up of quanta is of the same value. Murderers and saints are made up of quanta, and so are Rembrandt paintings and those of the Dadaists—but they have differing values.

Furthermore, Callicott's holism fails to provide ethical guidance. If all is one, like mere waves in the ocean or quanta that make up nature, nothing matters. That is, if I'm one with you, then why should you complain if I "steal" your wallet, your car, your computer, or your spouse? We're all one, so stealing is an illusion—all that happens is transfers of quanta-groupings. But ethics deals with larger wholes, with individuals as discrete but related units of consciousness. It assumes conflicts sometimes arise between your interests and mine, between your interests and those of animals, between your interests and those of future generations, between our interests and those of the environment (assuming that it makes sense to speak this way). But holism, in undermining or even denying *individuality*, dissolves this problem by putting blinders over its eyes.

In fact, by Callicott's logic, I would be justified in destroying the environment if I suddenly ceased to value my life. Sometimes people come to believe that they are worthless. So they may reason: I am worthless and might just as well commit suicide. I and the environment are holistically one, so I might just as well commit ecocide.

Is It Wrong to Intervene in Nature?

One problem for the ecocentric ethics, not much discussed, is the fact that nature itself undergoes radical changes. Earthquakes, glaciers, hurricanes, meteorites, and volcanoes do more to undermine or change the "stability, integrity, and beauty" of ecosystems than do humans—as the dinosaurs discovered to their grief some 70 million years ago. Nature and natural processes are always in flux. So why is it so bad if humans alter nature for their purposes—especially if they can improve on nature? Callicott follows Muir and Leopold in decrying domesticated animals as human creations, as "living artifacts, but artifacts nevertheless." They cannot survive in their original habitats—the wild. Sheep, cows, pigs, and chickens have "been bred to docility, tractability, stupidity, and dependency. It is literally meaningless to suggest that they be liberated." But by this logic, civilized humans are the most prominent of "living artifacts," for no animal has developed from his original habitat and lifestyle more than we have—needing extended levels of socialization (read "domestication"), education, medicine, sanitation, and technological support mechanisms (homes, clothes, electricity, and legal systems). Granted, it no doubt would be in our interest to recapture a deeper appreciation for nature and an ability to live in the wilderness, but it's doubtful whether there is anything immoral about creating culture, cities, and civilization—domesticating ourselves and animals. If we can "create" forms of animal life that flourish, that live longer and more meaningful lives than those in the wild, where life tends to be "red in tooth and claw," I see no reason to refrain from doing so. Granted we have bred some animals, such as pit bulldogs, whose quality of life may be very low, and granted that wild animals, such as wild dogs, can outsurvive their domestic counterparts; nevertheless, their domestic counterparts may live longer, without the threat of starvation, and have relations that are fulfilling. Our cavedweller ancestors could probably outsurvive most of us in the wild and demolish us in a fight, but that's hardly a reason for advocating a return to the cave. Granted, again, that we abuse both domesticated and wild animals, which no ethicist would condone, but we can work to correct this situation.

Suppose we could domesticate a volcano, would that be wrong? It is estimated that volcanic eruptions contribute vast amounts of carbon dioxide and particulates into the atmosphere, and these exacerbate the greenhouse effect. Suppose that through enhanced technological processes, we could greatly minimize these eruptions without creating some other bad side effects. We would be tampering with the ecosystem—domesticating it, as it were. But it would be for the good of humans and other animals. Or suppose modern technology discovered a means to transfer ozone in the lower atmosphere, where it is harmful to animals and humans, to the stratosphere, where it is needed to block harmful ultraviolet rays. Would it be wrong to domesticate nature? Similarly, building dams and creating landfills in order to build habitations for humans also "domesticates" nature, but are these always wrong? In Chapter 16 we will consider the question of whether we have obligations to leave nature alone ("nature knows best") or whether creating artificial forests, trees, and the like would be morally permissible. This leads to a final problem with holistic views of nature and the holists' loathing of human domestication of it. This problem involves demarcating the boundary between what is natural and what is artificial.

The Demarcation Problem

Every ethical theory must provide principles (or identify properties) that describe which objects have value and which do not. In doing so, they need to mark out a boundary between these two classes. This is called the problem of **demarcation.** Kant makes the demarcating property *rationality*. Classical utilitarianism makes it the ability to *experience pleasure and pain*. Contractualists make it an explicit or implicit *contract* between the relevant parties. Holistic environmentalists make it natural systems like species, mountains, and ecosystems—versus artificial objects like highways, buildings, cities, and domesticated animals. The difference is in how the two types come into being and continue to exist. Mountains, species, and ecosystems were created by natural processes, whereas buildings, cities, and domesticated animals are human contrivances. But the natural–artificial distinction begins to blur once we realize that organisms construct their environments in nature. As Elliott Sober notes,

> Organisms do not passively reside in an environment whose properties are independently determined. Organisms transform their environments by physi-

> cally interacting with them. An anthill is an artifact just as a highway is. Granted, a difference obtains at the level of whether conscious deliberation played a role, but can one take seriously the view that artifacts produced by conscious planning are thereby *less* valuable than ones that arise without the intervention of mentality? As we have noted before, although environmentalists often accuse their critics of failing to think in a biologically realistic way, their use of the distinction between "natural" and "artificial" is just the sort of idea that stands in need of a more realistic biological perspective.[19]

Anthills, beaver dams, beehives, birds' nests, and rat holes are all constructions of the respective animals. Bonobos and chimpanzees erect elaborate rituals for grooming, greeting, revenge, and forgiveness, which are very similar in design to human rituals of greeting, revenge, and forgiveness. It is difficult to see any difference in principle between natural and artificial. After all, we human animals are part of nature, only a more reflective part, which devises *memes* more extensively than bonobos and chimpanzees do. It's more a matter of degree than kind. I find it hard to discern any moral difference between the natural and the artificial.

Sober thinks the real difference is not between artificial and natural, but between natural objects and works of art:

> For both natural objects and works of art, our values extend beyond the concerns we have for experiencing pleasure. Most of us value seeing an original painting more than we value seeing a copy, even when we could not tell the difference. When we experience works of art, often what we value is not just the kinds of experiences we have, but, in addition, the connections we usually have with certain real objects. Routley and Routley have made an analogous point about valuing the wilderness experience: a "wilderness experience machine" that caused certain sorts of hallucinations would be no substitute for actually going into the wild. Nor is this fact about our valuation limited to such aesthetic and environmental contexts. We love various people in our lives. If a molecule-for-molecule replica of a beloved person were created, you would continue to love the individual to whom you actually were historically related. Here again, our attachments are to objects and people as they really are, and not just to the experiences that they facilitate.[20]

Sober points out that—like a work of art in a suitable frame and background setting—natural objects have appropriate contexts: the wild, the natural habitat rather than zoos. Furthermore, both environmentalists and aesthetes value *rarity*. From a moral point of view, rarity seems insignificant. Why should it matter whether there are 100 blue whales and 100 sperm whales rather than 200 blue whales and no sperm whales, any more than it matters that China is made up almost entirely of Asians rather than Caucasians and Africans? But from an aesthetic point of view, rarity does matter. If we discover that the Athens museum has the only Grecian urn of a certain period, the aesthetic value of that urn escalates. Viewed aesthetically, rare organisms may be especially valuable because they are rare.

The point is that this variety of environmentalist confuses aesthetic value with ethical value, taste with morality. But the two are not the same. It seems no more justified to kill a child who is disturbing a swamp environment than it is to kill a child because you like the taste of tender flesh.

Although the land ethic seems to go to an extreme in its one-sided holism, ignoring the claims of individual animals and humans and mistakenly eschewing all domestication of nature, there is something profoundly valid in the concern of the land ethicist that we realize we are all part of larger ecological systems. We cannot survive outside the complex web of the biotic community but are deeply dependent upon it. We need to take the insights of the land ethic into account as we work toward an adequate environmental ethic. For example, a preservationist could modify Leopold's theory to fit into an anthropocentric, prudential environmentalism, by saying, "It is important to recognize our interdependence with the ecosystem. We cannot exist without decomposers, solar energy, clean water, oxygen, and adequate nutrients. They all contribute to our well-being, so to act recklessly with regard to the ecosystems is tantamount to arbitrarily throwing unmarked mines in a field we and our children are about to walk through. It is massively imprudent. We should preserve the 'integrity, stability, and beauty' of the ecosystem because it is our life-support system, our spaceship—the only one we've got." Zoocentric sentientists could make similar adjustments.

Anthropocentric preservationists, zoocentric sentientists, and ecocentrists all recognize the need for a flourishing ecosystem. The question is, Who has the most cogent environmental philosophy?

THE GAIA HYPOTHESIS

Gaia, mother of all, I sing, oldest of gods,
Firm of foundations, who feeds all creatures living on Earth,
As many as move on the radiant land and swim in the sea
And fly through the air—all these does she feed with her bounty.
Mistress, from you come our fine children and bountiful harvests,
Yours is the power to give mortals life and to take it away.

—DONALD HUGHES[21]

One other holistic environmental philosophy which seems close to Leopold's land ethic is the Gaia hypothesis, first set forth by British scientists James Lovelock and Sidney Epton in the 1960s and developed by the American scientist Lynn Margulis and others.[22] "Gaia," as the poem that opens this section notes, is the ancient Greek word for mother. The idea of Earth as a *magna mater*, a "great mother" or "great soul," is not new. Plato speaks of the world soul in the *Timaeus*, and Lucretius believed that Earth was a mortal being who would grow old and finally die. The Hellenistic Jew Philo speaks of nature as a great "All-Mother," "most ancient and most fertile of mothers . . . by way of breasts, streams of rivers and springs, bestowing food for her children, to the end that both the plants might be watered and all animals might have abundance to drink."[23]

Lovelock and Epton ask us to consider whether Earth's living matter, air, oceans, and land surface form part of a giant system, a single biological superorganism. They present a scientific case for answering "yes" to this question. Through a series of considerations of climate patterns, atmospheric conditions, energy flow, and feedback loops, they show that Earth resembles a homeostatic organism that regulates itself, maintaining life in a vital balance of opposing forces. Lovelock, in studying Earth's atmosphere and then comparing it with those of Mars and Venus, discovered that our atmosphere is in a state of disequilibrium and that the hypothesis that Earth is a self-sustaining giant organism seems the only coherent way to account for it. That is, if Earth were governed simply by geochemical systems (rather than by biological systems) like Mars and Venus, its atmosphere would have been largely carbon dioxide, in which case life would have not appeared. The composition of Earth's atmosphere, which is about 21% oxygen and 78% nitrogen, differs from her neighbors Mars and Venus, whose atmospheres are about 90% carbon dioxide. Through various feedback systems, our planet functions in a mutually supportive, holistic manner so as to maintain a remarkable atmospheric equilibrium of gases in which life flourishes. Gaia has goals of her own and will survive without human beings, but, Lovelock muses, humanity may have been created by Gaia as the central nervous system to help protect her from catastrophe. Therefore, we should regard ourselves and all other things to be "parts and partners of a vast being who in her entirety has the power to maintain our planet as a fit and comfortable habitat for life."[24]

Although Lovelock doesn't exactly attribute conscious purpose to Gaia, he does suggest a limited kind of intelligence: "Some form of intelligence is required even within an automatic process, to interpret correctly information received from the environment. . . . The body's automatic temperature-regulating system is intelligent to the point of genius." The thesis begins to take on a religious aura:

> The evolution of *Homo sapiens,* with his technological inventiveness and his increasingly subtle communications network, has vastly increased Gaia's range of perception. She is now through us awake and aware of herself. She has seen the reflection of her fair face through the eyes of astronauts and the television

cameras of orbiting spacecraft. Our sensations of wonder and pleasure, our capacity for conscious thought and speculation, our restless curiosity and drive are hers to share. This new interrelationship of Gaia with man is by no means fully established; we are not yet a truly collective species, corralled and tamed as an integral part of the biosphere, as we are individual creatures. It may be that the destiny of mankind is to become tamed so that the fierce, destructive, and greedy forces of tribalism and nationalism are fused into a compulsive urge to belong to the commonwealth of all creatures which constitutes Gaia. It might seem to be a surrender, but I suspect that the rewards, in the form of an increased sense of well-being and fulfillment, in knowing ourselves to be a dynamic part of a far greater entity, would be worth the loss of tribal freedom.[25]

Powerful as she is, humans can harm their mother Earth, and ultimately themselves, by overpolluting, by destroying ecosystems, and by reducing the natural variety necessary for homeostasis. Lovelock calls on us to "make peace with Gaia on her terms and return to peaceful co-existence with our fellow creatures." This new harmony with Gaia will include preservation of species diversity and a rejection of the crude use of technology that regards nature only as a resource. It will mean a new biospheric ethics. Here is Lovelock's assessment of the meaning of Gaia:

> Art and science seem interconnected with each other and with religion, and to be mutually enlarging. That Gaia can be both spiritual and scientific, is, for me, deeply satisfying. From letters and conversations I have learnt that a feeling for the organism, the Earth, has survived and that many feel a need to include those old faiths in their system of belief, both for themselves and because they feel that Earth of which they are part is under threat. In no way do I see Gaia as a sentient being, a surrogate God. To me Gaia is alive and part of the ineffable Universe and I am part of her.
>
> When that great and good man Pope John Paul travels around the world, he, in an act of humility and respect for the Mother or Father Land, bends down and kisses the airport tarmac. I sometimes imagine him walking those few steps beyond the dead concrete to kiss the living grass; part of our true Mother and of ourselves.[26]

Mother Earth, the world's all-encompassing ecosystem is, if not a "surrogate God," a superorganism, but humans have developed technology to the point of being able to adversely affect the homeostatic balance of the Earth. The exacerbation of the greenhouse effect is one example of such an adverse effect; the depletions of the ozone layer and the rain forests are others.

The Gaia hypothesis is a renewal of ancient ideas dressed in the clothes of new science. The question is whether it is anything more than a new myth or suggestive metaphor. As James Kirchner and others have pointed out, the Gaia hypothesis is ambiguous.[27] When analyzed closely it can be shown to contain fragments of conflicting theories, which Kirchner labels *weak* and *strong*. The weak versions include assertions that the biota has a substantial influence over certain aspects of the abiotic world, such as the temperature and composition of the atmosphere. As D. Sagan and L. Margulis put it, "The Gaia hypothesis . . . states that the temperature and composition of the Earth's atmosphere are actively regulated by the sum of life on the planet."[28] A homeostatic process occurs in which the biotic elements influence the abiotic world, and does so in a way that is stabilizing. In other words, there are negative feedback loops, as occur in a properly functioning thermostat. For example, on a cold day, the thermometer in the thermostat causes the heat to come on until a predetermined warm temperature is reached, at which time the heat is turned off. Likewise the Earth's temperature and weather patterns may have built-in stabilizing features. Most scientists accept something like this weak version of the Gaia hypothesis. Another weak version of the Gaia hypothesis is that there is evolutionary reciprocity between the biotic and abiotic aspects of the world. The biota influences the abiotic environment, which in turn influences the biotic evolutionary process. Again, this seems plausible but hardly entails a super-living-being. The problem with the strong versions of Gaia, those which hold there is an inner telos, a purpose in nature (something that a theist would accept but not a Darwinian evolutionist), is that they are not scientifically testable. Kirchner challenges the Gaians to set forth the conditions which would falsify their theory. Unless the Gaians can give us positive reasons for accepting the superorganism view or tell us under what conditions they would surrender their hypothesis, it is scientifically worthless.

Of course, Gaians argue that critiques like Kirchner's are simplistic. Not everything is empirically testable. Perhaps we should exercise humility in the face of the mystery of our world, remaining open to the possibility that an inner purpose exists and acknowledging that we humans have the power to wreak havoc on the homeostasis and telos of our mother Earth. While the hypothesis cannot be proved, it is not without plausibility. Interestingly, some Protestant theologians, such as John Cobb of Claremont Graduate School, have embraced something akin to the Gaia hypothesis.[29] Their views are a version of *panentheism* discussed in this chapter: the view that God is *in* the world. Cobb and other theological panentheists hold that the world is God's body out of which His mind or soul emerges. So there is a progressive unfolding of God's purpose in the world's evolutionary development. Evolutionary history is His story. Hence, we ought to act humbly, reverently, and as good stewards of the Earth, since it is God's holy body. Holy Communion is a daily event. We should leave it to future children of God, our brothers and sisters, in a condition as good as it was when we found it.

COMMUNITY-BASED HOLISM

In a series of essays included in his book *Nature as Subject: Human Obligation and Natural Community*,[30] Eric Katz sets forth a pluralist version of holism, *community holism*, as opposed to the *organism holism* of the land ethicists and Gaians which we have been discussing in this chapter. The community metaphor is a better analogy to the environment than an organism because, unlike the parts of an organism, the parts of a community, as well as the community itself, have objective and intrinsic value. For the organicist only the whole has objective value. As Callicott says, "The effect upon the ecological systems is the decisive factor in determination of the ethical quality of action."[31] Katz compares a university community with the environment. Just as the members of the university—the students, faculty, administration, and support staff—are all autonomous in their own right and yet are members of an institution that is also autonomous, so the members of an ecosystem, as well as the ecosystem itself, are autonomous. Organic holism, like crude utilitarianism, would allow the sacrifice of any of its members as long as the "integrity, stability, and beauty" of the system is maintained. "Organic parts have no independent value—their value is derived from the entire functioning whole, the organism. The liver, muscles, and blood in a human body are important for what they do *for* the organic whole. They are not valuable in themselves, i.e., intrinsically, but only instrumentally in that they perform functions for other entities. Thus, the parts of the organism have value as parts of a system, just as an entity with instrumental value only possesses value through its functional relations with other entities."[32] Community holism would defend the individual's right as well as the right of the community as a whole. In this way Katz thinks his model avoids the charge of "ecological fascism" leveled against holism by philosophers like Regan. More broadly, he argues, correctly, that organic holism would have to allow substitutes for any of its parts as long as the system itself continued to function. But, if this is the case, organic holism would have to permit the substitution of artificial objects (such as plastic trees and rubber mountains) if the system would function as well with these. Similarly, it would have to accept the loss of species if they could be replaced by other entities.

Katz's community holism seems to go much farther in satisfying our intuitions about the importance of individuals in the world. His recognition that individuals are a locus of value is a needed improvement on other forms of holism, but at this point his theory seems undeveloped. Katz must do a better job of addressing two problems in particular:

1. He needs to provide an argument for his thesis that communities are autonomous and have objective value. The same criticisms we have noted

in the last two chapters haunt his theory. Normally, we think of individuals as being autonomous. Only individuals deliberate and make conscious decisions. Institutional decisions, such as those made by universities, are the result of individual political processes. Institutions are not conscious beings; they don't feel pain or experience love or hate. Perhaps Katz is misled by our metaphorical attributions to institutions—we say "Japan decided to devalue the yen" or "the University of Notre Dame mourns the death of Knute Rockne," but we don't mean this literally. The leaders of Japan decided to lower the value of the yen, and the alumni and members of the University of Notre Dame individually mourned the death of Knute Rockne. So Katz still has the task of defending holism itself. But there is a second problem, as well.

2. How do we prioritize our values when those of the community and those of the individual are in conflict? Katz considers John Passmore's suggestion that if we could enhance the wilderness experience, without seriously affecting the ecosystem's functioning, by eliminating flies (at least the biting kind), we should do so. Katz replies, "[T]here is something terribly wrong with this kind of modification. The best way I can express this is to say that human modifications harm the intrinsic value of the entities contained within natural systems. Individuals within natural systems have intrinsic value (among other reasons) by virtue of their existence in the *natural* world. Forcing these entities to conform to a human ideal, a human value, of what nature ought to be, would harm this intrinsic value. Thus, the modification of natural systems—even when the result is an increase in systemic well-being—is a violation of the intrinsic value of natural entities."[33] But how do we decide conflicts of interest? When parasites live symbiotically in the environment of my intestine, may I not destroy them? When is it appropriate to eliminate a predator or nuisance? May I swat a fly when it is about to bite me? May I take preemptive action and spray it with a ecologically benign fly-killer? Are all members of the community of equal intrinsic worth—the rocks, the shrubs, the trees, the rats, the domesticated animals, humans? If they are of equal worth, then we have no criterion to adjudicate conflicts of interest (except, perhaps, self-defense—which boils down to victory for the strongest).

But we are anticipating a larger discussion of egalitarianism to take place in the next chapter. So we must conclude that while Katz rightly recognizes deep problems in holism, his own community thesis account needs a better defense than he has provided.

Returning to the problem discussed at the beginning of this chapter—whether to dam the Hetch Hetchy—who is right? Holistic ecocentrists will argue that the artificial dam disrupts the stability and beauty of the biotic community and so is wrong. Anthropocentric conservationists will disagree. Some preservationists (the individualist variety) may also side with the holists, but for different reasons; they believe that (1) it is good to maintain the wilderness for aesthetic and recreational purposes and (2) it is not in our best interest to strain the natural ecosystem by locating so many people in the San Francisco Bay area, where water resources are not readily available. We must leave our discussion at this point, waiting until Chapter 16 to discuss it further. In the next chapter we will examine other versions of contemporary biocentrism, including more individualistic types.

Review Questions

1. Discuss the Hetch Hetchy project. Given what you know about it, would you say it was morally justified? Or was Muir right—we should not tamper this much with the wilderness but adjust our lives to living in accord with her?
2. Discuss Aldo Leopold's land ethic. What are its main features? Is it, as its critics allege, misanthropic?
3. From the perspective of the land ethic, death is good. What does the land ethicist mean by this? From the perspective of the individual, death is generally seen as evil. Can these two perspectives be reconciled?
4. What is our obligation toward ecosystems? Why should we care about them?
5. To what extent is it morally permissible to domesticate natural beings and objects? Do we have a duty to "improve" nature? Is it permissible?
6. Do environmentalists like those espousing the land ethic and Gaia hypothesis romanticize nature? Consider the critique offered by Aldous Huxley (author of *Brave New World*) of Wordsworth's romanticism about nature in his essay, "Wordsworth in the Tropics":

   > It is only very occasionally that [Wordsworth] admits the existence in the world around him of those "unknown modes of being" of which our immediate intuitions of things make us so disquietingly aware. Normally what he does is to pump the dangerous unknown out of nature and refill the emptied forms of hills and woods, flowers and waters with something more reassuringly familiar—with humanity, with Anglicanism. He will not admit that a yellow primrose is simply a yellow primrose—beautiful but essentially strange, having its own alien life apart. He wants it to possess some sort of a soul, to exist humanely, not simply flowerily. . . . But the life of vegetation is radically unlike the life of man.
   >
   > The jungle is marvellous, fantastic, beautiful, but it is also terrifying, it is also profoundly sinister. There is something in what, for want of a better word, we must call the character of great forests . . . which is foreign, appalling, fundamentally and utterly inimical to intruding man. . . .
   >
   > A few months in the jungle would have convinced [Wordsworth] that the diversity and utter strangeness of nature are at least as real and significant as its intellectually discovered unity.

Is Huxley's critique, as applied to ecocentric ethical systems, valid? Do eco-holists tend to glamorize nature, forgetting or glossing over its wildness, ferocity, and destructive aspects?

7. Discuss the difference between conservationists and preservationists. Note that they both disagree with consumptionists, who generally disregard future uses of natural resources. Which theory is the best?
8. Examine Lovelock's Gaia hypothesis. How plausible is it? Could it be used as a symbol to help us see our relationship to nature in a more accurate light?
9. What is the demarcation problem? Explain its significance.
10. Compare organic holism with community-based holism. What are the strengths and weaknesses of each position?

Suggested Readings

Callicott, J. Baird. "Animal Liberation: A Triangular Affair," *Environmental Ethics* 2:4 (Winter 1980).

———. *Companion to a Sand County Almanac*. Madison: University of Wisconsin Press, 1987.

Katz, Eric. *Nature as Subject: Human Obligation and Natural Community*. Lanham, Md.: Rowman & Littlefield, 1997.

Leopold, Aldo. *Sand County Almanac: Essays on Conservation from Round River*. New York: Oxford University Press, 1947.

Lovelock, James E. *Gaia: A New Look at Life on Earth*. New York: Oxford University Press, 1981.

Nash, Roderick. *Wilderness and the American Mind*. New Haven: Yale University Press, 1982.

Sober, Elliott. "Philosophical Problems for Environmentalism." In *The Preservation of Species*. Edited by B. G. Norton. Princeton: Princeton University Press, 1994.

Notes

1. Quoted in Roderick Nash, *Wilderness and the American Mind* (New Haven: Yale University Press, 1982). My discussion of the Hetch Hetchy episode is largely based on Nash's book.
2. Gifford Pinchot, *The Training of a Forester* (Philadelphia: Lippincott, 1914), p. 13.
3. Letter to Robert Johnson, 7 February 1909. Quoted in Nash, *Wilderness and the American Mind,* p. 167.
4. A third theory may be mentioned: *panpsychism*. This is the doctrine that each spatiotemporal object has a mental, or "inner," aspect. There may be varying degrees in which things have subjective or quasi-conscious aspects, some very unlike what we experience as consciousness. The motivation for panpsychism is that it is hard to see how consciousness can be caused by or be composed of purely nonmental objects. The hypothesis of the panpsychist is that mental qualities exist in embryo or potentially in physical things. While panpsychists believe that everything is psychic in nature, they need not also believe that everything is divine.
5. John Muir, *My First Summer* (Boston: Houghton Mifflin, 1911), p. 26. I have profited here from Max Oelschlaeger, *The Idea of Wilderness: From Prehistory to Age of Ecology* (New Haven: Yale University Press, 1991).
6. John Muir, *A Thousand Mile Walk to the Gulf* (Boston: Houghton Mifflin, 1916), p. 139.
7. John Muir, *The Wilderness World of John Muir,* ed. Edwin Way Teal (Boston: Houghton Mifflin, 1954), p. 312.
8. John Muir, *A Thousand Mile Walk to the Gulf,* p. 140.

9. John Muir, *Natural Parks* (Boston: Houghton Mifflin, 1916), p. 325.
10. Aldo Leopold, *Sand County Almanac: Essays on Conservation from Round River* (New York: Oxford University Press, 1947). Selections are reprinted in *Environmental Ethics*, 2nd ed., ed. L. Pojman (Belmont, Calif.: Wadsworth, 1998).
11. Alfred Tansley, "The Use and Abuse of Vegetational Concepts and Terms," *Ecologist* 16 (1935), quoted in Frank Golley, *A History of the Ecosystem Concept in Ecology* (New Haven: Yale University Press, 1993), p. 8.
12. Aldo Leopold, *Sand County Almanac,* p. 224.
13. J. Baird Callicott, "Animal Liberation: A Triangular Affair," *Environmental Ethics* 2:4 (Winter 1980). I note that when Callicott's article is reprinted this passage is sometimes omitted by the environmentalist editor (see for example, *Ethics and the Environment,* ed. Don Sherer and Thomas Attig, Englewood Cliffs, N.J., Prentice Hall, 1983, p. 65), but I think it is highly relevant to the discussion of the land ethic and biocentrism in general. The article is reprinted in *Environmental Ethics,* ed. L. Pojman.
14. Ibid.
15. Tom Regan, *The Case for Animal Rights* (Berkeley: University of California Press, 1983), pp. 361–62.
16. Callicott makes this comparison in "Animal Liberation."
17. J. Baird Callicott, "Intrinsic Value, Quantum Mechanics, and Environmental Ethics," *Environmental Ethics* 7 (1985), p. 274.
18. Ibid., p. 275.
19. Elliott Sober, "Philosophical Problems for Environmentalism," in *The Preservation of Species,* ed. B. G. Norton (Princeton: Princeton University Press, 1986), pp. 173–94. Reprinted in *Reflecting on Nature,* ed. Lori Gruen and Dale Jamieson (New York: Oxford University Press, 1994). My discussion in this section is indebted to Sober's essay.
20. Ibid.
21. Donald Hughes's poem "Gaia: An Ancient View of Our Planet," which is quoted in James E. Lovelock, *The Ages of Gaia: A Biography of Our Living Earth* (New York: W.W. Norton & Company, Inc., 1988), p. 203.
22. Lovelock, *Gaia;* James E. Lovelock and Lynn Margulis, "Homeostatic Tendencies on the Earth's Atmosphere," *Origin Life* 5 (1974): 93–103. James E. Lovelock and Sidney Epton, "In Quest for Gaia," *New Scientist* (Feb. 1975). This is reprinted in *Environmental Ethics,* ed. L. Pojman, pp. 190–93.
23. Philo, *On Creation,* sec. 133. The metaphor was rejected by the early Christian church as an idolatrous myth.
24. Lovelock, *Gaia,* p. 1.
25. Ibid., p. 148.
26. Ibid.
27. James W. Kirchner, "The Gaia Hypotheses: Are They Testable? Are They Useful?" in *Scientists on Gaia,* ed. Stephen Schneider and P. J. Boston (Cambridge: MIT Press, 1992). This article is reprinted in *Environmental Ethics,* ed. L. Pojman, pp. 194–201.
28. D. Sagan and L. Margulis, "The Gaia Perspective of Ecology," *The Ecologist* 13 (1983).
29. See Herman E. Daly and John B. Cobb, Jr., *For the Common Good* (Boston: Beacon Press, 1989).
30. Eric Katz, *Nature as Subject: Human Obligation and Natural Community* (Lanham, Md.: Rowman & Littlefield, 1997).
31. J. Baird Callicott, "Animal Liberation," p. 320.
32. Katz, *Nature as Subject,* p. 40.
33. Ibid., p. 45.

CHAPTER TEN

Contemporary Environmental Philosophy: Biocentric Egalitarianism

In this chapter we examine three of the most prominent contemporary environmental philosophies: Arne Naess's deep ecology, ecofeminism, and Paul Taylor's biocentric egalitarianism.

DEEP ECOLOGY

On September 3, 1972, the Norwegian philosopher Arne Naess (b. 1912) gave a lecture at the third World Future Research Conference in Bucharest, Romania, entitled "The Shallow and the Deep, Long-Range Ecological Movement," in which he compared two opposing views toward the environment.[1] He labeled the standard view of conservationists *shallow ecology* and characterized it as concerned with fighting pollution and resource depletion, which threatened the good of humanity. It was mainly an anthropocentric, individualistic, Western movement, concerned with the health and affluence of people in the developed countries.

The other view he called *deep ecology* or *Ecosophy T* (ecological wisdom—the *T* stands for "Tvergasten," Naess's rustic Norwegian retreat cabin, suggesting that ecophilosophy must begin with local concerns and simple living). It called for both a deeper questioning and a deeper set of answers to our environmental concerns. Specifically, it called into question some of the major assumptions about the consumerism, materialism, and individualism that govern our civilization. Deep ecology seeks to reestablish a set of principles, valued in our past but now in danger, that promote natural diversity, autonomy, decentralization, symbiosis, egalitarianism, and classlessness. It has philosophical roots in the Hindu idea of the *Atman is Brahman* (the self is [part of] God) and Spinoza's *Deus sive Natura* (God or nature, they are the same), showing a metaphysical interconnectedness between individuals and nature.[2] This movement delves deeper into the historical roots of our environmental crisis, arguing that shallow ecology is treating cancer with Band-Aids; it is treating symptoms, not the disease. The disease is our whole materialistic, consumer-oriented, technocentric, hierarchical, anthropocentric egoism. Unless we see this as constituting a set of false beliefs and misguided values and adopt a truer set of beliefs about our relationship to the environment, we will never solve these problems. We must

Table 10-1 Naess's Comparison of Shallow and Deep Ecology

Shallow Ecology	*Deep Ecology*
Natural diversity is valuable as a resource for us.	Natural diversity has its own (intrinsic) value.
It is nonsense to talk about value except as value for humankind.	Equating value with value for humans reveals a racial prejudice.
Plant species should be saved because of their value as generic reserves for human agriculture and medicine.	Plant species should be saved because of their intrinsic value.
Pollution should be decreased if it threatens economic growth.	Decrease of pollution has priority over economic growth.
Third world population growth threatens ecological equilibrium.	World population at the present level threatens ecosystems; the population and behavior of industrial states does so more than that of any others. Human population is today excessive.
"Resource" means resource for humans.	"Resource" means resource for living beings.
People will not tolerate a broad decrease in their standard of living.	People should not tolerate a broad decrease in the quality of life but in the standard of living in overdeveloped countries.
Nature is cruel and necessarily so.	Man is cruel but not necessarily so.

Reprinted from Arne Naess, "Identification as a Source of Deep Ecological Attitudes," in *Deep Ecology*, ed. Michael Tobias (Santa Monica, Calif.: IMT Productions, 1985).

move from the egocentric to the ecocentric! Naess provided a chart to mark the comparison (Table 10-1).

Like Leopold's land ethic, deep ecology is holistic, not individualistic. It attributes intrinsic and objective value to nature and the things in nature, not simply to humans. It seeks to live in harmony with nature, rather than view humanity as dominant over nature. It espouses ecological egalitarianism, evaluating all forms of life as equally valuable, with equal rights to flourish. In many ways deep ecology looks like a philosophical elaboration of the land ethic or Gaia hypothesis.

Naess highlights two features that separate shallow and deep ecology: self-realization and biocentric egalitarianism. Let us examine them in order.

Self-Realization

What is the self and where is the self? These are profoundly difficult questions, according to Naess, and Western philosophy has sloughed over them as though the answers were self-evident. He depicts traditional Western philosophy as positing the self as a soul or mind in a body. But this dualism is undermined both by our understanding of the way the brain works (consciousness—or mind—seems to be an emergent property of brain states) and by the fact that our self-understanding is deeply affected by culture, natural environment, chemical influences, and our relationships with others. There may be no separate self apart from others. Furthermore, the self changes over time and place. The "I" is not a fixed, individual item as many Westerners suppose.

Naess appeals to the Hindu idea of *Atman* for a richer understanding of the interconnected self. A relevant passage in the *Bhagavad Gita* (chapter 6:29) reads: "He whose self is harmonized by yoga sees the Self abiding in all beings and all beings in Self; everywhere he sees the same." Gandhi translated this passage, "The man equipped with *yoga* looks on all with an impartial eye, seeing *Atman* in all beings and all beings in *Atman.*" For Hindus the *Atman* (self) is *Brahman* (God). We are all in God, part of God as sparks are part of a grand fire. Self-realization "in its absolute maximum" is "the mature experience of oneness in diversity," as depicted in the verse just quoted. "The minimum is the self-realization by more or less consistent egotism—by the narrowest experience of what constitutes one's self and a maximum of alienation. As empirical beings we dwell somewhere in between, but increased maturity involves increase of the wideness of the self."[3]

Naess points out that Gandhi permitted poisonous snakes in his *ashram* and that antipoison medicines were frowned upon. Gandhi believed that "trust awakens trust, and that snakes have the same right to live and blossom as the humans":

> How do we develop this wider self? By identifying our self with others and with Nature. Identification is a spontaneous, non-rational, but not irrational, process through which the *interest or interests of another being are reacted to as our own interest or interests.* The emotional tone of gratification or frustration is a consequence carried over from the other to oneself: joy elicits joy, sorrow sorrow. Intense identification obliterates the experience of a distinction between *ego* and *alter* (other), between me and the sufferer. But only momentarily or intermittently: If my fellow being tries to vomit, I do not, or at least not persistently, try to vomit. I recognize that we are different individuals.[4]

We sense a solidarity with others. This does not eliminate conflicts of interest. Our vital interests involve killing some plants in order to live. Native Americans identified with the animals they were hunting and engaged in elaborate rituals to ask forgiveness of the animals. Naess appeals to ethologists who give examples of animals who identify across species. He cites the noted Austrian ethologist Konrad Lorenz, who claimed that a bird he was caring for once tried to seduce him, attempting to push him into its little home. We sympathize with suffering animals, with a stranded whale on a beach, with penguins in the Antarctic who might die out because of the effects of DDT upon the toughness of their eggs. Nature lovers often feel an identification with an animal, plant, mountain, river, or ocean. The extension of this identification to the whole of nature is the fulfillment of the idea of *atman*—the self that is in all things. The focus is on oneness and harmony with nature, not conflict and competition.

On the other hand, Western individualistic thinking depicts an antagonistic relationship between humans and nature ("man *against* nature"), emphasizing competition and alienation from nature. With this hostile attitude, is it any wonder that we ravage nature as we do?

Naess qualifies his notion of self-realization as identification with nature by stipulating that we may use nature to satisfy our *vital needs* and the *vital needs of those for whom we have special responsibility*. We may eat a plant or animal to sustain life, but we may not do so simply to live more affluently. We ought to live *simply* that others may simply *live*. Naess would have us decrease our population through birth control and gradually go back to simpler living. Hetch Hetchy would be restored to its earlier beauty and integrity.

Biocentric Egalitarianism

Two Californian philosophers, Bill Devall of Humboldt State University and George Sessions of Sierra College in Rocklin, have further developed the ideas of deep ecology.[5] They reiterate Naess's idea of "self-in-Self," where "Self" stands for organic wholeness, but they develop the idea of biocentric egalitarianism further.

> The intuition of biocentric equality is that all things in the biosphere have an equal right to live and blossom and to reach their own individual forms of unfolding and self-realization within the larger Self-realization. The basic intuition is that all organisms and entities in the ecosphere, as parts of the interrelated whole, are equal in intrinsic worth. . . . Mutual predation is a biological fact of life, and many of the world's religions have struggled with the spiritual implications of this. Some animal liberationists who attempt to side-step this problem by advocating vegetarianism are forced to say that the entire plant kingdom including rainforests have no right to their own existence. This evasion flies in the face of the basic intuition of equality. Aldo Leopold expressed this intuition when he said humans are "plain citizens" of the biotic community, not lord and master over all other species.
>
> Biocentric equality is intimately related to the all-inclusive Self-realization in the sense that if we harm the rest of nature then we are harming ourselves. There are no boundaries and everything is interrelated. But insofar as we perceive things as individual organisms or entities, the insight draws us to respect all human and nonhuman individuals in their own right as part of the whole

> without feeling the need to set up hierarchies of species with humans at the top.
>
> The practical implications of this intuition or norm suggest that we should live with minimum rather than maximum impact on other species and on the Earth in general. Thus we see another aspect of our guiding principle: "simple in means, rich in ends."[6]

All things, plants, animals, rivers, forests, and bacteria, have an equal right to exist, though we are permitted to kill to satisfy our vital needs.

Deep ecologists are concerned to create a gestalt shift in the way we think about ourselves and nature. They seek to have us see things holistically rather than through the economic individualist model. Naess writes:

> Confrontations between developers and conservers reveal difficulties in experiencing what is *real*. What a conservationist *sees* and experiences *as reality*, the developer typically does not see—and vice versa. A conservationist sees and experiences a forest as a unity, a gestalt, and when speaking of the *heart of the forest*, he or she does not speak about the geometrical center. A developer sees quantities of trees and argues that a road through the forest covers very few square kilometers compared to the whole area of trees, so why make so much fuss? And if the conservers insist, he will propose that the road does not touch the *center* of the forest. The *heart* is saved, he may think. *The difference between the antagonists is one rather of ontology than of ethics*. . . . To the conservationist, the developer seems to suffer from a kind of radical blindness. But one's ethics in environmental questions are based largely on how one sees reality.[7]

Deep ecology challenges our anthropocentric materialism, challenges us to a new vision of reality, challenges us to realize our oneness with nature—and then to transform our lives accordingly. Naess, Devall, and Sessions do not engage in close analytic argument for their position. They simply point out that there is another way of seeing things—and that it may be more accurate than the one we now have. If we compare world views, they are confident, we will agree that the vision offered by deep ecology is closer to the truth.

The "deep ecology" movement has received much criticism. For a start it has been noted that the dichotomous adjectives *deep* and *shallow* modifying the noun *ecology* are charged words. Using them as labels biases the discussion in one's favor. Calling your position "deep" and your opponent's position "shallow" is tantamount to wining the debate over theories by labeling your position "correct" and your opponent's "superficial." It's not the way to conduct a fair debate.

As you may expect, some of the same criticisms that we saw leveled at Leopold's land ethic have been leveled at ecosophy. It has been called environmental fascism, since it may permit harming individual humans (and animals) for the sake of the whole. Deep ecologist Sally Gearhart states:

> An extraterrestrial observing our polluted and diseased planet would have to conclude that *Homo sapiens*, the inventor of technology, was an evolutionary blunder and should now silently fold its wings and steal away. I agree with that cosmic observer. From the point of view of our fellow species and the earth itself, the best that can happen is that human beings never conceive another child, that the child being conceived at this very moment be the last human being ever to exist.[8]

A columnist for *EarthFirst!* wrote:

> I take it as axiomatic that the only real hope for the continuation of diverse ecosystems on this planet is an enormous decline in human population. . . . [I]f the AIDS epidemic didn't exist, radical environmentalists would have to invent one.[9]

Vice President Al Gore, a concerned environmentalist himself, attacked deep ecologists as equating humans with pathogens:

> [Humans are] a kind of virus giving the earth a rash and a fever, threatening the planet's vital life functions. Deep Ecologists assign our species the role of a global cancer, spreading uncontrollably, metastasizing in our cities and taking for our own nourishment and expansion the resources needed by the planet to maintain its health. . . . The Deep Ecology story considers human civilization a kind of planetary HIV virus, giving the earth a "Gaian" form of AIDS. . . . Global warming is, in this metaphor, the fever that accompanies a victim's desperate effort to fight the invading virus whose waste products have begun to contaminate the normal metabolic process of its host organism.
>
> The obvious problem with this metaphor is that it defines human beings as inherently and contagiously destructive, and deadly carriers of a plague upon the earth. And the internal logic of the metaphor points toward only one possible cure: eliminate people from the face of the earth. . . . Some of these who adopt this story as their controlling metaphor are actually advocating a kind of war on the human race as a means of protecting the planet.[10]

Some deep ecologists have engaged in eco-sabotage, pouring sand into the gas tanks of logging trucks and tractors, destroying power lines, and destroying animal research laboratories and spiking trees. Loggers have been injured when their saws hit spikes in these trees. The letter bombs sent by Unabomber Ted Kaczynski to destroy the "human pathogens" have been likened to the rhetoric of some deep ecologists. Perhaps this is unfair, but if so, deep ecologists need to explain their program better.

Deep Ecology has been called "mystical mush" because of its Eastern view of the self as Self—"If we are all one, then it doesn't matter if I take your wallet, does it?" And, of course, it faces the problem of justifying the idea that nature has objective value. I will deal with some of these issues later in the chapter when discussing Paul Taylor's work.

There is one problem for deep ecology that is almost never discussed in the literature—but I think it is important. Deep ecology rejects the dualism of "humanity-against-nature" for a holistic "humanity-in-nature." This has important implications. As far as we know, nature is deterministic or—if not completely deterministic—a process of blind chance and necessity. In either case, it is hard to see how *freedom of the will* can get a foothold in the system. If we're simply part of nature, then aren't we just as determined as nature itself? Freedom of the will—called *libertarian* freedom—is an idea that is rooted in our nature as agents who can act rationally in a countercausal manner. Our actions can be influenced but underdetermined by causal processes. That is, we are not completely determined by antecedent causes. Our wills can exercise independent control over our actions. So the idea of free will seems to presuppose just the kind of dualism that deep ecology (and all holistic natural systems like Spinoza's) rejects. Furthermore, if we are not free, how can we be truly responsible for our actions? But if we're not responsible for our actions, what is all the fuss about environmental responsibility—or even environmental ethics? Hence, if these inferences follow, and we are not free or responsible for our actions, then the price ecocentric holism asks us to pay for giving up the uniqueness of our humanity may be greater than these philosophers realize—or if they do realize it, they're not advertising it.

Of course, freedom of the will is a deep metaphysical problem, and many philosophers reject it in favor of a theory of free action that is compatible with determinism. That is, they accept determinism but point out that humans as rational agents act voluntarily and deliberately. Even if our voluntary actions (including our choices) are caused by antecedent events, the fact that we chose them is sufficient for us to be accountable for them. The doctrine is called *soft determinism* or *compatibilism*. It purports to reconcile free action with determinism, thus preserving personal responsibility. This may be correct. But if it is, then we still have located a significant difference between humans (perhaps some "higher" animals deliberate too) and the rest of nature. We can act rationally and so be held accountable for our actions. If we are unique in either the libertarian or compatibilist sense, then we enjoy a property that may give us greater worth, but certainly gives us greater responsibility—including responsibility for what we do with nature. But deep ecology with its biocentric egalitarianism and "self-in-Self" Spinozist worldview seems to lack any place for this feature.

ECOFEMINISM AND DEEP ECOLOGY

Ecofeminism, a term first coined by Françoise d'Eaubonne in 1972, is an umbrella term for an overlapping set of concerns related to the androcentric and patriarchal dominance of our culture. Karen Warren says that ecological feminism was born of an awareness of "women's potential for bringing about an ecological revolution" and a conviction that the illogic of "the logic of domination" must be exposed and the harms of the "dominations of women and nature" overthrown.[11] Warren claims that such evils as racism, classism, sexism, and environmental degradation are all caused by the oppressive, domineering relationship of patriarchy, which ecofeminism is committed to destroying. In place of the dominance model of androcentrism will arise an egalitarian ethics of caring, love, and peace with nature. Val Plumwood puts the message of ecofeminism this way:

> Ecofeminism has particularly stressed that the treatment of nature and of women as inferior has supported and "naturalized" not only the hierarchy of male to female but the inferiorization of many other groups of humans seen as more closely identified with nature. It has been used to justify for example the supposed inferiority of black races or indigenes (conceived as more animal), the supposed inferiority of "uncivilized" or "primitive" cultures, and the supposed superiority of master to slave, boss to employee, mental to manual worker. For Western society, which has particularly employed a stronger genderized concept of nature as a way of imposing a hierarchical order on the world, feminization and naturalization have been crucial and connected strands supporting pervasive human relations of inequality and domination both within Western society and between Western society and non-Western societies. The interwoven dualisms of Western culture, of human/nature, mind/body, male/female, reason (civilization)/nature, have been involved here to create a logic of interwoven oppression consisting of many strands coming together.[12]

Similarly, Carolyn Merchant, in *Radical Ecology,* declares the need for "an alternative vision of the world in which race, class, sex and age barriers have been eliminated and basic human needs have been fulfilled."[13] She urges us to adopt "alternative, nonpatriarchal forms of spirituality and alternative pathways within mainstream religions" as promoting a global ecological egalitarianism that replaces the patriarchy of dominance. As symbols of this new ethic, she points to the Chipko women—who led a tree-hugging movement active in India in the 1970s that resisted the incursions of agribusiness, lumber, and cash-crop companies planning to cut down the nation's sacred forests. These women's protests were nonviolent yet confrontational, involving song, marches, and tree-hugging in the company of their children.

Warren argues that feminism offers a different way of doing philosophy—first-person narrations tend to supplement abstract analysis. Such a process "gives voice to a felt sensitivity often lacking in traditional analytic ethical discourse, viz., a sensitivity to conceiving oneself as fundamentally 'in relationship with' others, including the nonhuman environment." An ethic of care replaces a relationship of domination.

She illustrates this with an account of her first rock-climbing experience. After she had a difficult time rock climbing, having viewed the rock as an obstacle to be conquered, she realized a precious relationship with the rock:

> I looked all around me—really looked—and listened. I heard a cacophony of voices—birds, trickles of water on the rock before me, waves lapping against the rocks below. I closed my eyes and I began to feel the rock with my hands—the cracks and crannies, the raised lichen and mosses, the almost imperceptible nubs that might provide a resting place for my fingers and toes when I began to climb. At that moment I was bathed in serenity. I began to talk to the rock in an almost inaudible, child-like way, as if the rock were my friend. I felt an overwhelming sense of gratitude for what it offered me—a chance to know myself and the rock differently, to appreciate unforeseen miracles like the tiny flowers growing in the even tinier cracks in the rock's surface and to come to know a sense of being *in relationship* with the natural environment. . . . I wanted to be with the rock as I climbed. Gone was the determination to conquer the rock, to forcefully impose my will on it; I wanted simply to work respectfully with the rock as I climbed. And as I climbed, that is what I felt. I felt myself caring for this rock.[14]

Warren comments that "*How* one climbs a mountain and *how* one narrates . . . about the experience of climbing are also *feminist issues*. In this way, ecofeminism makes visible why, at a conceptual level, environmental ethics *is* a feminist issue."

Margarita Levin criticizes Warren, not for her love of nature, but for lacking a clear theory. She comments on the preceding passage:

> It could equally well be argued that dental hygiene is a feminist issue. One can attack plaque, tartar, and cavities with the intention of destroying all traces of them, and while affirming the hierarchy of periodontism, reinforce the oppressive conceptual framework that sanctions sexism, racism, heterosexism, ageism, and classism. Or, one can approach one's mouth with a loving perception (and toothbrush) and an awareness of our relationship to our teeth and our gums that affirms our oneness with nature, presupposing and maintaining our difference from our teeth, and respecting our teeth as different from ourselves. The alarming thought that occurs is that Warren would probably agree with and welcome this account of toothbrushing if it appeared as a narrative in a feminist journal. . . . The focus of ecofeminist nature-love is not nature at all, but the nature-lover. It is, in other words, anthropocentric and not a little narcissistic.[15]

Warren argues that ecofeminism is the correct response to both sexism and naturism. It opposes the oppression of both women and nature. It replaces the male bias with a female one. She does not claim that there is anything *objectively* better about the feminine perspective. On the contrary, ecofeminism is contextual and relativistic: "[A] feminist ethic makes no attempt to provide an 'objective' point of view, since it assumes that in contemporary culture there really is no such point of view. As such it does not claim to be 'unbiased' in the sense of 'value-neutral' or 'objective.' However, it does assume that whatever bias it has as an ethic centralizing the voices of oppressed persons is a *better bias*—'better' because it is more inclusive and therefore less partial—than those which exclude those voices."[16]

This seems problematic. Where exactly does objectivity come in? If everything is contextual and relative, then isn't Warren's statement contextual and relative? If it's simply relative to culture, then the antifeminist culture is just as valid as Warren's ecofeminism. So Warren's relativistic ecofeminism is on the face of it in the same boat as patriarchy—simply another contextual option. True, she does give one criterion for assessing processes and worldviews—"inclusivity." But this seems inadequate and even false. Why is "more inclusive" necessarily *better*? If I want to be cured of cancer, I don't consult everyone in the town or country, but the experts. If I want to become a physicist, it's not the number of people I consult that matters but the quality of the instructor, his or her understanding of physics. I suspect that when we look more deeply into the matter, it will turn out that morality is objective (see my discussion in Chapters 2 and 3) and that inclusivity will play a role. As John Stuart Mill stated, diversity of ideas (not necessarily diversity of races or ethnicity or even gender) is what important for developing our worldviews. Diversity for diversity's sake is moral promiscuity, failing to differentiate between a Grecian urn and a chamber pot.

Ariel Salleh seems to have set off a debate about the relationship between ecofeminism and deep ecology in a 1984 article when she praised the ideas of deep ecology but criticized it for failing to oppose the evils of androcentrism and patriarchy.

> The deep ecology movement will not truly happen until men are brave enough to rediscover and to love the woman inside themselves. And we women, too, have to be allowed to love what we are, if we are to make a better world.[17]

Deep ecologists responded sharply to Salleh's manifesto on joining ecofeminism with deep ecology. Warwick Fox wrote that the ecofeminist fixation on androcentrism as "*the* real root of ecological destruction" is simplistic, replacing a one-sided androcentrism with an equally one-sided *gynocentrism:*

> It is actually perfectly possible to conceive of a society that is nonandrocentric, socioeconomically egalitarian, nonracist, and nonimperialist with respect to other human societies, but whose members nevertheless remain aggressively anthropocentric in collective benefit in ways that nonanthropocentrists would find thoroughly objectionable.[18]

Fox calls for a more inclusive, "transpersonal" transformation of consciousness—zoophilic, geophilic, Gaian—rather than a female-orientated transformation. In fact, might it not be the case that the identification of the feminine with nature is itself a mythic stereotype?

Michael Zimmerman has attempted to mediate the debate, arguing that ecofeminists are correct in pointing the finger at patriarchy as a destructive sociocultural practice. Both deep ecology and feminism have correctly diagnosed the "illness of Western culture . . . that women and men alike have been distorted by the effects of patriarchy" and that what is needed is a rediscovery of the sacred. "Perhaps . . . a postpatriarchal God/dess is necessary for women and men alike to develop a form of individuation that does not involve dissociation from the body, from nature, and from woman."[19]

I leave ecofeminism to your reflective judgment and move on to what many environmental philosophers consider the most sophisticated attempt to develop a deep ecological ethic—the work of Paul Taylor.

PAUL TAYLOR'S BIOCENTRIC EGALITARIANISM

> *I hold that a set of moral norms . . . governing human treatment of the natural world is a rationally grounded set if and only if, first, commitment to those norms is a practical entailment of adopting the attitude of* respect for nature *as an ultimate moral attitude, and second, the adopting of that attitude on the part of all rational agents can itself be justified. When the basic characteristics of the attitude of* respect for nature *are made clear, it will be seen that a life-centered system of environmental ethics need not be holistic or organist in its conception of the kinds of entities that are deemed the appropriate objects of moral concern and consideration. Nor does such a system require that the concept of ecological homeostasis, equilibrium, and integrity provide us with normative principles from which could be derived (with the addition of factual knowledge) our obligations with regard to natural ecosystems. The "balance of nature" is not itself a moral norm, however important may be the role it plays in our general outlook on the natural world that underlies the attitude of* respect for nature. *I argue that finally it is the good (well-being, welfare) of individual organisms, considered as entities having inherent worth, that determines our moral relations with the Earth's wild communities of life. . . .*
>
> *A life-centered system of environmental ethics is opposed to human-centered ones precisely on this point. From the perspective of a life-centered theory, we have prima facie moral obligations that are owed to wild plants and animals themselves as members of the Earth's biotic community. We are morally bound (other things being equal) to protect their good for* their *sake.*
>
> —PAUL TAYLOR[20]

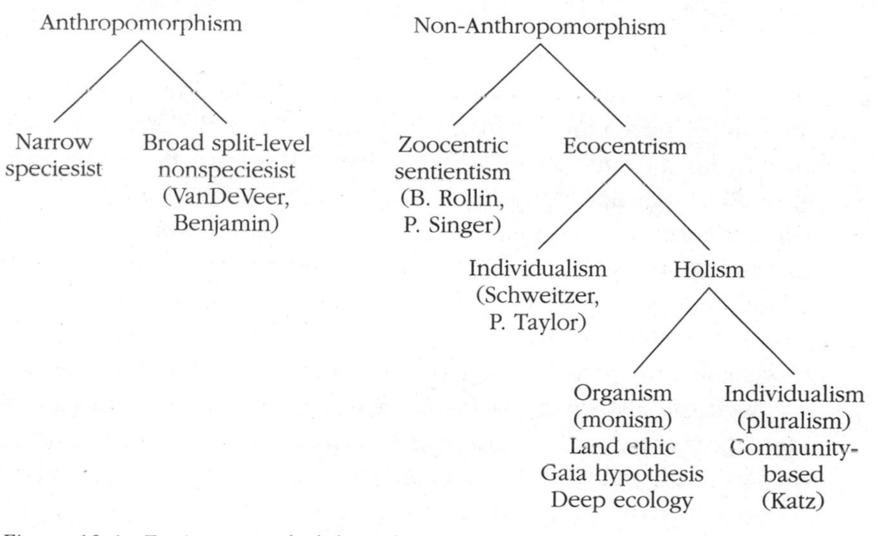

Figure 10-1 Environmental philosophies.

Not all biocentric ethical systems are holistic. Some treat individual members of species as morally considerable. Paul Taylor's biocentric egalitarianism, which appears in various forms in his publications, "The Ethics of Respect for Nature," "In Defense of Biocentrism," and *Respect for Nature,* represents the most sustained defense of ecological individualism in the literature.[21] While enormous attention has been given to Leopold's land ethics (see Chapter 9) and Naess's deep ecology (discussed earlier in this chapter), the philosophically more sophisticated work of Taylor has not received the critical attention it deserves.[22] More individualistic than either of its two biocentric relatives, Taylor offers a comprehensive, architectonic defense of biocentric egalitarianism, appealing to principles most people accept. Figure 10-1 shows Taylor's place in the scheme of environmental philosophies.

First, Taylor distinguishes *intrinsic value*, *inherent worth*, and *merit.*

1. An object has *intrinsic value* when the valuer values it "in and for itself." "We believe [it] should be preserved, not because of its usefulness or its commercial value, but simply because it has beauty, historical importance, or cultural significance."[23]
2. An object has *inherent worth* if it has value independent of valuers' valuing it, independent of anyone else's judgment of or desire for it.
3. Something, or someone, has *merit* by virtue of its satisfying certain standards.

The important idea is *inherent worth*, that which has value independently of anyone's valuing it. This is what Holmes Rolston and I have called *objective value.* Taylor says it entails "(1) that the entity is deserving of moral concern and consideration, . . . and (2) that all moral agents have a prima facie duty to promote or preserve the entity's good as an end in itself and for the sake of the entity whose good it is." Holding such a view is adopting an "ultimate moral attitude," one which meets all the criteria for a foundational moral principle in the sense that it functions as a strong presumptive guide to actions.

Taylor argues that every *living* being has inherent worth. He levels telling criticisms against, and distinguishes his position from, Leopold's land ethics (and, by implication, Naess's deep ecology) because he treats all and only living individual animals and plants as possessing inherent worth, whereas the two rivals hold that species, communities, and holistic systems possess such value.[24] Taylor sometimes gets into trouble by speaking of harming large numbers of a species as somehow worse than harming the sum of individuals harmed, but I think this problem is easily solved by dropping the notion of collective harm.

Taylor's argument takes the form of four propositions, each of which he defends with careful argument.[25]

1. Humans are not conquerors but simply plain citizens of the Earth's community. Their membership is evolutionarily on the same basis as that of all the nonhuman members.
2. The Earth's ecosystem is a complex web of interconnected elements, each sound, functioning element being mutually dependent on the rest.
3. Each individual organism is conceived of as a teleological center of life, pursuing its own good in its own way. Each living being has a particular role to play within a particular niche of the ecosystem, so that each may flourish in its own particular way.
4. Whether we are concerned with standards of merit or with the concept of inherent worth, the claim that humans by their very nature are superior to other species is a groundless claim, and in the light of propositions 1–3 must be rejected as nothing more than an irrational bias in our own favor. All living forms are of equal value.

From these we may conclude that

5. Therefore, all life forms are of equal inherent positive worth. The good of every individual organism has equal worth and deserves our equal moral consideration.

A brief explication of these premises is in order. I take premise 1 to mean that from the perspective of nature, humans are just one more result of the evolutionary process, no more important than any other aspect of nature. Anthropocentrists, especially religious ones, might challenge it, but ecocentrists accept it as basic.

Premise 2 states a well-known ecological principle that *everything is connected to everything else*, so that we never do just one thing. Our actions have wide-ranging consequences.

I take premise 3 to mean that each entity in nature has a good, seeks to fulfill its potential. Flowers grow upward toward the light; bees are drawn toward the sustenance of flowers; lions prefer the broad expanse of the wilderness to captivity in zoos; and birds fulfill themselves by soaring upward in the sky and eschewing cages.

Premise 4 is more controversial. Here is Taylor's case for it:

> In what sense are human beings alleged to be superior to other animals? We are indeed different from them in having certain capacities that they lack. But why should these capacities be taken as signs of our superiority to them? From what point of view are they judged to be signs of superiority, and on what grounds? After all, many nonhuman species have capacities that humans lack. There is the flight of the birds, the speed of a cheetah, the power of photosynthesis in the leaves of plants, the craftsmanship of spiders spinning their webs, the agility of a monkey in the tree tops. Why are not these to be taken as signs of their superiority over us?
>
> One answer that comes immediately to mind is that these capacities of animals and plants are not as *valuable* as the human capacities that are claimed to make us superior. Such uniquely human characteristics as rationality, aesthetic creativity, individual autonomy, and free will, it might be held, are more *valuable* than any of the capacities of animals and plants. Yet we must ask: Valuable to whom and for what reason?
>
> The human characteristics mentioned are all valuable to humans. They are of basic importance to the preservation and enrichment of human civilization. Clearly it is from the human standpoint that they are being judged as desirable and good. Humans are claiming superiority over nonhumans from a strictly human point of view, that is, a point of view in which the good of humans is taken as the standard of judgment. All we need to do is to look at the capacities of animals and plants from the standpoint of their good to find a contrary judgment of superiority. The speed of the cheetah, for example, is a sign of its superiority to humans when considered from the standpoint of a cheetah's good. If it were as slow a runner as a human it would not be able to catch its prey. And so for all the other abilities of animals and plants that further their good but are lacking in humans. In each case the judgment of human superiority would be rejected from a nonhuman standpoint.[26]

As Taylor himself recognizes, this is not a valid deductive argument as it stands, for nothing has been said about *positive value* in the first four premises. So Taylor needs an additional premise to follow the first one. What he offers is something like:

1* Every living thing that has a good is valuable.

That is, living things have interests. It is in the interest of the flower to flourish, in the interest of a tree to grow tall and spread its branches toward the sky, in the interest of an eagle to soar gracefully in the heavens, in the interest of a pig to eat in peace, being left alone by humans, and in the interest of humans to live healthy, cooperative, nonviolent, educated lives. Each one of us thrives in our own particular way. All living things who have interests are inherently good.

Even granting Taylor his four principles, the argument may commit the naturalist fallacy, the fallacy of deriving a value from a fact.[27] It is true that everything has a good, a way of flourishing (maybe many ways), but that says nothing about the value of the thing itself. *Having* a good and *being* good or possessing inherent value are separate concepts and the inference from one to the other needs justification. Satan may have a good without being good. Ivan the Assassin is *good at* assassinating tsars and commissars, but that doesn't give him any *worth*. Likewise, someone may be a *good* thief, and *have a good* (that is, be successful as a thief) as a thief, without being *good*. And a properly functioning human immunodeficiency virus (HIV) lodged snugly in a cell is flourishing, but it would be improper to say it has inherent worth.

Being good or having inherent value seems relative to having an interest. Consider the concept of good weather. Whether weather is good depends on whose interests are being considered. With regard to the farmer, rain may constitute good weather, whereas this may be bad weather for the sunbather or baseball player. A snowstorm may be bad weather for a person stranded in her car on an open highway, but good weather for the owner of a ski resort.

If value is relative to having an interest, then the notion of inherent objective worth is problematic. If Taylor holds that inherent objective worth is a coherent concept, he needs to provide us with a better argument. Incidentally, this same criticism applies to Holmes Rolston's discussion of value in Chapter 8. The natural locus of value seems to be the activity of evaluating, which conscious beings engage in.[28] To say that something is valuable is shorthand for the formula, "Some conscious agent A wants or commends some entity E relative to some standard S and for some purpose P."

So we value speedster Spike who runs the mile in less than 4 minutes relative to the standards of humans running the mile in races or while hunting for wild animals. Relative to the speed of a cheetah or leopard or sports car or light wave, Spike's talents are less impressive.

Could we not one-up Taylor's egalitarian equation of our worth with that of animals and plants? Could we not argue that Taylor is right to say that we all, plants, animals, and humans, are of equal value—but wrong in thinking that we have any positive value at all? We all are valueless, worthless (*de trop,* as Sartre would say). If so, Taylor's argument for biocentric egalitarianism seems to lead to this *nihilistic* conclusion.

Taylor responds to this point: "Yes, of course we *could* take an attitude of indifference toward all animals and plants and regard all of them as lacking any inherent worth (they could still be valued by human valuers as having instrumental value for this or that human purpose). But the question that must be answered if you are to refute my theory is: Would ideally rational, informed, and reality-aware valuers, *who accepted the belief-system of the biocentric outlook*, take such an attitude?"[29]

I agree that anyone who accepts Taylor's biocentric outlook might well be justified in holding the position that all life forms have positive value, but I don't think that the case has been made for taking that kind of biocentric outlook.

This leads to a related criticism. If all living beings are of equal worth, humans have no valid grounds for preferring the good, or even the survival, of other humans to that of other living beings. Call this the *speciesist critique of anthropocentrism.* Although Taylor allows for reasonable self-defense against threats to one's own life or health, biocentric egalitarianism would prevent us from intervening on behalf of other human beings in their struggle against wolves, sharks, bacteria, and viruses who are only seeking to realize their own good. If bacteria or a dangerous virus invades my body, I have a right to defend myself, but by what criterion is it permissible for doctors to kill these health- or life-threatening bacteria or viruses? Why prefer a human to an HIV's existence? The AIDS patient has a right of self-defense to destroy the HIV infecting him or her, but what grounds are there for third parties, like experimenting scientists or doctors, to take sides and engage in viral destruction? Are our scientists and physicians well-paid assassins, hired guns knocking off innocent enemies? On Taylor's logic, the HIV virus occupies a niche in the ecosystem and has a good of its own, and as such it is just as valuable as we are.[30] Even more troubling, if the virus is only innocently threatening our health—say, with a mild case of flu—Taylor's *principle of proportionality* (harming in proportion to the harm threatened) would seem to prohibit the use of antivirus drugs.[31]

While Taylor may bite the bullet here and simply insist that morality sometimes calls on us to make great sacrifices, most reasonable people will take this implication as a reductio ad absurdum of his egalitarianism. Asking precisely why we reject Taylor's position may help us discover the nature of moral principles.

But I think there is an even more serious problem with Taylor's egalitarianism, one which drives it even further toward nihilism. If racism, sexism, and speciesism are unjustified, discriminatory prejudices, isn't *biocentrism* also a prejudice? biocentric chauvinism? Why ascribe intrinsic value only to living things and not to nonliving things like robots, computers, cars, missiles, watches, rocks, air, helium atoms, feces, and dorm rooms? On what basis can we say that an amoeba, but not a robot, has a telos, a good? What is this mysterious substance that gives all and only living things inherent worth? Watches are as telic as protozoa, it would seem, and thermostats with feedback circuits and self-correcting missiles are as self-regulating as plants.

If living things are completely made from chemicals, aren't these chemicals the true possessors of inherent worth? But the chemicals are composed of atoms, and the atoms of electrons, protons, and neutrons—aren't these then the ultimate centers of value? Living entities are systems, but there are other nonliving systems that can prosper, that have a "good." Dorm rooms and houses can flourish and become run-down just like living entities. Why is it not immoral to interfere with the flourishing of a dorm room or to smash a rock or destroy a computer or robot as it is to kill an amoeba or fruit fly or cockroach who lives in the dorm room? But if the idea of a good or inherent worth can be extended to things as small as atoms and as complicated as systems and artifacts, how shall we make moral distinctions? Are two rats not worth more than one baby? two termites more than one house? two molecules more than one human? The expanding circle of moral considerability expands until nothing is left outside it, and so ends up an undifferentiated pantheism beyond good and evil. To make everything equally valuable is to end up without the kind of action-guiding directives that the notion of inherent worth was supposed to provide. If this is so, pantheism becomes nihilism.

Do you agree with this reasoning? Does endowing everything with equal value end in nihilism, the death of all value? Or do your intuitions reject this kind of reductio ad absurdum argument?

But there are other problems for one to consider. Taylor's ultimate attitude of respect for nature seems similar to G. E. Moore's version of the *good.* For Moore "good" refers to an unanalyzable, simple property like the color yellow, only, unlike yellow, it is nonnatural. It's not empirically identifiable, but known through the intuitions. Moore asks us to "imagine one world exceedingly beautiful. Imagine it as beautiful as you can . . . and then imagine the ugliest world you can possibly conceive. Imagine it simply one heap of filth." Even if there were no conscious beings who might derive pleasure or pain in either world, Moore avers, the beautiful one should exist. It would be a good thing for it to exist and a bad thing for the ugly one to exist. For Taylor, apparently, it would be better for trees to exist in peace than for both humans and trees to exist if humans unjustly went about cutting down trees, say, for the purposes of making paper for books—even books about environmental ethics.

Taylor goes so far as to advocate affirmative action in terms of reparations for offended species. We must atone for our former crimes and give special treatment to wetlands and wildflowers. But how do we pay back the bees for all the honey we have stolen from them or the salmon for all the fish we have eaten or all the flukes or roundworms or bacteria for all of their species that we have annihilated—we genocidal monsters! What infinite compensation do we owe nature? How can we possibly repay the termites that we have exterminated in favor of our lifeless property?

In the end Taylor's position becomes a moral misanthropy. Killing a wildflower is tantamount to killing a human and may be worse. Since humans are *the* conscious perpetrators of antibiocentric behavior, their demise would not be a bad thing. It would follow that if the destruction of the human race would promote the integrity, stability, and beauty of the biotic community, human genocide would be justified. Similar thoughts are expressed elsewhere in the environmental literature. In his book *Desert Solitaire,* Edward Abbey, referring to humanity's anti-environmentalism, says that he would sooner shoot a man than a snake. Similarly, Aldo Leopold's famous dictum, discussed in the last chapter, "A thing is right when it tends to preserve the integrity, stability, and beauty of the biotic community. It is wrong when it tends otherwise," might justify killing humans if that would preserve the integrity of the biotic community. Taylor develops his thesis in great detail. It goes as follows. The well-being of humans is dependent on the well-being of the biosphere (or ecosystem) rather than vice versa. From the point of view of the ecosystem, humans are unnecessary, gratuitous, spongers, and parasites.

> Every last man, woman, and child could disappear from the face of the Earth without any significant detrimental consequence for the good of wild animals and plants. On the contrary, many of them would be greatly benefitted. The destruction of their habitats by human "developments" would cease. The poisoning and polluting of their environment would come to an end. The Earth's land, air, and water would no longer be subject to the degradation they are now undergoing as the result of large-scale technology and uncontrolled population growth. Life communities in natural ecosystems would gradually return to their former healthy state. Tropical forests, for example, would again be able to make their full contribution to a life-sustaining atmosphere for the whole planet. The rivers, lakes, and oceans of the world would (perhaps) eventually become clean again. Spilled oil, plastic trash, and even radioactive waste might finally, after many centuries, cease doing their terrible work. Ecosystems would return to their proper balance, suffering only the disruptions of natural events such as volcanic eruptions and glaciation. From these the community of life could recover, as it has so often done in the past. But the ecological disasters now perpetrated on it by humans—disasters from which it might never recover—these it would no longer have to endure.[32]

The solution seems simple. The moral thing to do is to eliminate human beings.

Something has gone wrong with Taylor's reasoning. What? I think it is Taylor's confusion of being a moral patient with moral agency. As I argued in Chapter 7, moral capacities are *special*, giving their possessors higher value. Actualizing the capacity in moral character and action is actualizing objective value. Only humans and perhaps some other animals—such as some primates (chimpanzees, for example) and dolphins—have the capacity for moral deliberation and can actualize objective value in their lives, so only humans and these other beings have the high value this capacity endows. As M. Schonfeld puts it, "It is irrelevant for moral standing that only humans, but no birds, can do math, or that only birds, but no humans, can fly, but it is *not* irrelevant that only humans, but no birds, can be moral agents in the terrestrial biosphere at present."[33] Schonfeld goes on to apply this point to Taylor's biocentric egalitarianism:

> Taylor, however, denies this claim as well. Since only moral agents can be said to be good or bad, Taylor argues, the issue of superiority is applicable only to them, but not to a comparison of humans with nonhumans. [Taylor writes:] "It is not false, but simply confused to assert that humans are moral superiors of animals and plants." This rebuttal, it seems to me, is guilty of a fallacy of equivocation. "Superior," in Taylor's argument, refers both to the comparative meritoriousness of moral performances and to the comparative larger amount of morally relevant features. Although it is certainly confused to take humans as being superior to animals in terms of the comparative merits of their performance . . . , it is not confused to take humans as being superior to animals in terms of the comparative larger amount of morally relevant features they possess. Hence, one can assign humans superiority in terms of moral capacities, which are, at present on earth, uniquely their own. Most humans are potential and actual moral agents, whereas animals can never be more than just moral patients. This additional moral capacity of most humans constitutes their superior moral standing, and gives their interests overriding force in cases of conflicts. Thus, we could formulate the following rule: *The interests of potential and actual moral agents can override the interests of mere moral patients provided the interests involved are comparable.* The rule is justified by factual differences in moral capacities which are of immediate relevance for differences in moral standing.[34]

If this is correct, if the capacity for and actual exercise of moral qualities gives one greater value, then we can avoid the reductio ad absurdum argument contained in Taylor's theory. We can conclude that it is rational to afford human interests more weight than animals or plants.

Taylor's brilliant but problematic biocentric system is valuable for what it tells us about values and ethics. It corrects the holism of the land ethic, rightly identifying the individual as the primary locus of value. It is more philosophically sophisticated than deep ecology or contemporary ecofeminism. Perhaps his strongest arguments are against the two major ethical traditions, Kantianism and utilitarianism. Kantian deontologism, based on an anthropocentric valuation of *rationality,* and utilitarianism, with its bias toward *sentience,* are now seen as problematic. But Taylor's ark runs aground on the shoals *of inherent worth* (objective value). In the end his position, like those of ecocentrists and deep ecologists, boils down to an appeal to our intuitions—to the ultimate attitude that living things have inherent value. His strategy is to get us to admit that some things, namely, us, have inherent value. If we accept that premise, he leads us down a **slippery slope**—each showing that there is no relevant difference between the "higher" form of life and the "lower"—to conclude that every living thing must have it. If some things (such as humans) have inherent value, then every living thing must.

There are several ways to block Taylor's move—supposing we accept his criticisms of mainstream Kantian and utilitarian ethics. We have already discussed three of them. The first is to accept a religious position on the matter. God has endowed humans alone (or humans and X—fill in the blank according to your religion) with inherent value. Taylor believes that such positions are fraught with insurmountable metaphysical and epistemological difficulties, but he hasn't shown them; and many will disagree. When the Judeo-Christian tradition flourished, anthropocentrism did as well. On the Judeo-Christian model, such as we described in Chapter 6, an anthropocentric environmental ethic will be built on a stewardship model of human responsibility for nature.

The second alternative to Taylor's program is to give up the idea of inherent worth altogether and base morality on a less metaphysically ambitious basis: on sociobiological principles or contractualism, as outlined in Chapter 4. In the latter case, rational creatures see that it is in their interest to adopt certain rules and inculcate them in their children and society at large. No appeal to inherent value is necessary. Of course, this sociobiological move takes the glamour out of ethics and seems to leave us with a shrunken anthropomorphism, one even more narrow than most of us began with.

The third alternative, the nonutilitarian but consequentialist, core morality, which I have defended in Chapters 2, 3, and 7, accords special status to human beings on the basis of their greater ability to deliberate and act morally, but also accords greater responsibility for ameliorating suffering and promoting good—including the good of distant and future people and animals. It includes the split-level principle that warrants special treatment to humans over other organisms when it comes to basic needs, but requires sacrifices of humans regarding their nonbasic needs when the basic needs of animals are concerned. It also recognizes our symbiotic relationship with nature. There is a glory about being human which sets us above nature; while, at the same time, because we are a part of nature, we can alter nature, even our own nature. But the moral law must be the criterion whereby we realize our potential. Exactly how we do this with regard to the environment will be the subject of the second part of this book.

The virtues of Taylor's system are that it compels us to rethink our entire basis for morality in the light of environmental considerations and that it carries its biocentric premises out to their logical conclusion. But our worry is that it is an inadequate system: It ends in a moral misanthropy and ultimate nihilism.

Review Questions

1. Here are Bill Devall and George Session's list of principles of deep ecology. Read them over and discuss them.

   BASIC PRINCIPLES

   1. The well-being and flourishing of human and nonhuman life on earth have value in themselves (synonyms: intrinsic value, inherent value). These values are independent of the usefulness of the nonhuman world for human purposes.
   2. Richness and diversity of life forms contribute to the realization of these values and are also values in themselves.
   3. Humans have no right to reduce this richness and diversity except to satisfy *vital* needs.
   4. The flourishing of human life and cultures is compatible with a substantial decrease of the human population. The flourishing of nonhuman life requires such a decrease.
   5. Present human interference with the nonhuman world is excessive, and the situation is rapidly worsening.
   6. Policies must therefore be changed. These policies affect basic economic, technological, and ideological structures. The resulting state of affairs will be deeply different from the present.
   7. The ideological change is mainly that of appreciating *life quality* (dwelling in situations of inherent value) rather than adhering to an increasingly higher standard of living. There will be a profound awareness of the difference between big and great.
   8. Those who subscribe to the foregoing points have an obligation directly or indirectly to try to implement the necessary changes.

2. Consider the criticism that deep ecology lacks a notion of free will. Is this necessarily the case? How might deep ecologists respond to this charge? Would they say that free will exists in nature? or that it is an illusion? How can we be both free and a part of God?
3. Examine the quotation by Sally Gearhart: "An extraterrestrial observing our polluted and diseased planet would have to conclude that *Homo*

sapiens, the inventor of technology, was an evolutionary blunder and should now silently fold its wings and steal away. I agree with that cosmic observer. From the point of view of our fellow species and the earth itself, the best that can happen is that human beings never conceive another child, that the child being conceived at this very moment be the last human being ever to exist."

Is this a fair assessment of the issue? Consider the criticism that deep ecology (and the land ethic) is misanthropic. Do you agree with this criticism? Or is this too harsh a statement? Would it be better to say that deep ecology (and the land ethic) simply put humans in their proper ecological place as plain citizens, rather than lords and masters of the Earth?

4. Discuss ecofeminism's basic ideas. What are its strong and weak points? Consider Karen Warren's contextualism. Does her relativism undermine her theory? Explain.
5. Consider Paul Taylor's biocentric egalitarianism. Does it avoid many of the criticisms of the land ethic and deep ecology? How so? What problems do you see it having? Write an essay showing the strengths and weaknesses of Taylor's theory.
6. Compare deep ecology and Taylor's biocentric egalitarianism with "enlightened" anthropocentric environmental ethics. Taylor and the deep ecologists claim that anthropocentrism does not have the resources to solve our environmental problems because it is palliative and does not go to the heart of the matter. What we need, they allege, is a biocentric or holistic worldview, a new vision of our relationship with nature, which anthropocentrism fails to provide. Do you think this is true?
7. Which approach to environmental ethics seems the most promising at this point?

Suggested Readings

Attfield, Robin. *The Ethics of Environmental Concern*. New York: Columbia University Press, 1983.

Bookchin, Murray. "Social Ecology Versus Deep Ecology." In *Environmental Ethics*. Edited by L. Pojman. Belmont, Calif.: Wadsworth, 1998.

Devall, Bill. *Simple in Means, Rich in Ends: Practicing Deep Ecology*. Salt Lake City: Peregrine Smith, 1985.

Devall, Bill, and George Sessions. *Deep Ecology: Living As If Nature Mattered*. Salt Lake City: Peregrine Smith, 1985.

Levin, Margarita. "A Critique of Ecofeminism." In *Environmental Ethics*. Edited by L. Pojman. Belmont, Calif.: Wadsworth, 1998.

List, Peter C., ed. *Radical Environmentalism*. Belmont, Calif.: Wadsworth, 1993.

Naess, Arne. *Ecology, Community and Lifestyle: Outline of an Ecosophy*. Translated by David Rothenberg. Cambridge: Cambridge University Press, 1989.

———. "The Shallow and the Deep Long-Range Ecological Movement." *Inquiry* 16 (1973). Reprinted in *Environmental Ethics*. Edited by L. Pojman. Belmont, Calif.: Wadsworth, 1998.

Salleh, Ariel. "Deeper than Deep Ecology: The Ecofeminist Connection." *Environmental Ethics* 6 (1984).

Sterba, James. "Environmental Justice: Reconciling Anthropocentric and Nonanthropocentric Ethics." In *Environmental Ethics*. Edited by L. Pojman. Belmont, Calif.: Wadsworth, 1998.

Taylor, Paul. *Respect for Nature*. Princeton: Princeton University Press, 1986.

Warren, Karen J. "Feminism and Ecology: Making Connections." *Environmental Ethics* 9 (1987).

———. "The Power and the Promise of Ecological Feminism." *Environmental Ethics* 12 (1990). Reprinted in *Environmental Ethics*. Edited by L. Pojman. Belmont, Calif.: Wadsworth, 1998.

Zimmerman, Michael. "Feminism, Deep Ecology, and Environmental Ethics." *Environmental Ethics* 9 (1987).

Notes

1. Arne Naess, "The Shallow and the Deep, Long-Range Ecological Movement," *Inquiry* 16 (Spring 1973). Reprinted in *Environmental Ethics,* 2nd ed., ed. L. Pojman (Belmont, Calif.: Wadsworth, 1998), pp. 134–36. See also his "Ecosophy T: Deep Versus Shallow Ecology," also reprinted in *Environmental Ethics,* ed. L. Pojman, pp. 137–43. The best single work (anthology) on deep ecology is *Deep Ecology for the Twenty-First Century,* ed. George Sessions (Boston: Shambhala Press, 1995). The terms *deep* and *shallow* seem gratuitously question-begging, but Naess uses them to indicate his sense that traditional anthropocentric environmentalism doesn't go to the heart of the matter—a whole reorientation of our worldviews.

2. Deep ecologists draw ideas and inspiration from Spinoza, but it is doubtful whether they should rely on him for so much. He was, after all, an anthropocentrist, writing "Besides man, we know of no particular thing in nature in whose mind we may rejoice, and whom we can associate with ourselves in friendship or any sort of fellowship; therefore, whatsoever there be in nature besides man, a regard for our advantage does not call on us to preserve, but to preserve or destroy, according to its various capabilities, and to adapt to our use as best we may." *Ethics*, part 4: 26, trans. R. Elwes (New York: Dover Press, 1955).

3. Naess, "Ecosophy T: Deep Versus Shallow Ecology," p. 139.

4. Ibid.

5. Bill Devall and George Sessions, *Deep Ecology: Living As If Nature Mattered* (Salt Lake City: Peregrine Smith, 1985). The selection I quote from is reprinted in *Environmental Ethics,* ed. L. Pojman, pp. 137–43.

6. Ibid., p. 146.

7. Arne Naess, *Ecology, Community and Lifestyle: Outline of an Ecosophy*, trans. David Rothenberg (Cambridge: Cambridge University Press, 1989), p. 66.

8. Sally Gearhart, "An End to Technology," *in Machina Ex Dea: Feminists Perspectives on Technology,* ed. Joan Rothschild (New York: Pergamon Press, 1983), p. 183.

9. Quoted in Carolyn Merchant, *Radical Ecology: The Search for a Livable World* (New York: Routledge, 1992), p. 175.

10. Al Gore, *The Earth in Balance* (New York: Houghton Mifflin, 1992), pp. 216–17.

11. Karen Warren, "The Power and the Promise of Ecological Feminism," *Environmental Ethics* 12 (1990), pp. 125–46. Reprinted in *Environmental Ethics,* ed. L. Pojman.

12. Val Plumwood, "Ecosocial Feminism as a General Theory of Oppression," in *Ecology,* ed. Carolyn Merchant (Atlantic Highlands, N.J.: Humanities Press, 1994), p. 211.

13. Carolyn Merchant, *Radical Ecology* (New York: Routledge, 1992), pp. 235–36. See also her *Ecological Revolutions: Nature, Gender, and Science in New England* (Chapel Hill: University of North Carolina Press, 1989).

14. Karen Warren, "The Power and the Promise of Ecological Feminism."

15. Margarita Garcia Levin, "A Critique of Ecofeminism," in *Environmental Ethics*, ed. L. Pojman, pp. 183–88.

16. Ibid.

17. Ariel Salleh, "Deeper than Deep Ecology: The Eco-feminist Connection," *Environmental Ethics* 6 (1984), pp. 339–45.

18. Warwick Fox, "The Deep Ecology–Ecofeminist Debate and Its Parallels," in *Deep Ecology for the Twenty-First Century*, ed. George Sessions.

19. Michael Zimmerman, "Deep Ecology and Ecofeminism: Emerging Dialogue," in *Reweaving the World: The Emergence of Ecofeminism*, ed. I. Diamond and G. F. Orenstein (San Francisco: Sierra Club, 1990).

20. Paul Taylor, "The Ethics of Respect for Nature," *Environmental Ethics* 3 (Fall 1981). Reprinted in *Environmental Ethics*, ed. L. Pojman. I have italicized "respect for nature" in the quotation to highlight it.

21. Paul Taylor, "The Ethics of Respect for Nature," *Environmental Ethics* 3 (Fall 1981); "In Defense of Biocentrism," *Environmental Ethics* 5 (Fall 1983); *Respect for Nature* (Princeton: Princeton University Press, 1986).

22. The critiques known to me, Gene Spitler's "Justifying a Respect for Nature," *Environmental Ethics* 4 (1982), and John Kleinig's *Valuing Life* (Princeton: Princeton University Press, 1991), chap. 4, raise important issues on this issue.

23. Taylor, *Respect for Nature*, p. 73f.

24. Ibid., p. 118.

25. They are found in "The Ethics of Respect for Nature," *Environmental Ethics* 3 (1981) and *Respect for Nature*, p. 99f.

26. Taylor, *Respect for Nature*, p. 129ff.

27. Taylor rejects any logical entailment between the "is-statement that a being has a good of its own" and "the ought-statement that it is worthy of moral consideration and hence ought to be treated in a certain way." In a letter to me he writes, "This logical gap is the reason why *commitment* to a moral principle (or the *taking* of a moral attitude) is needed to get from the assertion that living things have a good of their own to the claim that they possess inherent worth." But the question remains, Why should one take this attitude and not some other? (Paul Taylor, letter to author, 19 March 1993.)

    When Taylor replies that his system is justified by its coherence, I can only reply that an infinite set of systems, including fairy tales and Satanic systems, can be justified in that manner. Coherence may generally be a necessary condition for justification, but it is not sufficient.

28. Taylor has replied to this point: "The question, Are all values dependent on human values, their standards and purposes? is unclear. What is meant by the *noun* 'value'? I think the real issue has to do with the *truth* of value judgment. It is the issue of moral realism vs. antirealism. When I was writing *Respect for Nature* (RN), I left the metaethical assumptions ambiguous. My present metaethical views, which I believe are consistent with the moral theory presented in RN, are: an externalist moral psychology combined with a *moral realist* metaphysics (regarding the truth of moral judgments) and a *coherentist* moral epistemology (regarding the justification of moral judgments)." (Paul Taylor, letter to author, 19 March 1993.)

29. Ibid.

30. Taylor does list five principles for adjudicating conflicts between human and nonhuman interests: self-defense, proportionality, minimum harm, distributive justice, and restitutive justice (restoring ill-gotten gains). The basic needs of humans and nonhumans are to be decided impartially (via distributive justice), but the basic needs of nonhumans should override the nonbasic needs of humans. Some philosophers have pointed to these considerations to support the thesis that Taylor's views are not antihuman. But this seems

dubious. Perhaps I have a right to defend myself against a virus, but what right does a third party—say, a physician—have to enter in on my behalf against the virus? If all are equal, the virus has just as much a right to live as I have—and the physician violates the equality principle by so intervening to harm the virus. It's hard to avoid hierarchies.

31. Taylor responds to this point: "Your 'second criticism' is important. . . . The problem is complex, and I am not sure how to solve it. . . . I have yet to see a good argument (that doesn't rely on our 'intuitive' judgments, which are simply embedded in our culture's anthropocentrism) for unequal degrees of 'moral significance' attached to different species. What has to be shown, it seems to me, is that the *intrinsic* (inherent) *value* (worth) of organisms varies with what species they belong to. Although this *seems* to be obviously true, the more it is critically examined the more it turns out to be just an assumed belief that is part of the value-conditioning we all receive from our culture. It has the same status as racial bias in a racist culture, or a sexist bias in our culture!" (Paul Taylor, letter to author, 19 March 1993, p. 7.)

32. Paul Taylor, "The Ethics of Respect for Nature."

33. M. Schonfeld, "Who or What Has Moral Standing?" *American Philosophical Quarterly* 29 (1992), p. 357.

34. Ibid., pp. 357–58.

CHAPTER ELEVEN

Population: General Considerations

The happiness of a country does not depend, absolutely, upon its poverty or its riches, upon its youth or its age, upon its being thinly or fully inhabited, but upon the rapidity with which it is increasing, upon the degree in which the yearly increase of food approaches to the yearly increase of an unrestricted population.

—THOMAS MALTHUS, 1798[1]

In February 1966, in the plaza in front of a train station in Calcutta, India, I saw dozens of people lying motionless on the ground in broad daylight. Each person was covered with a thin cloth from head to toe. Here and there a skeletal brown arm or foot spilled out from under its dust-gray cloth. Sometimes half a face, eyes shut, was bared to the sun. Passengers with trailing clouds of beggars, porters, hawkers, and taxi drivers streamed past without concern, stepping over or around the motionless bodies. I asked my Bengali host whether the people lying on the plaza were dead or sleeping. He answered that Calcutta had perhaps half a million people whose only homes were the city's streets and sidewalks. As for these people, he tried to reassure me, "If a person does not move for three days, the station-master has him carted away."

—JOEL E. COHEN, demographer[2]

THOMAS MALTHUS: PROPHET OR VILLAIN?

No philosopher has been vilified more, "refuted" more, than the English parson, philosopher, and economist Thomas Malthus (1766–1834). Named after him, the label *Malthusian* is tantamount to *misanthrope* in many circles. On the face of it, Malthus was an unlikely candidate for such controversy—the neighbor of Jane Austen and friend of the economist David Ricardo, known for his "spotless integrity" and "sweetness of temper and tenderness of heart," a benevolent and pious man, foregoing fame and fortune. A frugal father, friendly clergyman, circumspect scholar, and moderately talented philosopher, he lived an exemplary, if quiet, life. But he did one unspeakable, politically incorrect thing. In an essay written two hundred years ago (1798), he dared to assert that you can have too much of a good thing.[3] The Earth, or a nation, can have too many people. He argued that although human beings have inherent value, unrestricted population growth can bring disaster.

Malthus's theory is simple. It starts with two premises: (1) that food is necessary to human existence and (2) "that the passion between the sexes is necessary, and will remain nearly in its present state." He rejected the claims of the utopian anarchists—his fellow Englishman, Richard Godwin (1756–1836), and the French aristocrat, Marquis de Condorcet (1743–94)—who believed in the perfectability of man, in the ability of humanity to remove all obstacles to progress, which they saw as including the sexual passions.[4] Malthus was skeptical of utopian visions and especially those that included the desexing of humanity. He wrote that "those who from coldness of constitutional temperament have never felt what love is, will surely be allowed to be very incompetent judges, with regard to the power of this passion to contribute to the sum of pleasurable sensation in this life." Godwin, who publicly rejected marriage as a deprecation of freedom but secretly married the feminist philosopher Mary Wollstonecraft, never forgave Malthus for making him look like a frozen fish.

Since food is necessary and the power of sexual passions likely to remain constant throughout history, Malthus reasoned, "the power of population is definitely greater than the power in the Earth to produce subsistence for man. Population when unchecked increases in a geometrical ratio. Subsistence increases only in an arithmetical ratio. A slight acquaintance with numbers will show the immensity of the first power in comparison with the second." Malthus surveyed the census figures of numerous nations and supported his thesis with a host of impressive statistics.

Since the increase of food is arithmetical and the increase of population tends to be exponential (or geometrical), population will tend to outrun its supply of food, so that we are faced with the alternatives of curtailing birth rates or facing the four horsemen of the Apocalypse: hunger, disease, rampage, and war (Figure 11-1).[5] Nature will balance population with resources in her ruthless way if we do not act, but we have the choice of limiting births, as a more benign method. As an eighteenth-century clergyman, Malthus rejected artificial contraception and abortion and instead advocated delaying marriage and abstaining from sex until marriage. He was the first philosopher in history to call on the government to inaugurate a population policy to encourage abstinence, late marriages, and small families. Although he believed in charity and temporary help for the unemployed, he opposed the welfare policies of his time, the Poor Laws, arguing that they encouraged poor adolescents to have children they could not afford. No one, he argued, has an inalienable right to subsistence. No one has a right to bring children into the world whom he or she cannot afford to raise. His rejection of welfare was bitterly opposed as cruel, heartless, and inhuman; he was seen as a modern Cassandra who prophesied doom,

Figure 11-1 Albert Dürer, *The Four Horsemen of the Apocalypse*

"And I saw, and behold, a white horse, and its rider had a bow; and a crown was given to him, and he went out conquering and to conquer. When he opened the second seal, I heard the second living creature say, 'Come!' And out came another horse, bright red; its rider was permitted to take peace from the earth, so that men should slay one another; and was given a great sword. When he opened the third seal, I heard the third living creature say, 'Come!' And I saw, and behold, a black horse, and its rider had a balance in his hand; and I heard what seemed to be a voice in the midst of the four living creatures saying, 'A quart of wheat for a day's wages, and three quarts of barley for a day's wages; but do not harm oil and wine!' When he opened the fourth seal, I heard the voice of the fourth living creature say, 'Come!' And I saw and behold, a pale horse, and its rider's name was Death, and Hades followed him; and they were given power over a fourth of the earth to kill with sword and with famine and with pestilence and by wild beasts of the earth." (*The Revelation to John* 6:2–8) gloom, and disaster.[6] The poet Samuel Coleridge was scandalized by Malthus's views on population growth, writing, "I solemnly declare that I do not believe that all the heresies and sects and factions, which the ignorance and the weakness and the wickedness of man have ever given birth to, were altogether so disgraceful to man as a Christian and as a philosopher as this abominable tenet." Malthus's contemporary Karl Marx labeled him the "principal enemy of the people," while Pierre Leroux said "Malthusians propose an annual massacre of the innocent."[7] On the other hand, Malthus advocated free medical aid to the poor, opposed child labor, and called for universal education, so that every citizen could have the means to find gainful employment.[8] He complained that his critics had grossly misinterpreted him: "It is an utter misconception of my argument to infer that I am an enemy to population. I am only an enemy to vice and misery, and consequently to that *unfavorable proportion* between population and food which produces these evils." And again, "In the desirableness of a great and efficient population, I do not differ from the warmest advocates of increase."[9]

Was Malthus correct about the inevitability of population outrunning food supply? Critics point out that he overlooked the power of technology to enhance our productive capacity.[10] Actually, Malthus did take this into account, writing that "if the capacity of the soil were at all times put properly into action, the additions to the produce would, after a short time and independently of new inventions, be constantly decreasing till, in no very long period, the exertion of an additional laborer would not produce his own subsistence."[11] He acknowledged that technological innovation could extend our power to produce food, but he discounted its long-term efficacy. Such improvements are only temporary and actually lead to greater misery, for when the overshoot of the carrying capacity finally occurs, the disaster will be all the greater. The higher technology takes us in increasing population beyond the natural carrying capacity, the deeper the fall when the crash comes. Sooner or later innovation will succumb to population growth rates.

Is there evidence for Malthus's theory? There is. Shortly after Malthus's death, Ireland experienced a tragic famine that corroborated his theory. In the late seventeenth century, the population of Ireland was about 2 million people, mostly living in misery. Then a technological revolution occurred, as the potato was introduced from America, enabling the Irish to raise much more food per acre than ever before. The result of this benevolent technological improvement was an increase in population from 2 million to 8 million by 1845. Between then and 1847 the potato blight, a previously unknown disease, destroyed the crops, causing a nationwide famine, which was followed by typhus. Two million Irish died of starvation and disease;

another 2 million emigrated; and the remaining 4 million learned a sharp lesson that has still not been forgotten. Ireland's population has been roughly stable since that date, in spite of the fact that the nation, with a predominantly Roman Catholic population, forbids the use of artificial contraceptive materials. The stability has been achieved by an extraordinary increase in the age at marriage—and the fact that many people enter the priesthood or become nuns.

Similar stories could be told of countries in Africa, such as Ethiopia, Somalia, Sudan, Senegal, and Mauritania, where modern medicine saved lives and significantly lowered infant mortality; but the resulting population increases led to insufficient food resources during lean years.

CASSANDRA MEETS DR. PANGLOSS

Optimist: This is the best of all possible worlds!
*Pessimist: I'm afraid so.**

In 1980 a famous bet took place. An economist named Julian Simon from the University of Illinois bet three doomsdayer ecologists, Paul Ehrlich, John Holdren, and John Harte, $1,000 that the market price of five metals (copper, chrome, nickel, tin, and tungsten) would fall between 1980 and 1990. In 1990 the three ecologists paid up. All five metals had fallen in price. The publicity of the bet did much to enhance the reputation of Julian Simon and advance the cornucopian cause. Simon, who died while I was writing this book (February 12, 1998), held that human ingenuity can solve every problem we are facing. Using an impressive number of graphs and statistics showing economic growth over the last several decades, he argues in his major work, *Ultimate Resource,* that on average life is getting better for the vast majority of humanity. Where life is bad—say, in China, the former Soviet Union, or sub-Saharan Africa—it is because of lack of political and economic freedom. Free-market forces have the ability to respond to problems with ingenious solutions. Indeed, necessity is the mother of invention. Rather than despair of our future, as Simon claims the doomsdayers do, we should be optimistic. He points to the fact that we are living longer, are healthier and wealthier, are able to produce more food, and are able to substitute synthetic materials for more expensive metals. There is no scarcity of energy supplies. We have actually reduced our pollution of

* Dr. Pangloss is Voltaire's caricature of the philosopher Leibniz, who taught that this is the best of all possible worlds. Cassandra, in Greek mythology, always prophesies doom and gloom.

water and air. One hundred years ago if your child had diabetes, you would watch helplessly as he went blind and died. Today medicine affords him a healthy life. Centuries ago you had to give up reading as your eyes got dim around the age of 40. Now eyeglasses or contact lenses allow us to read into old age.

Simon relates how he was converted to cornucopianism: "One spring day about 1969, I visited the U.S. AID office on the outskirts of Washington, D.C., to discuss a project intended to lower fertility in less-developed countries. I arrived early for my appointment, so I strolled outside in the warm sunshine. Below the building's plaza I noticed a road sign that said *Iwo Jima Memorial*. There came to me the memory of reading a eulogy delivered by a Jewish chaplain over the dead on the battlefield at Iwo Jima, saying something like, How many who would have been a Mozart or a Michelangelo or an Einstein have we buried here? And then I thought: Have I gone crazy? What business do I have trying to help arrange it that fewer human beings will be born, each one of whom might be a Mozart or a Michelangelo or an Einstein—or simply a joy to his or her family and community, and a person who will enjoy life?"[12]

Population, according to Simon, is not a problem. "The ultimate resource is people—skilled, spirited, hopeful people—who will exert their wills and imaginations for their own benefit as well as in the spirit of faith and social concern."[13] Human beings are our greatest resource. The more we have the better. He argues that if doomsdayers like Ehrlich say "that their lives are of value to themselves, and if the rest of us honor that claim and say that our lives are of value to us, then in the same manner the lives of additional people will be of value to those people themselves. Why should we not honor their claims, too?"[14] He rejects the idea that people in either developing countries or the developed countries should be coerced to practice birth control. High fertility leads to the survival of the group. Families should have exactly as many children as they want to have, no more, no less. In sum, our science and technology, products of human creativity and intelligence, can save us, if we let them:

> We now have in our hands—in our libraries, really—the technology to feed, clothe, and supply energy to an ever-growing population for the next 7 billion years. Most amazing is that most of this specific body of knowledge developed within the past hundred years or so, though it rests on knowledge that had accumulated for millennia, of course. Indeed, the last necessary additions to this body of knowledge—nuclear fission and space travel—occurred decades ago. Even if no new knowledge were ever invented after those advances, we could go on increasing forever, improving our standard of living and our control over our environment. The discovery of genetic manipulation certainly enhances

> our powers greatly, but even without it we could have continued our progress forever.[15]

And again, "Even if no new knowledge were ever invented . . . we would be able to go on increasing our population *forever* while improving our standard of living."[16] Simon doesn't mean that there is an infinite amount of oil or copper in the Earth, but that, whenever those resources expire, human ingenuity can invent substitutes. He has deep faith in the free-enterprise system to solve our material problems. Yes, problems arise: Resources become scarce in the short run, and prices rise; but the higher the prices rise, the more incentive there is for inventors and entrepreneurs to search for new solutions. Necessity being the mother of invention, new and cheaper substitutes are found, and progress is made. "I do not suggest that nature is limitlessly bountiful. Rather, I suggest that the possibilities in the world are sufficiently great so that with the present state of knowledge—even without the additional knowledge that human imagination and human enterprise will surely develop in the future—we and our descendants can manipulate the elements in such fashion that we can have all the raw materials that we desire at prices ever smaller relative to other goods and to our total incomes. In short, our cornucopia is the human mind and heart, and not a Santa Claus natural environment. So it has been throughout history, and therefore so is it likely to be in the future."[17]

Ironically, more than 130 years before Simon wrote his major work, Karl Marx and Friedrich Engels attacked Malthus and the doomsdayers in much the same vein. In 1844 Engels gave the basic critique that Simon develops:

> Has it been proved that the productivity of the land increases in an arithmetical progression? The extent of land is limited—that is perfectly true. But the labor power to be employed on this area increases along with the population; and even if we assume that the increase in yield due to this increase does not always rise in proportion to the labor, there remains a third element—which the economists, however, never consider important—namely science, the progress of which is just as unceasing and at least as rapid as that of populations. . . . Science increases at least as fast as population. The latter increases in proportion to the size of the previous generation. Science advances in proportion to the knowledge bequeathed to it by the previous generation and thus under the most ordinary conditions grows in geometrical progressions—and what is impossible for science?[18]

In other words, technology, the offspring of science, also grows exponentially, and it can keep pace with human population.

Who is right: the Malthusian doomsdayers or the Engels-Simon cornucopians? In this chapter we look at the evidence. You will need to decide where the truth resides.

WHAT ARE THE FACTS? HOW BAD IS THE POPULATION PROBLEM?

The Universal Declaration on Human Rights describes the family as the natural and fundamental unit of society. It follows that any choice and decision with regard to the size of the family must irrevocably rest with the family itself and cannot be made by anyone else.

—U THANT, former Secretary-General of the United Nations[19]

In South India, at a certain festival, it was customary to roll out a giant ceremonial chariot in honor of Lord Vishnu. The chariot was wheeled but no provision was made for steering it. Pulled and pushed by devotees in religious ecstasy, it crushed everything in its path: people, animal, buildings. It was called the Juggernaut.

Population growth has something of that same terrible appearance of inevitability. However, like the Juggernaut, it is driven—not by fate—but by humans who are unconcerned about the consequences of its progress. If we can bring those driving the population Juggernaut to consider the consequences of what they are doing perhaps we can turn it from its present destructive course.

—LINDSEY GRANT[20]

Is the world overpopulated? How serious is the increasing growth of the world's population? How does this impact on the environment?

To find the annual rate of population change demographers (population specialists) compare the annual birth rate with the annual death rate. When the birth rate of a region is greater than the death rate, population is growing. When the death rate is greater, population is declining, and we have negative population growth (NPG). When the birth rate equals the death rate, we have zero population growth (ZPG). The formula is

$$\text{Annual rate of population change (\%)} = \frac{\text{Birth rate} - \text{death rate}}{10}$$

For example, in 1991 the birth rate for the world was 27 (per 1,000 people) and the death rate 9. Thus the world's population in 1991 grew at rate of 1.8%.

$$1.8\% = \frac{27 - 9}{10}$$

The result was 97 million more people that year.

Recall our discussion of exponential growth in Chapter 1. In evaluating population growth one must keep in mind that such growth increases *exponentially* rather than *linearly*. Linear growth increases by adding one unit to the sum: 1, 2, 3, 4, 5, and so forth. Exponential growth increases by a fixed percentage of the whole over a given time. The doubling pattern goes

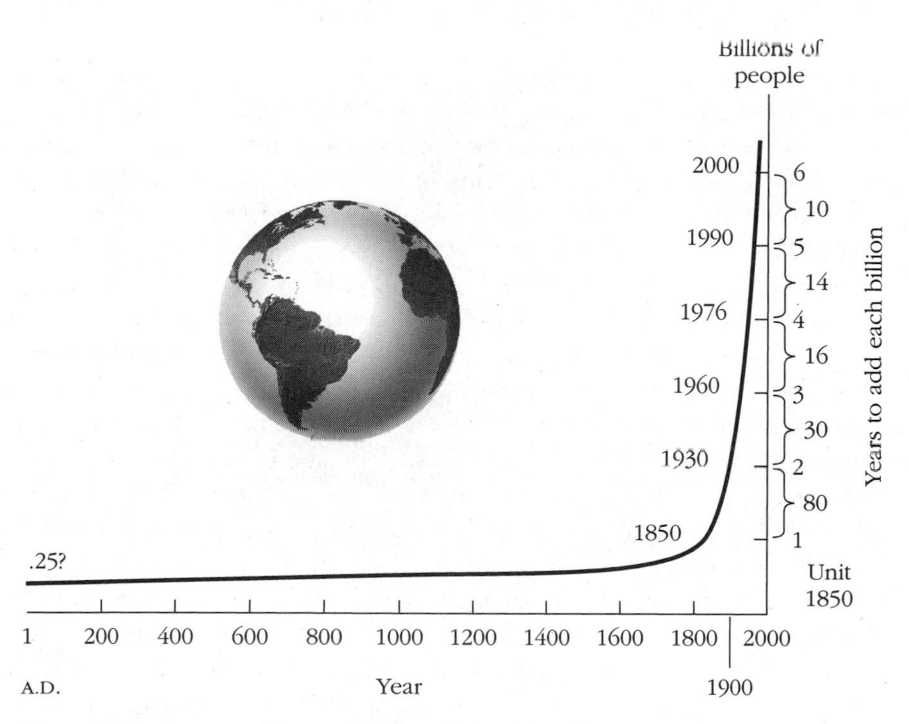

Figure 11-2 J curve of the world's population growth.

1, 2, 4, 8, 16, 32, 64, and so forth. Recall that we used *the rule of 70* to calculate the growth of a particular population. If a sum increases at 1% per annum, it will double in size in 70 years. So a community of 1,000 people growing at 1% per annum will reach 2,000 in 70 years. If we increase the growth rate, say to the present 1.7% rate of the U.S. population, the doubling time is 41 years ($70/1.7 \approx 41$). If it increases at a 2% rate per annum, it will double in 35 years. An 8% rate, such as occurred in Rwanda before its recent civil war, caused a doubling of the population every 8½ years.

Figure 11-2 depicts the world's population growth from the time of Christ, when the world's population was estimated to have been around 250 million, until today, when it is 6 billion.

Crowded conditions prevail in many parts of the world. Famines have worsened in areas of Africa and Asia. Today the global population is growing not by 70 million per year but by 90 million (1995). The growth rate is about 1.7%, translating into a doubling time of 41 years, so that, if the present trend continues, the world's population will reach about 12 billion in 2040, doubling again to 24 billion by 2081. (See Figure 11-2.)

There is evidence that the growth rate is slowing down. Demographers report that the population of some sub-Saharan African countries is leveling off. Unfortunately, the causes are famine, disease, and war, such as the civil war in Rwanda. In some countries, such as Sudan and Zimbabwe, AIDS is wreaking havoc in the population. A recent news report stated that 25% of the adults in Zimbabwe were infected by AIDS. Even so, projections prepared by the United Nations' demographers indicate that if human populations continue to increase at the present rate, the population will increase more than 130 times in 160 years, from 5.3 billion in 1990 to about 694 billion in 2150. As I noted in Chapter 1, this humongous population would require more water than falls from the skies to grow the grain needed to feed this number of people.

Compare this with the negative population growth of Germany and slower growth of the United States (1.7%). The present U.S. population is up from about 70 million in 1900 to around 270 million today. About 1.2 million people per year are added to our population through immigration. Demographers project that at current trends we will have a population of 397 million by the year 2050. If we are straining our resources and producing pollution now, the prognosis for the future should give us pause.[21]

A slight increase in the world's growth rate would result in the world's population density becoming that of present-day New York City by the year 2300. Because a disproportionate number of people in the less developed countries are under 15 years of age (averaging 39%—and in Kenya 50%), a tremendous **demographic momentum** will cause continuing exponential population growth for a long time to come *even if* we inaugurate a zero-growth global policy tomorrow. For example, if over the next 33 years the average family size of India dropped from its present 3.9 to the replacement rate of 2.2, India's present population of over 900 million would still soar to about 2 billion by 2100.

These figures are projections, not hard predictions. No one knows exactly what will happen. A prediction is categorical. It says "X will happen." A projection is a *hypothetical prediction*. For example, if you see me driving my car at 80 mph toward the edge of a canyon, you could say, "If you continue on the present course, you'll go over the cliff and very likely die." Of course, I can change course. It may be that your hypothetical prediction makes me realize the danger of my act. So when we *project* that the world will have a population of 11.2 billion by the year 2050 or India will have 2 billion people by 2100, we are saying that *if* the growth rate continues at the present rate, these will be the consequences. Although projections are not hard predictions, they are worth taking seriously, for they describe general tendencies and, as with the car speeding toward the canyon's edge, show the direction in which we are heading.

We can explain resource consumption and pollution growth in terms of exponential growth by factoring in population increase. If a population grows and does not reduce its resource consumption or pollution, resource consumption and pollution will exponentially increase. Our present environmental problems, including population growth, resource use, extinction of species of plants and animals, destruction of ecosystems, depletion of the ozone layer, and pollution, are intertwined and examples of exponential development. For example, if you have $1 million and invest it at 10% per year, you can spend up to $100,000 without depleting your capital, but if you spend more than your investment allows, your capital will eventually be gone, and you will become bankrupt. If you spend $200,000 per year, you will deplete your capital in less than seven years.

The Earth has many wonderful resources, our precious capital, but we are spending that capital at a dangerously fast rate, depleting soil, aquifers, the ozone layer, rivers, atmosphere, forests, petroleum deposits, and so forth.

How serious is population growth? The more people there are, the more food, water, and energy is needed, and the more pollution is produced.

HOW MANY PEOPLE CAN THE EARTH SUSTAIN?

A recent United Nations report says that the problem of population growth is urgent because of "the extent of urban growth, the extent of environmental damage, the impending food crisis in many developing nations, the extent of infant and maternal mortality, and the continuing low status of women." Fresh water and topsoil are disappearing, and cities are crammed—packed with masses of unemployed people. The recent famines and wars in Ethiopia, Angola, Mauritania, and Somalia have been attributed to rapid population growth in those countries. The per capita production of grain was down in 1995 and is expected to continue to decline further, while the cost of importing grain rose—making hunger even more likely in third world countries. Meat production increased as more people throughout the world, especially in China, demanded more meat in their diet. This is a dangerous trend, since meat production involves considerable loss of nutrient energy (see Chapter 7). As more people adopt middle-class lifestyles, they tend to use more resources and produce more pollution, resulting in more environmental degradation. As we noted in Chapter 1, this is sometimes referred to as *consumption overpopulation* (CO) as opposed to *people overpopulation* (PO). Ecologists speak of *environmental impact* (I) as the product of *population* (P) times its per capita *resource consumption* (R) times its per capita *pollution output* (O):

$$P \times R \times O = I$$

People overpopulation occurs when the number of people in a geographical area is simply too great for the carrying capacity of the land. Consumption overpopulation occurs when the people in an area consume more resources than are necessary for survival and produce large quantities of pollution. Many demographers believe that the world already has reached its population carrying capacity and that we are experiencing people overpopulation in several areas of the world, especially in India, Bangladesh, China, and many countries in Africa.

Let column P = population, R = resource use per unit of resource, O = pollution per unit of resource, and I = environmental impact:

$$P \times R \times O = I$$
$$\text{PO } 100 \times 1 \times 2 = 200 \text{ units}$$
$$\text{CO } 10 \times 10 \times 5 = 500 \text{ units}$$

Currently 90% of the world's births take place in the third world, so these countries have larger populations than do the Western industrialized nations and Japan, but we have a far greater environmental impact (I) due to our consumption habits. In the example just given, we see a community of 10 having an environmental impact 2½ times that of a poorer community of 100. A member of CO has 25 times as great an impact as a member of PO. The total environmental stress on our planet is increasing 5.5% per year. As of 1990 it had increased to 450 times that of 1880 and it is still growing. Recently (October 1, 1998) ecologists Paul Ehrlich and Stephen Schneider declared that overconsumption was beginning to rival overpopulation as the most serious environmental problem of our age. They pointed out that the average American's negative impact is several times higher than that of the average person in a third world country like India, Sudan, or Rwanda. In some cases our negative impact is as much as 100 times that of individuals in these countries. As already noted, with 4.5% of the world's population we use about 34% of its mineral resources, 30% of its food, and 25% of its nonrenewable energy; and we create about 36% of its pollution. Our cars and airplanes are wasteful consumers of fuel. We also use 33% of the world's paper, cutting down 1,000 acres of forest each day to produce our paper. The Sunday edition of the *New York Times* requires the cutting down of about 150 acres of forest. But, as the living standards of third world nations rise, they consume more and hence tend to add to environmental degradation.

WHAT SHOULD WE DO?

Population growth presents a challenge to us all. The subject stirs the deepest passions, and hence some environmental groups try to ignore it, lest it tear them apart. As I write this chapter (summer 1998), the Sierra Club, to which I belong, is bitterly split over population and immigration. Should it make reduction of population, including both lower birth rates and lower immigration rates, a vital part of its environmental strategy, or should it concentrate on the root causes of population and immigration growth? Although the Sierra Club members voted to keep national population and immigration questions separate from its main task of promoting mainline environmental issues, demographic questions are surely relevant to these traditional issues. Essentially, we have four options:

1. *Increase the carrying capacity of the environment.* In other words, make a bigger pie.[22] That is, use technology to produce greater yields of food and water, recycling water and other resources. Necessity is the mother of invention. This is the solution of the cornucopians who, in the spirit of Condorcet, see population as a challenge to be inventive.

2. *Reduce the number of people.* In other words, allow fewer people at the table. That is, cut the birth rate to zero growth. This is the solution of the neo-Malthusians, like Ehrlich and Hardin. By reducing population size we lessen the burden on the carrying capacity, so that we may muddle through tough times and may even achieve sustainable lifestyles.

3. *Live simpler lives.* Turn from energy-inefficient meat diets to more efficient vegetarian diets, reduce our use of fossil fuel, reject the model of conspicuous consumption, and live closer to nature. This is the message of the radical environmentalists, like Henry Thoreau, John Muir, Dave Foreman, Murray Bookchin, and Barry Commoner. Environmentalists like Commoner, Bookchin, and Foreman, not to mention Unabomber Ted Kaczynski, would have us reject the technological fix as a pseudosolution. Their motto is "Live simply that others may simply live."

4. *Reform and redistribute our resources.* Learn good table manners at Earth's table. Help poor nations to develop small technology; compensate developing countries for preserving their rain forests. Educate men and women in the developing countries so that they can make wise decisions, utilizing their resources for greater prosperity. Liberate women from oppressive situations where childbearing is seen as a perpetual duty. Promote global social justice, so that the rich do not dominate the poor and oppress nature. This is the solution of the Marxists, socialists, and political liberals.

Which is the correct strategy? Let us examine each of these strategies.

Increase the Carrying Capacity

The cornucopians urge us to let technology solve our ecological problems by making new inventions. We have already noted Julian Simon's statements, "We now have in our hands—in our libraries, really—the technology to feed, clothe, and supply energy to an ever-growing population for the next 7 billion years," and "Even if no new knowledge were ever invented . . . we would be able to go on increasing our population *forever* while improving our standard of living."[23] Here is a letter that appeared in the *New York Times* in 1998:

> The debate on how many people this country can and should contain assumes that there is a limited supply of natural resources and some maximum number of people that can share them. Both food resources and population are potentially unlimited. For anthropologists, increases in population are the measure of progress. History testifies that the invention of new and better ways of doing things allows a greater number of people to survive, and survive longer. Over the past two centuries in this country, the number that can be supported by a square mile of land has risen from 6.1 in 1800 to 25.6 in 1900 and reached 64 people per square mile in 1980. It makes no sense to assume that resources are limited and that population is cancerous. (Myles J. Duffy, Somerville, Massachusetts; March 13, 1998)

As we noted in our discussion of Simon at the beginning of this chapter, cornucopians point to the new strains of grains, corns, and wheat that agronomers have produced, with several times the yield of previous strains; to fertilizers and pesticides that have increased productivity; to nuclear energy, which is both cheap and powerful; to medical technology, especially vaccines, which has drastically lowered infant mortality and led to higher living standards and longer lives throughout much of the world. In this century, technology has given us the airplane, television, the computer, the automobile, air conditioning, refrigerators, heart transplants, birth control pills, safe abortions, and penicillin and other wonder drugs.

All this is true. We have much to thank science and technology for. But with almost every blessing has come a potential curse. The new petrochemical herbicides, fertilizers, and pesticides run off into and pollute the water supply. DDT and dieldrin are carcinogenic. Pesticides and herbicides also cause new and more powerful strains of pests and bacteria to emerge, which in turn call for ever greater pesticides and herbicides. Nuclear energy may be cheap and powerful, but it may also cause radiation sickness, especially when mistakes occur as they did in the tragic Chernobyl accident. It also creates a problem of disposing of nuclear waste, which has a half-life of hundreds of thousands of years. Medical science has lowered the infant mortality rate throughout the world; but if other conditions are not changed, the infants grow up malnourished in overpopulated areas, only to die during famine or pestilence. Statistics from the International Monetary Fund show that the world's annual economic growth rate is slowing down considerably. The number of people in absolute poverty is now 1.2 billion, with 400 million in semistarving conditions.[24] The automobile is an amazing means of transport, but it produces enormous amounts of harmful pollutants, especially nitrogen oxides, carbon monoxide, and carbon dioxide, which contribute to smog, which in turn causes emphysema and lung cancer. On April 7, 1998, the Centers for Disease Control announced that bronchial asthma had doubled in the United States over the past decade. Automobile pollution also exacerbates the greenhouse effect. The automobile also demands roads and parking lots, which take large amounts of prime farmland out of the economy. Refrigeration and air conditioning have made the Sun Belt inhabitable, but they demand huge amounts of electricity and send CFCs into the stratosphere, breaking up the ozone layer. Television and the computer can serve us as good tools, but they also tend to become addictive, creating couch potatoes and virtual reality junkies, robbing us of social skills and human relations. Furthermore, technology has created atomic, chemical, and biological weapons and other instruments of mass destruction that could destroy human life on the planet.

Few would deny that technology has contributed to the improvement of our lives, but two other caveats must be mentioned. One is that we are in danger of worshipping our tools and toys, rather than wisely using them. Recall the a story of a man who discovered his shadow (see Chapter 1). Becoming dazzled by its shifting motions and its ubiquitous presence, he gradually treated it with more and more deference. His affection grew into a consuming passion. He slept with the light on so as not to obliterate his shadow's life. He always walked in such a way that his shadow had full presence. He took such pains to give pride of place to his shadow that eventually the man became the shadow of his shadow. Technology is analogous to our shadow. We may become the shadow of our shadow, mistaking our projections for reality.

The second caveat is that we are not always wise enough to understand the long-term effects of our inventions. Sometimes *nature knows best*—and we at least should be ready for unexpected negative consequences of our technological interventions. Here are a couple of examples of this principle. In 1955 an island off North Borneo was experiencing a large number of cases of malaria. Attempting to ameliorate the predicament, the World Health Organization introduced the synthetic chemical pesticide, dieldrin, into the infected regions to kill the malaria-carrying mosquitoes. The treatment was successful and greatly reduced malaria in the area. Unfortunately, there were unintended side effects. The dieldrin killed other insects, including flies and cockroaches that lived in the island houses. But small lizards that also lived in the houses and that ate the flies and cockroaches began to die from the poisoned flies and cockroaches. The cats that ate the lizards died of the poison in the lizards. Without their natural predators around, the rats who inhabited the adjacent jungle had a field day. They overran the villages, bringing with them typhus carried by the fleas that resided on them. But there was more. The dieldrin killed wasps and other insects that fed on a particular caterpillar that escaped the effects of dieldrin. With their predators dwindling in number, the caterpillars munched their way through the thatched roofs of the houses. The thatched roofs began to cave in. Eventually, the situation was brought under control, but the lesson is that interfering with nature often has unpredictable consequences.

In the early 1960s the former Soviet Union scored a tremendous public relations victory by building the billion-dollar Aswan Dam on the Nile River. The dam was to provide five benefits to Egypt: prevent severe flooding of the Nile; store water behind the dam in a gigantic 2,000-square-mile reservoir, Lake Nasser; provide drinking water for the people of the region; release water as needed to increase flood control and provide irrigation for the plains below the dam; and provide electricity for the inhabitants of Cairo. The goals were more or less achieved. Some severe flooding was prevented; irrigation added 1 million acres of arable land to Egypt, although the crops are of a poor quality. Egypt's rice and cotton crops were saved during the droughts of 1972 and 1973. Irrigation also allows farmers to harvest three crops per year instead of one. The dam provides 50% of Egypt's electricity.

But there are some deficits. Before the dam was built, rich silt deposits in the valley provided a natural fertilizer for crops and nutrients for fish, and also washed salts out of the soil. With the flood waters now held in the reservoir, most of the silt sediment sinks to the bottom of Lake Nasser. Without the annual supply of fertilizing silt, the land below the dam has lost most of its fertility. Hence, large amounts of synthetic fertilizer, inferior to silt, have been imported into the region, and much of the electricity produced is used by fertilizer plants. Buying fertilizer costs one-fifth of a small farmer's income. About 400,000 acres of land have been ruined by the mineral salt buildup that ordinarily would have been washed away by the annual floods, so that three-fourths of the land made arable by the dam has been ruined. A billion-dollar drainage system is being built to correct this.

Now that the Nile's nutrient-loaded silt no longer reaches the waters at the river's mouth, Egypt's sardine, mackerel, shrimp, and lobster industries—the country's dominant sources of protein—have been destroyed. Slowly moving irrigation canals caused the spread of schistosomiasis, a chronic parasitic disease carried by blood flukes; it causes severe stomach cramps, chronic inflammation of the bladder, and damage to lungs, liver, and heart. The resulting pain and exhaustion reduce the workers' ability to farm, often to a maximum of 3 hours per day. Millions of rice farmers in the Nile Delta are dying of schistosomiasis, which has lowered their life expectancy to about 26 years. On balance the Aswan Dam has turned out to be an ecological and economic disaster for Egypt, bringing it to the brink of bankruptcy—but for the aid of the United States. Why doesn't the government of Egypt destroy the dam? It doesn't do so for one reason: electricity.

The lesson to be derived from these examples is not that we should abandon science and technology and go back to a primitive life without electricity, pesticides, airplanes, or computers, although simplifying our lifestyles would be good (I will discuss this in Chapter 18). The lesson is that we are not omniscient, that we never do just one thing; but, because of the interconnectedness of nature, our actions often have unexpected consequences. Joel Cohen has proposed a law of action: "It is difficult to do just what you intended to do."[25]

Cornucopians like Julian Simon and Myles J. Duffy believe that resources and population are potentially unlimited. This is sheer nonsense. The Earth is of limited size and is subject to the laws of physics.

Consider population alone. As mentioned earlier, the U.N. demographers state that at present growth rates the world will have 694 billion people by the year 2150, which will require more water for growing food than falls from the skies. Even if you think desalination of the ocean's water could save the situation, a time would come when even this would be used up. Paul and Anne Ehrlich have asked us to consider how long it would take for the 1994 population of 5.6 billion to reach a size where there were *"ten human beings for every square meter* of ice-free land on the planet."[26] At the 1994 growth rate of 1.6% (a doubling time of 43 years), it will only take 18 doublings or 774 years to bring the Earth's population to that point. After 1,900 years at that growth rate, the mass of the human population would be equal to the mass of the Earth; and after 6,000 years it would equal the mass of the universe. Any population that grows by any positive number, be it only .01% per year, will eventually exceed any fixed size. Granted, with ingenuity and luck we could put more people on the planet than now exist, but it is likely that the quality of life will suffer. At present no nation is living within its means, on its renewable resources. We are all using up our natural capital at an alarming rate. As the Ehrlichs put it, *"Homo sapiens* is collectively acting like a person who happily writes ever larger checks without considering what's happening to the balance of the account."[27] If 6 billion people are depleting our supply of farmland and water supply, causing ozone depletion and an exacerbation of the greenhouse effect, what will 30 or 40, not to mention 694 billion, do to it? If crowded conditions exacerbate social malaise, what will more densely populated cities bring? The examples given—the destruction of malaria-carrying mosquitoes in North Borneo and the building of the Aswan Dam in Egypt—should make us hesitate before putting so much faith in a technological fix. Technology alone doesn't seem to be the answer.

Reduce the Growth of Population

We don't know exactly how many people the Earth or the United States can hold without diluting the quality of life, but we can see how difficult it is to educate, feed, and offer good health to the nearly 6 billion people now living. The more people, the more resources are used; the more resources are used, the more pollution is created; the more pollution is created and the fewer resources there are, the worse off will be members of future generations to whom we have duties. If you consider that the kinds of lifestyles we are presently living in the West use an enormous amount of resources and produce an enormous amount of pollution—and that this is the kind of lifestyle the developing countries are aiming at—then you can sense a deep problem.

Doomsdayers tend to support China's coercive one-child policy. Government sometimes must provide sanctions and incentives to keep down the birth rate if it doesn't want a high death rate. Hugh LaFollette claims that we don't have a natural right to propagate, but may have an obligation not to do so, especially in light of the danger of overpopulation. Society normally regulates potentially harmful activities, such as practicing law and medicine, selling drugs and driving vehicles, by requiring a license to do these things. But having children is as important and as potentially harmful as any of these activities. LaFollette asks, Shouldn't we require parents to qualify for having children by meeting minimal standards?[28]

The doomsdayers are correct in announcing that the Earth has limits, but when they specify population limits, they invariably end up embarrassed. Perhaps, since we don't know how many people can live lives of quality on Earth, we ought to play it safe, aim at zero or even negative growth, at least for the moment, and aim at creating a good life for all of the Earth's present inhabitants. This leads us to the next strategy.

Live More Simply

Radical ecologists—including the classical environmentalists Henry David Thoreau and John Muir, but also unlikely contemporary bedfellows such as Dave Foreman, Wendell Berry, Barry Commoner, and Murray Bookchin, who hold this view—reject the technological solution of the cornucopians. Technology is more part of the problem than it is the solution. We don't need bigger nuclear power plants; we need to live closer to nature, sharing our resources. But some of these ecologists, like Commoner and Bookchin, also reject what they regard as "antihumanism" in doomsdayer thinking. People are good and don't need coercive governmental pressure to limit family size. They believe that if you educate people, liberate women, and get closer to nature, families will naturally become smaller. But the key word is *simplify:* We need to develop sustainable agriculture, reducing our use of petrochemical fertilizers and pesticides, and adopt a vegetarian diet. We need to live sustainable lifestyles, use less electricity, ride bicycles and trains instead of automobiles and trucks, use solar energy instead of coal-burning power plants, and develop new methods of sewage disposal. The problem is that we can't (or won't) go back to primitive conditions. We desire the comforts of civilization: electricity, computers, airplanes, air conditioning, refrigeration, medicine, and the automobile. And the developing nations would like to have these things too. Simplicity is relative. Still, we can use our ingenuity to educate people, especially the young, who are our hope for the world, to renounce the idols of a consumption-oriented society and embrace one that is based on love, justice, reason, and moral values.

Redistribute Resources

Socialists and liberals call on us to work for global social justice, interpreted as respecting universal rights and creating conditions in which people everywhere can reach their full potential as rational agents. This would call for universal laws regarding pollution control and resource use. It would mean that the wealthy nations would reform their overconsumption and aid the poorer ones. Countries like Brazil, Malaysia, and Indonesia would not have to bear the entire burden of preserving the rain forest, but would be aided by the more technologically advanced nations in developing sustainable economic uses for the rain forest (such as fruit growing, rubber tapping, and the production of herbal medicine) and would be compensated for any economic sacrifice they would be making by not exploiting the rain-forest ecosystems. We need the best minds of our time to work on a global ethic that is joined with international law. A global tragedy of the commons threatens us all, as each nation deems it in its interest to engage in self-interested activities that deplete total resources and create overwhelming problems of pollution. There must be mutually agreed upon, mutually coercive rules to limit our mutually destructive behavior. Either we all hang together in solving our global problems or we will each *hang* alone.

Which of these four is the correct strategy? Perhaps all four have some merit and can be reconciled to one another in a single global ethic—though getting everyone to agree on a combined strategy may take a long time—perhaps longer than we have time for. The problem is so large that we must use every legitimate means at our disposal to solve it. So while we must use technology wisely to produce more goods, we must be careful not to become a slave of our technology, as the radical environmentalists correctly caution us; we need to realize that social justice must be done and *be seen to be done* in distributing goods equitably. If we are to ask Brazil to save its rain forest, we should compensate it for the economic loss, at least by helping Brazilians to develop the rain forest into a source of fruits and medicinal herbs. Similarly, we must learn to live simpler lifestyles, turning to vegetarian diets, smaller, electric cars, and public transportation. We must preserve our topsoil and aquifers and develop sustainable agriculture. But none of this is likely to be sufficient unless we cut the population rates. We don't know how many people the Earth can sustain, but if we are to live lives of quality, without fear of a dangerous greenhouse effect and a severely damaged ozone layer, and with space for hiking, sufficient topsoil for crops, clean air to breathe, and clean water to drink, we must halt the exponential rate of population growth.

Meanwhile, we must become ecology-conscious, learning to think globally but act locally. We need to adopt simpler lifestyles, practicing sustainable agriculture and economics. Our forests must be renewed and we must promote social justice here and, as much as possible, abroad. Wilderness ought to be preserved as a place for renewal and recreation and increasing the quality of life. What should be done about immigration? The United States is the number one recipient of immigrants in the world. Cornucopians like Simon and Jacqueline Kasun believe that there is plenty of room for a lot more immigrants and that immigrants enhance the economy. Socialists, leaders of the Sierra Club, and Zero Population Growth believe that we should attack the root causes of overpopulation, ignorance and the oppression of women in poor countries, rather than slam the door on those who are knocking on our doors, longing to breathe free.

Ecologists like Ehrlich, Hardin, and Grant believe that we are already overcrowded. They believe that in order to send a message of responsibility to other countries, as well as to keep our own population down, the United States should greatly reduce immigration rates (from the current estimate of 1.2 million legal and illegal immigrants) and promote birth control both here and abroad. The United States, Canada, and Australia once provided for the overflow of people from crowded countries. But now these nations are reaching the limits of their carrying capacity, so that each nation must take responsibility for stabilizing its population within the context of a global ethic. There is also the issue of fairness. Many of us would love to have large families, but we restrain ourselves, limiting ourselves to one or two children. But why should we sacrifice our love of a large family, if the government is going to allow in millions of immigrants, many of whom have or will have larger families than we do? Statistics show that many immigrant families have five or six children. Having children must be seen as a privilege with deep moral responsibility. No one should bring children into the world whom he or she cannot support. Our goal is for every child, regardless of gender or race, to have the opportunity for a good life.

Some environmentalists object to this kind of thinking, arguing that twenty or fifty poor people have less environmental impact that one suburban American. This is true; and, as I will argue in subsequent chapters, we all, and especially the affluent, must learn to live more simply, more in harmony with the environment. But, at the same time, we must have morally responsible natalist and immigration policies. Both lowering our consumptive habits and keeping population rates under control are necessary for justice.

So was the doomsdayer Malthus right? In a way he was. Although he underestimated the ability of humanity to increase the Earth's carrying capacity by means of science and technology, as cornucopians rightly argue, he saw that there are limits to growth, that overpopulation brings the four horsemen of the Apocalypse: poverty, disease, war, and death. Even if the cornucopians are right and we are not in serious danger, we should bet with the doomsdayers, reduce global population, simplify our lives, using less resources and creating less pollution. It's a cardinal rule of gambling not to bet more than you can afford to lose, and we cannot afford to lose our planet, Earth. We do not have a spare planet to hop onto if Simon's side loses. Furthermore, as we will see in our discussion of pollution, especially the greenhouse effect and depletion of the ozone layer, there is a time lag before the bad effects of our policies are fully experienced. By the time the full price of our folly hits us, it may be too late to stave off disaster.

In sum, even if Malthus and his doomsdayer followers exaggerate the situation, prudence would counsel us to play it safe—to act as though our planet were in danger—by living within our means and at the same time reducing the population, so that those who are living and will be born in the future may have the opportunity to live lives at least as good as our own.

The point about the quality of life leads to a final reflection. Even if the cornucopians are right about the Earth's ability to feed billions more people, it would still invite a tragedy, for we would have to convert ever greater tracts of wilderness to farmland and thus destroy habitats and further imperil already endangered species. An ever larger share of the world's population would have to live in ever closer quarters on a planet whose surface would have to be transformed to accommodate our growth; we would see the conversion of forests and open spaces to urban, industrial, and agrarian areas and the conversion of the oceans to food factories.[29] Whether the optimism of the cornucopians or the pessimism of the doomsdayers is right is not the only issue: The prospect of an overcrowded world, filled with compact high-rise apartment buildings, with dwindling forests, deserts, and open spaces, is nightmarish.

Review Questions

1. Read over Malthus's theory on population growth. Was he a villain or a prophet? Should we commemorate the bicentennial of his essay on population? Do you think he was right or wrong?
2. Consider the quotations by Julian Simon ("We now have in our hands—in our libraries, really—the technology to feed, clothe, and supply energy to an ever-growing population for the next 7 billion

years") and Myles Duffy on the Earth's unlimited resources and ability to provide for an unlimited population. How do you interpret their assertions? Are they correct?

3. Compare the four strategies for dealing with global population growth. Which, or which combination, seems the best strategy? Explain.
4. What should we be doing about population growth in our own nation? How serious is the problem? Illegal immigration is thought to account for a large percentage of total U.S. immigration—perhaps as much as 400,000 each year. At this moment the Sierra Club, one of the largest and most influential environmental groups in the United States, with 550,000 members, is in the throes of an acrimonious debate over whether to take a public stand on immigration. In February 1996 the Club's board of directors adopted a resolution to support efforts to reduce global population but to take no stand on U.S. immigration levels, whereas grassroots members believe that the Club should take a comprehensive stand in favor of the reduction of U.S. population, which includes lowering immigration rates as well as birth rates. The board accuses the anti-immigrationists of racism. "It's offensive to people of color, even those who think we should limit immigration," said Carl Pope, the executive director of the club. "It says to the world, we in the U.S. can enjoy a lifestyle that is a threat to the planet. We will not let you come here; we have no intention to get off this lifestyle." The board argues that the Sierra Club should attack the root causes of global population problems, including championing the right of all families to maternal and reproductive health care and the empowerment and equity of women. Those who support a moratorium on immigration dismiss the charge of racism as an attempt to slander them. Immigration restrictions are color-blind. They argue that we cannot consider issues of the wilderness, resource consumption, and pollution separately from issues of population growth, including immigration. What is the correct course for the Sierra Club to take? What would you recommend and why? (See "An Uncomfortable Debate Fuels a Sierra Club Election," *New York Times,* 5 April 1998.)
5. Should the government get involved in the activity of bringing down population growth? What do you think of LaFollette's idea of licensing parents? Explain your answer.
6. Read over Joel Cohen's description of poverty in Calcutta, quoted at the beginning of this chapter. What can be done to ameliorate the kind of suffering described in this passage? What has it got to do with population? Compare it with the statements made by Simon and Duffy.
7. Examine Simon's argument that because Ehrlich values his life and we all value our lives, we ought to value the potential lives of those who could be born under a generous pro-fertility policy. Is this a good argument for an enlarged population?

8. Discuss the cornucopian versus doomsdayer wager. How should we bet? How should we live in the light of these alternatives?
9. America is said to be a nation of immigrants. Some people believe it is the most important reason for our success. Consider the poem by Emma Lazarus (1849–87) inscribed on the Statue of Liberty in New York Harbor:

   Give me your tired, your poor,
   Your huddled masses yearning to breathe free,
   The wretched refuse of your teeming shore,
   Send these, the homeless, tempest-tost to me:
   I lift my lamp beside the golden door.

   Discuss the meaning of this poem. Do you agree with its message? Explain your answer.

Suggested Readings

Bouvier, Leon, and Lindsey Grant. *How Many Americans?* San Francisco: Sierra Club Books, 1994.

Cohen, Joel E. *How Many People Can the Earth Support?* New York: Norton, 1995.

Ehrlich, Paul. *The Population Bomb*. New York: Ballantine, 1968.

Ehrlich, Paul, and Anne Ehrlich. *Betrayal of Science and Reason*. Washington, D.C.: Island Press, 1996.

Grant, Lindsey. *Juggernaut: Growth on a Finite Planet*. Santa Ana, Calif.: Seven Locks Press, 1996.

Hardin, Garrett. *Living Within Limits: Ecology, Economics, and Population Taboos*. New York: Oxford University Press, 1992.

Kasun, Jacqueline. *The War Against Population*. San Francisco: Ignatius, 1988.

Malthus, Thomas. "On Population." 1830. In *Three Essays on Population: Thomas Malthus, Julian Huxley, and Frederick Osborn*. New York: Mentor, 1956.

Myers, Norman, and Julian Simon. *Scarcity or Abundance? A Debate on the Environment*. New York: Norton, 1994. A debate between Norman Myers and Julian Simon.

Simon, Julian. *The Ultimate Resource 2*. Rev. ed. Princeton: Princeton University Press, 1996.

Notes

1. Thomas Malthus, *Essay on the Principle of Population As It Affects the Future Improvement of Society* (London: 1798).
2. Joel E. Cohen, *How Many People Can the Earth Support?* (New York: Norton, 1995), p. 8.
3. Malthus, *Essay on the Principle of Population*. During the bicentenary of this famous work, in 1998, there were few, if any, commemorations, let alone celebrations.
4. Richard Godwin (1756–1836) published *Enquiry Concerning Political Justice* (1793), in which he announced that a utopian time would come when humans would not need to propagate, since they would live for ages, if not forever. "The men therefore whom we are supposing to exist, when the earth shall refuse itself to a more extended population, will probably cease to propagate. The whole will be a people of men, and not of children. Generation will not succeed generation, nor truth have, in a certain degree, to recommence her career every thirty years. Other improvements may be expected to keep pace with those of health and longevity. There will be no war, no crimes, no administration of justice, as it is called, and no government. Besides this, there will be neither

disease, anguish, melancholy, nor resentment. Every man will seek, with ineffable ardor, the good of all." He was the husband of the feminist philosopher Mary Wollstonecraft and father of Mary Shelley, the creator of Frankenstein. Marquis de Condorcet was a French nobleman who had a firm hope that humanity could be perfected. He supported the French Revolution only to become its victim. He was imprisoned in Bourg-La-Reine, where he was found dead. It is not clear whether he took poison or was poisoned by others. It was remarked by the French lawyer Pierre Vergniaud before he was guillotined in 1793 that the "revolution eats its own children." It seems to be the case with Condorcet.

5. The Apocalypse, referred to in modern Bibles as "The Book of the Revelation," is the last book of the New Testament. In the sixth chapter John describes the end of the world, wherein four horsemen wreak destruction on the Earth. See the painting by the German artist Albert Dürer (Figure 11-1).
6. In Greek mythology Cassandra was the daughter of King Priam of Troy, who tricked Apollo into giving her the gift of prophecy. Gaining revenge, Apollo decreed that no one should believe her prophecies, which were all doom, gloom, and disaster.
7. Quoted in Garrett Hardin, *Living Within Limits: Ecology, Economics, and Population Taboos* (New York: Oxford University Press, 1992), p. 27. My discussion is indebted to Hardin's book as well as to Jane Soames Nickerson's *Homage to Malthus* (London: Kennikat Press, 1975).
8. William Peterson, "Marx Versus Malthus: The Men and the Symbols," *Population Review* 1 (1957), pp. 26–27.
9. Malthus, quoted in Hardin, *Living Within Limits,* p. 27.
10. Joel E. Cohen appreciates but is critical of Malthus. "Malthus has been wrong for nearly two centuries because he did not foresee how much people can alter the human carrying capacity of the Earth. Will Malthus continue to be wrong for the next two centuries?" (p. 429). Cohen argues that "to believe that no ceiling to population size or carrying capacity is in prospect, you have to believe that nothing will stop a sufficient proportion of additional people from increasing the Earth's carrying capacity by more than, or at least as much as, they consume" (p. 445). In other words, you have to have the kind of unlimited utopian faith that Condorcet and Godwin had in the ability of humanity to overcome all obstacles to population growth. (Cohen, *How Many People Can the Earth Support?*)
11. Thomas Malthus, "On Population" (1830), in *Three Essays on Population: Thomas Malthus, Julian Huxley, and Frederick Osborn* (New York: Mentor, 1956), p. 32.
12. Julian Simon, *Ultimate Resource 2* (Princeton: Princeton University Press, 1996), p. xxxi.
13. Ibid., p. xxxviii.
14. Ibid., p. xxxii.
15. Norman Myers and Julian Simon, *Scarcity or Abundance? A Debate on the Environment* (New York: Norton, 1994), p. 65.
16. Julian Simon, *The State of Humanity* (Oxford: Blackwell, 1995), p. 26. See *Ultimate Resource 2,* chap. 3 for a discussion of this thesis.
17. Simon, *Ultimate Resource 2,* p. 67.
18. Friedrich Engels, quoted in Kenneth Blaxter, *People, Food and Resources* (Cambridge: Cambridge University Press, 1986), p. 102.
19. Secretary-General of the United Nations U Thant, *International Planned Parenthood News* 168 (February 1968), p. 3.
20. Lindsey Grant, *Juggernaut: Growth on a Finite Planet* (Santa Ana, Calif.: Seven Locks Press, 1996), p. 2.
21. For a discussion of the population growth of the United States, see Leon Bouvier and Lindsey Grant, *How Many Americans?* (San Francisco: Sierra Club Books, 1994). See also Lindsey Grant, *Juggernaut: Growth on a Finite Planet.*

22. I am indebted to Joel E. Cohen, *How Many People Can the Earth Support?* p. 17, for this metaphor and for this way of framing the options.

23. The two quotations are from Julian Simon in Norman Myers and Julian Simon, *Scarcity or Abundance?* p. 65, and Simon's *The State of Humanity,* p. 26.

24. Although Simon regularly challenges his opponents to bet high amounts with him on the future, he declined Paul Hawken's offer to bet $100,000 that living systems will continue to deteriorate over the next 10 years. Simon has also turned down an offer from Paul Ehrlich and Stephen Schneider to bet $1,000 per trend that 15 continental and global indicators "pertaining to material human welfare" will worsen over the next decade. The trends include global warming, more carbon dioxide and nitrous oxide in the atmosphere, more AIDS-related deaths, lower sperm count in human males, and an increased gap in wealth between the richest 10% and lowest 10% of humanity.

25. Cohen, *How Many People Can the Earth Support?* p. 75.

26. Paul and Anne Ehrlich, *Betrayal of Science and Reason* (Washington, D.C.: Island Press, 1996), p. 66.

27. Ibid., p. 68.

28. Hugh LaFollette, "Licensing Parents," *Philosophy and Public Affairs* 9:2 (1979). Reprinted in *Environmental Ethics,* ed. L. Pojman.

29. I am indebted to an anonymous reviewer for forcefully reminding me of this point.

CHAPTER TWELVE

Population and World Hunger

[Everyone has] the right to a standard of living adequate for the health and well-being of himself and his family, including food.

—*United Nations Declaration on Human Rights,* 1948

Feeding the hungry in some countries only keeps them alive longer to produce more hungry bellies and disease and death.

—JOSEPH FLETCHER, "Give If It Helps, Not If It Hurts"

More than one-third of the world goes to bed hungry each night.* Ten thousand people starve to death each day, and 1.3 billion people (nearly a quarter of the human race) live in absolute poverty, with incomes less than one dollar a day. We know that 841 million people are chronically undernourished. The United Nations Food and Agriculture Organization (FAO) estimates that almost half of these are children. In Africa the *number* of hungry people has more than doubled, and the *proportion* of population that is hungry has increased 13% in the past 25 years. The FAO predicts that if current trends continue, 265 million Africans (the size of the present population of the United States) will suffer hunger in the year 2010, an increase from 148 million in 1981 and 215 million in 1992. While famines have ravaged parts of Africa and Asia, another third of the world, the industrialized West, lives in relative affluence, wasting food or overeating. The rich get richer and the poor get poorer.

World hunger is one of the most intractable problems facing humankind today. Some environmentalists deny that it should be discussed in a book on environmental ethics, but I disagree. It is an environmental issue because it is tied to population growth, deforestation, soil erosion, and the just distribution of resources. Poor farmers in the rain forest cannot afford to worry about saving endangered species or the rain forest itself, because they are compelled to work to feed their families, even at the expense of the environment. World hunger is a global environmental issue. What can be done about it? What obligations, if any, do we in the affluent West have to distant, needy people, those who are hungry or starving? To what extent

---

* Hunger is a condition in which people do not get enough food to provide nutrients they need for fully productive, active, and healthy lives. Malnutrition, which impairs both physical and mental health, results from undernutrition, inadequate consumption of one or more nutrients.

should population policies be tied to hunger relief? These are the questions discussed in this chapter.

I will discuss four responses to these questions: (1) the neo-Malthusian response set forth by Garrett Hardin; (2) the liberal response, exemplified by Peter Singer and Richard Watson; (3) the conservative (or libertarian) response represented by Thomas Hobbes, Robert Nozick, and others; and (4) a moderate alternative position, taking into consideration the valid insights of each of these other positions.

FOUR ETHICAL RESPONSES TO WORLD HUNGER

The contrast between neo-Malthusians and liberals can hardly be imagined to be greater. Liberals assert that we have a duty to feed the hungry in famine-ridden areas either because the hungry have a right to food or because of utilitarian reasons maximizing welfare or happiness. Neo-Malthusians deny such a right and assert that we have an opposite utilitarian duty to refrain from feeding the hungry in famine-ridden areas. Conservatives take the middle road in this debate and assert that while we do not have a duty to feed the hungry, it is permissible and praiseworthy to do so. It is an act of supererogation, an act going beyond the call of duty. The moderate position accepts some tenets of conservatism and some of liberalism. We begin with the neo-Malthusians.

Neo-Malthusianism

In the last chapter we discussed the theory of Reverend Thomas Malthus (1766–1834) that population size tends to outrun food production, leading to misery, until war, disease, famine, and other disasters restore a natural balance. Partly due to modern agricultural technology and the spread of birth control devices, Malthus's predictions haven't been universally fulfilled. In the United States and Canada, for example, food production has been substantially greater than what is needed for the population. Neo-Malthusians are those ecologists who accept Malthus's basic thesis but modify it in the light of technological innovation. A nation that is not maintaining the proper food-to-population ratio should not be helped from the outside by increments of food, they believe. To feed such sick societies is, to quote Alan Gregg, former vice president of the Rockefeller Foundation, like feeding a cancer: "Cancerous growths demand food; but, as far as I know, they have never been cured by getting it."[1]

The most prominent neo-Malthusian today is Garrett Hardin, emeritus professor of human biology at the University of California at Santa Barbara, who in a series of articles set forth the idea of lifeboat ethics.[2] Hardin's position can be succinctly stated through three metaphors that he has made famous: "lifeboat," "tragedy of the commons," and "the ratchet." Let us examine his use of each of these.

1. *Lifeboat.* Hardin compares the world to a sea in which a few lifeboats (the affluent nations) are surrounded by hordes of drowning people (the populations of the poor nations). Each lifeboat has a limited carrying capacity, such that it cannot possibly take on more than a tiny fraction of the drowning people without jeopardizing the lives of its passengers. Besides, need of a safety factor always dictates that we ought to leave a healthy margin between the actual number on board and the maximum possible number. The optimum population is somewhat below the maximum population. According to Hardin, the affluent nations currently are right around that optimum figure, probably beyond it, so that it is self-destructive to take the world's poor on board our lifeboats. Not only must we adhere to a population policy of zero-growth, but we must also have stringent immigration policies that prevent immigrants from swamping our boat.
2. *The tragedy of the commons.* Recall our discussion in Chapter 4. There is a public field (a "commons") where shepherds have been grazing their sheep for centuries. Because of the richness of the field and the poverty of the shepherds, the field is never overgrazed. Now there comes a time when the carrying capacity of the field is reaching its limit. At this point the short-term rational self-interest of each farmer is to add one more sheep to the commons in spite of its limitations. The farmer reasons that by grazing yet one more sheep he will be reaping a positive factor of 1 (the value brought on by the extra sheep) and losing only a fraction of the negative unit 1, the loss of the field's resources, since all the herdsmen share that equally whether they participate in overgrazing or not. So it is in each shepherd's interest to overgraze. But if too many shepherds act in their short-term self-interest in this way, it soon will be against their interest, for the pasture will be ruined. Hence the tragedy of the commons! A similar tragedy is occurring in our use of natural resources. We are in danger of depleting the world's resources through wanton overuse. To prevent such a tragedy, we must have mutually agreed upon, mutually coercive laws governing population increase, overgrazing, overfishing, deforestation, pollution, and the like. Each nation must manage its own commons, and where one fails to do so, it must be left to its own misery. Benevolent intervention on the part of misguided do-gooders is likely only to increase the overall misery. This leads to the next metaphor.
3. *The ratchet.* Hardin argues that there is a natural relationship in ecosystems so that once a species has overshot the environment's carrying capacity, nature takes care of the situation by causing a dieback of the population of the species and eventually restoring a balance (Figure 12-1). For example, a serious decline in the natural predators of deer in

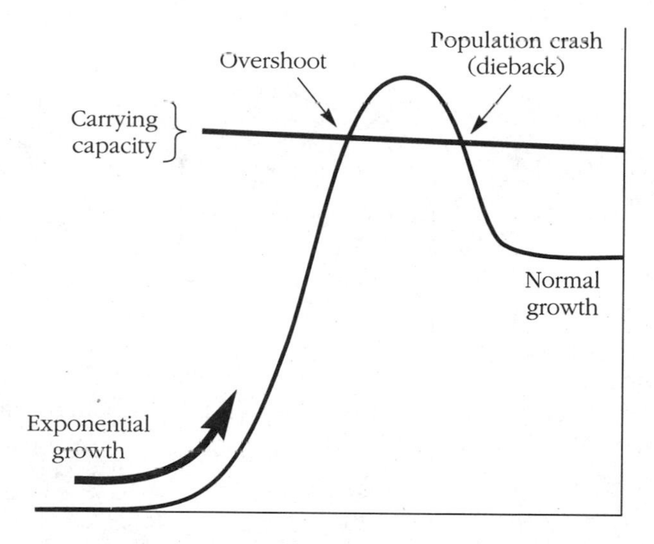

Figure 12-1 The relation of the carrying capacity of an environment to exponential population growth.

an area causes the deer population to increase exponentially until it overshoots the carrying capacity of the land for deer. The land cannot provide for this increase of deer, so they begin to starve, causing a dieback in their population until conditions are such that they can again increase at a normal pace. Likewise with human population systems. Once people in a given area have exceeded the carrying capacity of the environment, there will come a period of scarcity, resulting, à la Malthus, in famine, disease, and war over scarce resources; what ensues is a dieback or lowering of the population below the level of the carrying capacity. Nature will take care of the tragedy of the commons. Where people refuse to constrain their procreative instincts, nature intervenes and does it for them.

Now let some well-meaning altruists intervene in order to thwart nature's iron law. The altruists send food to the starving, fending off the effects of famine for a time. But what happens? The people procreate and the population increases even further beyond the carrying capacity of the land, so that soon there are even more people starving, so that another even greater altruistic effort is needed to stave off the worsening situation. And so a herculean effort is undertaken, and the population is saved once again. But where does this process lead? Only to an eventual disaster. The ratchet effect keeps raising the level of the population without coming to terms with the natural relation of the population to its environment, and that is where Malthus's law is valid (Figure 12-2). Sending food to those who are not taking steps voluntarily to curb their population size is like feeding a cancer.

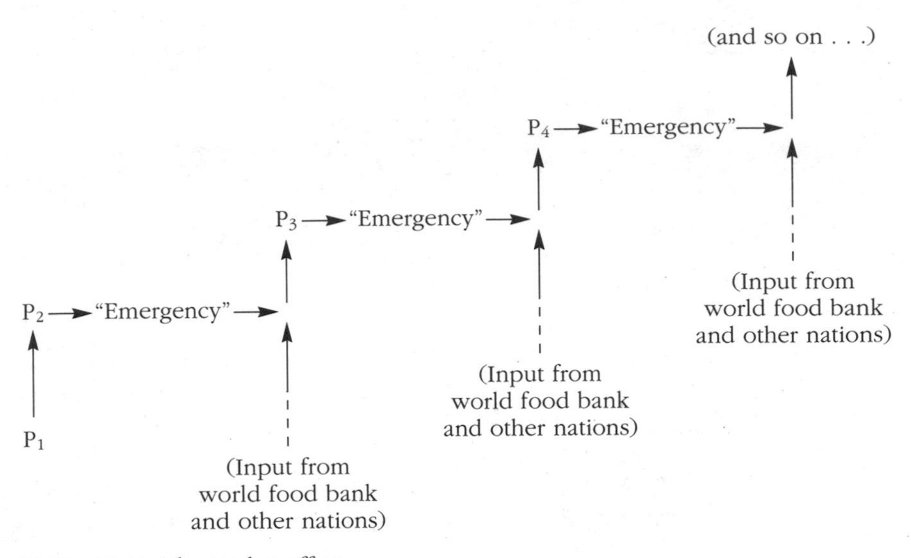

Figure 12-2 The ratchet effect.

For Hardin, it is wrong to give aid to those who are starving in overpopulated countries because of the ratchet effect. It only causes more misery in the long run. "How can we help a foreign country to escape overpopulation? Clearly the worst thing we can do is send food. . . . Atomic bombs would be kinder. For a few moments the misery would be acute, but it would soon come to an end for most of the people, leaving a very few survivors to suffer thereafter."[3]

Furthermore, we have a natural duty to our children and posterity to maintain the health of the planet as a whole. By using resources now for this short-term fix, we rob our children and future generations of their rightful inheritance. The claims of future people in this case override those of distant people.

In summary, Hardin has three arguments against giving aid to the poor in distant lands: (1) It will threaten our lifeboat by affecting the safety factor and causing our carrying capacity to become strained; (2) it will only increase the misery due to the ratchet effect; and (3) it will threaten the welfare of our descendants to whom we have prior obligations. For all of these reasons we are morally required not to give aid to the hungry.

What can be said about this kind of reasoning? Is Hardin right about the world's situation? Let us examine the arguments more closely. Consider the lifeboat argument. Is the metaphor itself appropriate? Are we really so nicely separate from the poor of the world? Or are we vitally interdependent, profiting from the same conditions that contribute to the misery of the underdeveloped nations? Haven't colonialization and commercial arrangements worked out to increase the disparity between the rich and the poor nations of the earth? We extract cheap raw materials from poor nations and sell them expensive manufactured goods (such as radios, cars, computers, and weapons) instead of appropriate agricultural goods and training. The structure of tariffs and internal subsidies discriminates selectively against underdeveloped nations. Multinational corporations place strong inducements on poor countries to produce cash crops such as coffee and cocoa instead of food crops needed to feed their own people. Besides this, the United States and other Western nations have often used foreign aid to bolster dictatorships—like the Somoza regime in Nicaragua and the military juntas in Chile and El Salvador—which have resisted social change that would have redistributed wealth more equitably. For example, in 1973 when President Allende of Chile requested aid from the United States, not only was he turned down but our government also aided in bringing his reformist government to ruin. When the military junta that replaced Allende took power and promised to maintain American business interests, eight times the amount of aid Allende had asked for was given to that government.

Hardin's lifeboat metaphor grimly obscures the fact that we have profited and are profiting from the economic conditions in the third world. Perhaps a more apt metaphor than "lifeboat" might be "octopus"—a powerful multinational-corporation octopus with tentacles clutching weapons and reaching out into diverse regions of the globe. Our nation protects, encourages, and even intervenes in the affairs of other nations on the basis of its relations to these corporations. But if that is the case, how can we dissociate ourselves from the plight of people in these countries? Keeping the poor out of our lifeboats might be permissible if we hadn't built the boats out of rubber taken from them in the first place. The fact is, even if you can justify the commercial dealings we have with the rest of the world, we are already involved with the hungry of the world in a way that the lifeboat metaphor belies.

The question of distributive justice haunts Hardin's argument. He admits that survival policies are often unjust; but he argues, as a utilitarian, that survival overrides justice, that it is better for some to survive unjustly than to be just and have everyone perish: "complete injustice, complete catastrophe." But this fails to consider a whole middle range of possibilities where justice could be at least a contributing factor to a solution that would take the need for one's own survival into consideration. Justice would demand that some attention be given to redistributing the world's wealth. At present the United States, which constitutes less than 4.5% of the world's population, consumes some 35% of its food (much of it thrown into garbage cans or rotting in storage silos) and 38% of its energy and is responsible for creating 33% of the world's pollution. If Hardin is so concerned about preserving the world's purity and resources for posterity, justice would require that we sacrifice the overfed, overweight, overnourished, overconsuming, overpolluting, greedy Americans who throw into garbage cans more food than some nations eat.

Regarding Hardin's views on the carrying capacity and the ratchet effect, several objections are in order. First, how does Hardin know which nations have exceeded their carrying capacity? The very notion of the carrying capacity, given our technological ability to produce new varieties of food, is a flexible one. Perhaps experts can identify some regions of the Earth (such as its deserts) where the land can sustain only a few people, but one ought to be cautious in pronouncing that Bangladesh or India or some regions of the Sahel in Africa are in that condition. Too many variables abound. New agricultural or fishing methods or cultural practices may offset the validity of technical assessments.

Second, Hardin is too dogmatic in proclaiming the lawlike dictum that to aid the poor is to cause the escalation of misery. Granted, we can make things worse by merely giving food handouts, and a population policy is needed to prevent the ratchet effect that Hardin rightly warns against. But there are intelligent ways to provide aid—such as providing agricultural instruction and technological know-how to nations committed to responsible population policies, promoting mutually beneficial international trade agreements, and seeking ways to eradicate those conditions that cause famine and malnutrition. We should also set a good example of what a just, disciplined, frugal society should be. But to dismiss these options out of hand and simply advocate pushing people off our lifeboat is as oversimplified as it is cruel.

Finally, Hardin's food-population theory ignores the evidence that, contrary to the ratchet projections, population growths are affected by many complex conditions other than food. Birth rates can fall quite rapidly, sometimes before modern birth control devices are available. Specifically, a number of socioeconomic conditions can be identified that cause parents to have fewer offspring. These conditions include parental security and faith in the future, the improved status of women in the society, literacy, and lower infant mortality. Procuring these conditions requires agricultural reform, some redistribution of wealth, increased income and employment opportunities, better health services, and fresh expenditures on education. Evidence suggests that people who perceive the benefits of a smaller family will act prudently. The theory that favorable socioeconomic conditions will cause a natural decrease in their birth rate is called the **benign demographic transition theory (BDT).** Although it is controversial, it may give us some reason to hope that population growth will level off. How much hope? Later in the chapter, we will discuss that issue in more detail. We must now turn to the second philosophical theory on world hunger, liberalism.

Liberalism

The liberal position on world hunger is that we have a duty to help the poor in distant lands. There is something inherently evil about affluent people failing to come to the aid of the poor when they could do so without great sacrifice. Liberal theorists on this issue come in several varieties. Some are utilitarians who argue that sharing our abundance and feeding the poor will very likely maximize utility or happiness. Some are deontologists who argue that we have a fundamental duty to use our surplus to aid those less well off. Some deontologists simply find it self-evident that the needy have a right to our resources. Witness the report of the Presidential Commission on World Hunger: "Whether one speaks of human rights or basic human needs, the right to food is the most basic of all. . . . The correct moral and ethical position on hunger is beyond debate."[4] Others appeal to the principle of justice, arguing that the notion of fairness requires that we aid the most needy in the world. Still others are radical egalitarians—perhaps the label *liberal* doesn't strictly apply to these—who argue that the principle of equality overrides even the need for survival so that we should redistribute our resources equitably even if it means that all of us will be malnourished and risk perishing. I think that we can capture most of what is vital to the liberal program if we examine Peter Singer's theory, which covers the first two types of liberalism, and Richard Watson's theory, which is a version of radical egalitarianism.

Peter Singer's article "Famine, Affluence, and Morality," written on the eve of the 1971 famine in Bangladesh, sets forth two principles, either of which would drastically alter our lifestyles and require us to provide substantial assistance to distant, poor, and hungry people.[5] The *strong principle* states that "if it is in our power to prevent something bad from happening without thereby sacrificing anything of *comparable* importance, we ought morally to do it." While this has similarities to utilitarian principles, it differs from them in that it does not require maximizing happiness, simply the amelioration of suffering through sacrifice of our goods to the point where we are just about at the same place as the sufferer. The idea behind this principle is utilitarian: diminishing marginal utility, which states that transferring goods from those with surplus to those with needs generally increases total utility. For example, if you have $100 for your daily food allowance and I have no allowance at all, your giving me some of your money will actually increase the good that the money accomplishes; for the gain of, say, $10 by me will enable me to survive, thus outweighing the loss you suffer. But there will come a point where giving me that extra dollar will not make a difference to the total good. At that point you should stop giving. If we were to follow Singer's strong principle, we would probably be giving a vast proportion of our GNP (gross national product) to nonmilitary foreign aid instead of the present 0.21% (that's two tenths of one percent) or the 0.7% advocated by the United Nations as a fair share for rich countries.

Singer's *weak principle* states that we ought to act to prevent bad things from happening if doing so will not result in our sacrificing anything *morally significant*. He asks you to suppose that you are walking past a shallow pond and see a child drowning. You can save the child with no greater inconvenience than wading into the water and muddying your suit or dress. Should you not jump into the pond and rescue the child? Singer thinks it is self-evident that nothing morally significant is at stake in the sacrifice.

While Singer prefers the strong principle, he argues that the weak principle is sufficient to ground our duty to aid needy, distant people, for what difference does it make if the drowning child is in your hometown or in Africa or Asia? "It makes no moral difference whether the person I can help is a neighbor's child ten yards away or a Bengali whose name I shall never know, ten thousand miles away." He or she is still a human being and the same minimal sacrifice is required. Furthermore, the principle makes no distinction between cases in which I am the only person who can do anything and cases where I am just one among many who can help. I have a duty in either case to see that what is needed is accomplished. Call this the "no-exception proviso."

Singer's two principles have generated considerable debate and many ethicists have accepted one or both of them, but there are problems with each. John Arthur has noted in his critique of Singer that the strong principle is too strong and the weak principle is too weak.[6] On the strong principle our right to our property and lifestyles is too easily overridden by the needs of others. For example, if I meet a stranger going blind and I could prevent her becoming completely blind by giving her one of my eyes, I should take steps to have my eye removed—even, according to the no-exception proviso, if there are others on whom she has a greater claim to some assistance than she has on me. Likewise, if I meet a man about to lose his kidneys or lung functions, I have a prima facie duty to give him one of my kidneys or lungs, a duty that can be overridden only by finding someone on whom he has a stronger claim who will donate his or her organ. Woe the person who meets someone in need of all three of these organs—an eye, a lung, a kidney! If no one else is doing his or her duty, you are left with the responsibility of yielding your organs—even if this results, as it surely will, in a severe change in your lifestyle. As long as you have not reduced your lot to the level of the other person's, you must go on sacrificing—even for strangers.

Richard Watson's position on our responsibility to distribute our resources with the hungry is more severe than Singer's.[7] From a deontological perspective, he argues that the principle of equal worth of individuals calls for the food of the world to be distributed equally. "All human beings are moral equals with equal rights to the necessities of life. Differential treatment of human beings thus should be based only on their freely chosen actions and not on accidents of their birth and environment." It is our sacred duty to share scarce resources with every needy person even if this means that we all will be malnourished, that no one will get sufficient food, and that ultimately everyone will perish. Equality trumps survival, even the survival of the human race.

Watson's equality-absolute has problems. The equal absolute right of each person to life's necessities needs a defense, which Watson fails to afford. I don't see why I am obligated to sacrifice the lives of my children and myself simply because there is not enough food to feed all of us. Suppose you and your family and friends (50 people in all) work hard and grow enough food to feed 100 people for the next six months. If you feed only your 50 people, you will have enough food for a year, after which a new crop can be harvested. But there are 200 people who need food. If you share your food with all 200, you will all die. If you share it with 50 others beside your family (100 in all), you all can survive for the next six months, but must hope for outside aid after that. What is the morally correct thing to do? (1) Feed only your community (50 people who did the work)? (2) Feed these 50 and another 50 outsiders? (3) Feed all 200 and perish together? (4) Draw lots to determine who should get the food and live? I think we are morally permitted to opt for (1), because we have a right to try to survive and flourish, as long as we are not unjustly harming others. As we argued in the early chapters of this book, we have a prima facie right to the fruits of our labor, so that we need not divest ourselves of life's necessities in order to help others.

An adequate moral theory must make room for self-regarding reasons. I am required to make *reasonable* sacrifices for others, but not at the cost of what would severely detract from the quality of my own life. Watson explicitly rejects the notion of reasonableness in morality. Morality is often unreasonable, according to him. But I see no reason to accept that verdict. If morality were truly unreasonable, rational people would be advised to opt out of it and choose a more rational quasi-morality in its place. My thesis throughout this book has been that moral principles are reasonable requirements. In general, they are in our long-term interest.

Of course, you are free to go beyond the call of duty and donate your organs to strangers. It is certainly noble of you to volunteer to do so. But such supererogatory acts are not duties as such. Extreme utilitarians and absolutist egalitarians confuse morality with extreme altruism or saintliness.

We turn to Singer's weaker principle. If we can prevent an evil by sacrificing something not morally significant, we should do so. This seems closer to the truth, but John Arthur has argued that it is too weak, for what can be morally significant varies from person to person. For example, my record collection or collection of rare pieces of art might be a significant part of what makes life worth living for me, so that to sacrifice them for the poor would be of moral significance. Is giving up owning a television set or possessing a nice car or having a nice wardrobe morally significant? For many people they are. Nevertheless, there are occasions where sacrificing these things for the poor or needy might be morally required of me. Even in the case where a child is drowning in the pond, you could refuse to jump in the water to save the child, using Singer's weak principle, for you could argue that having clean, unspoilt clothes is morally significant for you. Of course, it would really have to be the case that they were morally significant—but for some people they truly are.

Singer could respond that Arthur's objection fails because he is making moral significance overly relative. There is an objective truth to the matter of whether something really is morally significant. Singer would need to qualify his principle of what is morally significant by relational terms: In situation S object O is morally significant to person P (whether he or she realizes it or not). Compared to saving someone's life, wearing clean, unspoilt clothes is not morally significant whatever the misguided dandy might think to the contrary. Not every supposedly morally significant trait is really so. If I believe that burning witches is the way to save our nation from the devil and go around burning those who fit my description, I am simply misguided. Likewise, if I think that my baseball card collection is more important than saving my best friend's life, I have a bad set of priorities—friends really are more valuable than baseball cards; and if I fail to realize this, I am missing a deep moral truth.

While Singer's weak principle, suitably qualified, can survive the kind of attack that philosophers like Arthur hurl at it, it may not be good enough to get him the hunger relief that he advocates. Other factors need to be addressed. For example, do needy strangers have rightful claims on my assets even though I have done nothing to cause their sorry state? Do the starving have rights to my property? We turn now to the conservative position.

Conservatism

Conservatives on world hunger argue that we have no duty at all to give aid to distant needy people. Representative of the view in question are such libertarians as John Hospers, Robert Nozick, and Ayn Rand, and contractualists like Thomas Hobbes and, more recently, Gilbert Harman, Howard Kahane, and William Nelson.[8] Typically, conservatives, in the minimalist sense I am using the term, reject the notion that we have positive rights that entail others to come to our aid or promote our good unless there is a contractual agreement between us. The one right we have is that of freedom. the right not to be interfered with, the right to possess our property in peace. As long as I have a legitimate claim on my property (that is, I have not acquired it through fraud or coercion), no one may take it from me, and I may refuse to share it regardless of how needy others are.

We may not positively harm others, but we need not help them either. No moral duty obligates you to dirty your clothes by trying to save the child drowning in the muddy pond. Of course, it shows bad character not to save the child, and we should endeavor to be charitable with our surplus and support good causes; however, these are not strictly moral duties but rather optional ideals. So it follows that if hungry Esau is prompted to sell his birthright to that feisty chef, Jacob, so much the worse for Esau. And if a poor African country decides to contract with a Western corporation to shift from growing a high-protein crop to a cash crop like coffee, as long as no external force was used in the agreement, the contract is entirely just; and the corporation need feel no guilt when the poor nation undergoes a famine and finds itself unable to supply its people with adequate protein. No rights have been violated. The country simply made a foolish choice.

If you believe that the contractual approach to ethics, which we examined in Chapter 4, is the correct approach, the conservative position will appeal to you. It may have considerable merit, but it also has certain weaknesses. Unless it is supplemented with a theory of natural duties, contractualism limits moral obligations to agreements made. If a poor country agrees to accept hazardous waste from a rich country, there is nothing more morality has to say about the matter. But, as I have suggested in Chapter 2, morality, in large part, has to do with the promotion of human flourishing and ameliorating suffering, so that we have some duties to help others, even when we have not contracted ahead of time to do so. On the contractualist model the 38 people who for 35 minutes watched Kitty Genovese get beaten to death, who did not lift a finger to call the police or lift their windows to shout at the assailant, did nothing wrong. But if we have a duty to promote human flourishing and ameliorate suffering, these onlookers did have a duty to do something on Ms. Genovese's behalf, and they are to be faulted for not coming to her aid.

It would seem, then, that there are positive duties as well as negative ones. Both utilitarian and deontological theories are better at recognizing positive duties to help others and promote human flourishing, even where no contract is in force.

A Moderate Alternative

There is a moderate position between the liberal and the conservative that accepts part of each position but rejects other parts. It goes as follows.

Morality originates in group living, tribes. People discover that certain rules are necessary for survival and happiness. These are rules against killing each other, against promise breaking, against violating property rights, and against lying and cheating; there are also rules promoting justice, cooperation, respect for others, and beneficence. Those living in a society implicitly agree to live by this core morality (discussed in Chapter 2). They resolve their conflicts of interest through compromise or impartial third parties—the primitive origins of law. But they notice that other groups do not respect their lives or property, and there is no way to resolve differences through impartial review. The Other is the enemy toward whom the rules do not apply. Indeed, it is only by not respecting the enemy's life and property that one can survive and flourish.

Eventually, the two groups learn to accept an intertribal core morality. They begin with a mutual nonaggression pact and start to respect each other as equals, cooperate instead of fight with each other, and subject their differences to an impartial review. Nevertheless, in many situations, members of a tribe will feel a greater responsibility to aid members of their own family and tribe than to aid members of the neighboring one. If my child, a neighboring child, and a child of another tribe all need a pair of shoes, and I have only enough resources to procure one pair, I will feel a duty to give them to my child. If I can procure two pair easily, I will give the first pair to my child, but sell the second pair at a low price to my neighbor. If I go into the shoemaking business, I will still be likely to give favored treatment to my neighbor over the person from the other tribe. Greater opportunity for reciprocity arises with my neighbor than with the family of another tribe, so it makes sense to treat that family as special.

Moderate moral theory recognizes special responsibilities to family, friends, and neighbors. This is the reason Singer's drowning child example is misleading. I can do only so much good. I can save only so many drowning children. While I may have a duty to give of my surplus to help save drowning children in a distant land, I have a stronger duty to help those with whom I have intimate or contractual ties.

This said, the other side of the coin needs to be turned over and the liberal program acknowledged; for another aspect of morality is the enlargement of the circle of benevolence and flourishing, the utilitarian aspect of maximizing good. We need to expand the small circle of moral consideration and commitment from family, community, and country to include the whole world. We need to do this for two reasons. First, it just is good to do so. Helping as many people (and animals) as possible without harming yourself is part of the meaning of the idea of promoting the flourishing of sentient beings. Second, it is in our self-interest to do so. Unless we learn to live together on this small planet, we may all perish. Humanity is no longer innocent. Technology—through its inventions of atomic weapons, biological weapons, poisonous chemicals, and so forth—is available to destroy all sentient life. One nation's adverse environmental impact can affect the rest of the world. If one nation pollutes the air through spreading sulfur dioxide or carbon dioxide, the rest of the world suffers the effects. If the Brazilian or Peruvian farmers cut down large segments of the Amazon rain forest, we all lose oxygen and ecological diversity. We are all in each other's debt. If we don't hang together, we *hang* alone.

So the same considerations that led to mutual cooperation between our original tribes must inform our global policies. A rational core morality must reign internationally. While we will still have priority commitments to family and friends, we cannot allow selfishness to hinder generous dealings with the rest of the world. So Hardin's metaphor of nations as lifeboats has only limited applicability. It may justify careful immigration policies that prevent overcrowding, but it should not prevent assistance to other nations who may be helped and who someday may be able to help us. In a sense the whole earth is one great lifeboat in which we'll sink or float together.

But something must be said in Hardin's behalf. He points to a crucial problem that deserves our concentrated attention: population policy. Even if we finally opt for the benign demographic transition theory, many situations may not wait for that policy to take effect. The facts are that the population of the world is multiplying at an exponential rate. Since 1930 the earth's population has increased from 2 billion to the present 5.9 billion, heading for 6 billion by 1999. In 1968 Paul and Anne Ehrlich wrote *Population Bomb,* warning that the population of the world (then 3.5 billion) was growing exponentially at a rate of 70 million per year and that if strong measures were not taken, we would likely have Malthusian problems of famine and disease. Critics pejoratively labeled such cautioners as "doomsdayers," but they have been proved correct.

As we noted in Chapter 11, crowded conditions prevail in many parts of the world. Famines have become worse in areas of Africa and Bangladesh. Today the global population is growing not by 70 million per year, but by 90 million. As we also noted in Chapter 11, the growth rate is 1.7%, which means the Earth's population is likely to double in 41 years and is projected to rise to about 11.2 billion by 2050, unless there is a significant dieback or policies of population control are implemented. A slight increase in the growth rate would result in the world's population density becoming that of present-day New York City by 2300. Add to this the following: The innovative technological development of food has leveled off; topsoil is being depleted; pesticide-resistant strains of crop destroyers are appearing throughout the Earth's agricultural areas; and there is evidence of changing weather patterns, probably brought on by the greenhouse effect, causing diminished farm production.[9] And, as though this were not serious enough, the ocean fish harvests are declining.

We noted in our critique of Hardin that the benign demographic transition theory (BDT) holds that we should concentrate on the root causes of population growth and let the process solve this problem. How does this theory work? It idealizes population changes in four stages. In stage 1, the *preindustrial stage*, the severe living conditions give rise to a high average birth rate and a high average death rate, resulting in very little, if any, population increase. In stage 2, the *mortality transition stage*, the death rate falls, while the birth rate remains high, so that the population increases. In stage 3, the *fertility transition* (or *industrial stage*), the average birth rate decreases, due to availability of birth control, the improved status of women, lower infant mortality, general education, and the rising cost of raising children. In stage 4, the *postindustrial stage*, the average birth and death rates are both low, tending toward zero growth. The population has leveled out but is much larger than it was at stage 1. There is some evidence for the BDT. Separate studies by Revelle, Brown, Eberstadt, and Rich have shown that several countries that have progressed in these areas have cut their birth rate dramatically.[10] Some demographers point to the fact that in the 1970s China brought its birth rate down from 40 per 1,000 people to 30 in about five years' time and that Cuba brought its birth rate down to 27 per 1,000. The conclusion is that we should go to the causes of overpopulation (through those activities mentioned in stage 3) and not punish countries for their "overpopulation."

But others have argued that the benign demographic transition theory has severe problems. Demographer Joel Cohen points out that some countries reduced their birth rates prior to industrialization or educational development. In other cases, improved public health measures in developing countries caused an enormous rise in life expectancy without significantly reducing the birth rate.[11] Furthermore, after China and India made significant progress toward cutting fertility rates in the 1970s and despite the growing benefits of industrialization, their fertility rates increased in the 1980s. Referring to the 1980s and 1990s, Cohen notes that population growth rates are rising. Seventy-eight million people live in countries with a total fertility rate above seven children per woman, and 708 million live in countries with six or more children per women. Why did the trend reverse? There are several reasons, one being China's softening of its austere one-child policy and another being India's relaxation of its family-planning policy.[12] The point is that the BDT is too idealized a model. It may indicate a tendency in industrialization to lead to population stabilization, but there is no lawlike necessity about it. People must understand that it is to their advantage to have fewer children, and effective techniques of fertility reduction must be available.[13] Even with the benefits of industrialization and the availability of birth control, the average birth rate may continue for a long time, so that the population increases at a very high rate, putting a stress on the ecosystem and exceeding the carrying capacity of the land. Famines in the coming years could have devastating effects.

We have cause for alarm, and if the shouts of neo-Malthusians like Hardin and Ehrlich are needed to wake us up, let us thank them for waking us up—even as we work for kinder, more just solutions to the problem. The point is that hard choices will have to be made, and food aid should be tied to responsible population control for the survival and well-being of humanity on this planet.

A proposal that improves on Hardin's lifeboat ethics is the triage approach first set forth by Paul and William Paddock in *Famine—1975* and advocated by Joseph Fletcher.[14] The term *triage* (French word for "sorting") comes from wartime medical policies. With scarcity of physicians and resources to cope with casualties of battle, the wounded were divided into three groups: those who would probably survive without medical treatment, those who would not survive even with treatment, and those for whom treatment would make the decisive difference. Only this last group would receive medical aid. The Paddocks and Fletcher urge us to apply the same policy to world hunger. Given limited ability and scarce resources to help, we should not aid those nations who will survive without our aid or those who will not be able to sustain themselves even with our help. We should direct all of our attention to those people and nations for whom our input could make a decisive difference. The aim should be to enable these people to become self-sufficient, responsive to the carrying capacity of their environment. As the Chinese proverb says, "Give a man a fish today and he will eat it for a day. Teach him how to fish, and he will eat for the rest of his days."

As repulsive as the triage strategy may seem to some, it should not be dismissed out of hand. Perhaps at present no nation is hopeless, so that we still have time to forestall the nefarious effects of overpopulation. We should give the benign demographic theory a chance to work, supporting social reform at home and abroad with our actions and our pocketbooks; but if we cannot effect global changes, the time for triage may soon be upon us. The doomsdayers are to be taken seriously. They may not be correct, but their warnings should be heeded.

Meanwhile, the broad-based core morality outlined in the first three chapters of this book would seem to require contributing personally to hunger relief organizations and urging that national policy provide appropriate agricultural know-how and technological assistance to nations in dire need. At the same time, we should support family-planning programs both here and abroad, aiming to provide each human being with as high a quality of life as is possible. The option is not food *or* population control, but food *and* population control. The world must see these as two sides of the same coin, a coin that is the entrance payment to a better future for all people.

If the preceding discussion is correct, we do have a duty to give aid to the needy both in our own country and in other lands. It is a duty to exercise benevolence in order to ameliorate suffering and promote human flourishing. We are not required to bring ourselves to the same level of poverty as those we help—as Watson and Singer's strong principle advocates—but we should be giving more than most of us are. No one can tell another person just how much to donate, but each of us must consult his or her conscience. Besides that, we should be living as personal examples of ecological responsibility, as good stewards of Earth's resources; and we should call upon our leaders to increase nonmilitary aid to underdeveloped countries where the need is greatest.

Review Questions

1. Discuss the neo-Malthusian views, especially those of Garrett Hardin, on the relationship between population and food production. Evaluate his three arguments against giving aid to starving people in distant places.
2. What is the "benign demographic transition theory"? How strong is it and what might be said against its use in policy development?
3. Examine Peter Singer's argument in favor of helping famine victims. Evaluate his principle, "If it is in our power to prevent something bad from happening, without sacrificing anything else of comparable moral importance, we ought morally to do it." If we followed this principle, what would be the likely results? What moral considerations would outweigh our obligation to give aid to famine victims or anyone else in need?
4. How would a contractualist deal with the problem of aiding famine victims? Evaluate this position.
5. Someone who reviewed the manuscript of this book suggested this problem: Suppose a farmer in the rain forest says, "My principal duty is to feed my children, not to worry about the global effects of cutting down rain forests. You Americans are happy to ignore such special relations when you say that *I* should worry about the rain forest because it is a treasure to the world. Yet you rely on special relations when you think my country ought to worry about feeding its poor, and your country ought to worry about feeding your poor." How would you respond to this farmer? What is the link between hunger and environmental degradation?
6. Discuss the notion of triage. How does it apply to the problem of world hunger?
7. Sometimes we treat hunger as an abstract idea, forgetting just how terrible it is. I suggest reading a report or essay on the subject. You might want to read Knut Hamsun's *Hunger,* one of the most poignant portrayals of the personal agony of the psychology of starvation.

Suggested Readings

Aiken, William, and Hugh LaFollette, eds. *World Hunger and Moral Responsibility*. Englewood Cliffs, N.J.: Prentice Hall, 1977.

Arthur, John. "Rights and Duty to Bring Aid." In *World Hunger and Moral Responsibility*. Edited by William Aiken and Hugh LaFollette. Englewood Cliffs, N.J.: Prentice Hall, 1977.

Brown, Lester. *Tough Choices: Facing the Challenge of Food Scarcity*. New York: Norton, 1996.

Cohen, Joel E. *How Many People Can the Earth Support?* New York: Norton, 1995.

Fletcher, Joseph. "Give If It Helps, Not If It Hurts." In *World Hunger and Moral Responsibility*. Edited by William Aiken and Hugh LaFollette. Englewood Cliffs, N.J.: Prentice Hall, 1977.

Hamsun, Knut. *Hunger*. Translated by Robert Bly. New York: Bard Books, 1967.

Hardin, Garrett. "Lifeboat Ethics: The Case Against Helping the Poor." *Psychology Today* (1974).

———. "Living on a Lifeboat." *BioScience* (1974).

———. *The Limits of Altruism*. Bloomington: Indiana University Press, 1977.

Lappe, Francis, and Joseph Collins. *World Hunger: Twelve Myths*. New York: Grove, 1986.

O'Neill, Onora. *The Faces of Hunger*. London: Allen & Unwin, 1986.

Paddock, Paul, and William Paddock. *Famine—1975*. Boston: Little, Brown, 1968.

Singer, Peter. "Famine, Affluence, and Morality." *Philosophy and Public Affairs* (1972).

Notes

1. Alan Gregg, "A Medical Aspect of the Population Problem," *Science* 121 (1955).
2. Garrett Hardin, "Lifeboat Ethics: The Case Against Helping the Poor," *Psychology Today* (1974); "Living on a Lifeboat," *BioScience* (1974); *The Limits of Altruism* (Bloomington: Indiana University Press, 1977).
3. Garrett Hardin, "The Immorality of Being Softhearted," *Stanford Alumni Almanac* (January 1969).
4. Presidential Commission on World Hunger, *Overcoming World Hunger: The Challenge Ahead* (Washington, D.C.: Government Printing Office, 1980).
5. Peter Singer, "Famine, Affluence, and Morality," *Philosophy and Public Affairs* (1972).
6. John Arthur, "Rights and Duty to Bring Aid," in *World Hunger and Moral Responsibility*, ed. William Aiken and Hugh LaFollette (Englewood Cliffs, N.J.: Prentice Hall, 1977).
7. Richard Watson, "Reason and Morality in a World of Limited Food," in *World Hunger*, ed. William Aiken and Hugh LaFollette.
8. John Hospers, *Libertarianism* (Los Angeles: Nash, 1971); Robert Nozick, *Anarchy, State, and Utopia* (New York: Basic Books, 1974); Thomas Hobbes, *Leviathan* (1651); Gilbert Harman, "Moral Relativism Defended," *Philosophical Review* (1975); Howard Kahane, "Making the World Safe for Reciprocity," in *Reason and Responsibility*, ed. Joel Feinberg (Belmont, Calif.: Wadsworth, 1989); and William Nelson, *What's in It for Me?* (Oxford: Westview Press, 1991). Ayn Rand, *The Virtues of Selfishness* (New York: New American Library, 1964) also holds this position from an explicitly egoistic perspective.
9. Statistics are based on the *Population and Vital Statistics Report of the United Nations Statistical Office* and the *U.N. Demographic Yearbook*, 1990. Information is available through the Population Reference Bureau, 777 14th St., N.W., Suite 800, Washington, DC 20005. See also the journal *Population and Environment*, especially 12:3 (Spring 1991).
10. See Roger Revelle, "The Ghost at the Feast," *Science* 186 (1974); W. Rich, "Smaller Families Through Social and Economic Progress," *Overseas Development Council Monograph* #7, Washington, D.C., 1973; Nick Eberstadt, "Myths of the Food Crisis," *New York Review of Books*, 19 February 1976; Lester Brown, *In the Human Interest* (New York: Norton, 1974), p. 119, and *World Without Borders* (New York: Praeger, 1974), p. 140f; and M. S. Teitelbaum, "Relevance of Demographic Transition Theory in Developing Countries," *Science* 188 (1975).

11. Joel Cohen, *How Many People Can the Earth Support?* (New York: Norton, 1995), chap. 4.
12. Ibid., pp. 64–65.
13. Ashley Coale lists three general prerequisites for declining fertility: (1) Fertility must be within the calculus of conscious choice. Potential parents must consider it an acceptable mode of thought and form of behavior to balance the advantages and disadvantages before deciding to have another child. (2) Reduced fertility must be advantageous. Perceived social and economic circumstances must make reduced fertility seem an advantage. (3) Effective techniques of fertility reduction must be available. Procedures that will in fact prevent births must be known, and there must be sufficient communication between spouses and sufficient sustained will, in both, to employ them successfully. One of Coale's students summed this position up by saying "potential parents must be ready, willing, and able to control their fertility." Quoted in Cohen, *How Many People Can the Earth Support?* p. 62. Cohen says the most significant factor in reducing the birth rate is the availability of birth control.
14. Paul and William Paddock, *Famine—1975* (Boston: Little, Brown, 1968) and Joseph Fletcher, "Give If It Helps, Not If It Hurts," in *World Hunger,* ed. William Aiken and Hugh LaFollette.

CHAPTER THIRTEEN

Air Pollution, the Greenhouse Effect, and Ozone Depletion

We brush our teeth with fluoride compounds, rub on propylene glycol deodorants, clothe ourselves in rayon and nylon or treated cotton and wool, drive cars filled with the products of a liver carcinogen called vinyl chloride, talk on plastic phones, walk on synthetic tiles, live within walls coated with chemical-laden paint. Our food, kept fresh in refrigerators by heat-absorbent refrigerants, contains preservatives and chemical additives. And, of course, it has been grown with the aid of chemical fertilizers and insecticides. The detergents and the medicines, the foam rubber and the floor cleaners—all had their underside of waste which is about to make a dramatic public reentry.

—LESTER BROWN

Climate is an angry beast, and we are poking it with sticks.

—WALLACE BROECKER

Nature has been besieged. It was unprepared for this bombardment of artificial chemicals. Nowhere in Earth's soil, the sky's air, or the ocean's waters is there the ability for dismantling these complex new substances bonded together by the ingenuity of humans. Many of the chemicals, such as chlorinated hydrocarbons (DDT, aldrin, endrin, and dieldrin) are based on long molecular chains that are solid enough to withstand even degradation by sunlight and dissolution by water or breakdown by acids. There are hundreds, if not thousands, of new artificial chemicals that nature doesn't know how to handle—hatched in the cauldron of technological witches' brew—all of which threaten the lives of humans and animals over the globe. Hence, the urgent and global problem of pollution.

POLLUTION: A GENERAL CONSIDERATION

On March 24, 1989, the oil tanker *Exxon Valdez* ran aground off the Alaskan coast, spilling 1.26 million barrels of crude oil into Prince William Sound. It was the worst oil spill in history. The pristine beauty of the Alaskan coast, with its wealth of birds, fish, and wildlife, was degraded. Five hundred square miles of the Sound were polluted. Millions of fish, birds, and wildlife were killed. The fishing industry, which earns $100 million annually in Prince William Sound, ground to an abrupt halt, and many people lost their means of livelihood. The Exxon Corporation was unprepared for an accident of this magnitude. It had only 69 barrels of oil dispersant on hand in Alaska, when something like 10,000 were needed to clean up the spill. The damage caused by the spill, some which can still be seen, was reckoned to be $8.5 billion. The ship's captain, Joseph Hazelwood, was found guilty of negligence and operating the tanker under the influence of alcohol—testimony reported that he had drunk 14 glasses of vodka; Exxon was fined $100 million. But Greenpeace placed an advertisement in newspapers, showing Joseph Hazelwood's face, with the caption: "It wasn't his driving that caused the Alaskan oil spill. It was yours. The spill was caused by a nation drunk on oil. And a government asleep at the wheel."

Modern industrial society has provided enormous benefits, making our lives far more comfortable and freeing us from backbreaking drudgery. From electric lighting, washing machines, air conditioners, and gas-driven automobiles to modern medical miracles, supersonic airplanes, and microchip computers, our lives have been enriched with possibilities that a little more than a century ago existed only in dreams or in science fiction. Yet with these life-enhancing, technological wonders has come waste and pollution. This pollution in turn jeopardizes our vital supply line, nature, threatening to undo the benefits technology has brought us.

Pollution may be broadly defined as any unwanted state or change in the properties of air, water, soil, or food that can have negative impact on the health, well-being, or survival of human beings or other living organisms. Many pollutants are undesirable chemicals produced as by-products when a resource is converted into energy or a commodity. Types of pollution include contaminated water, chemically polluted air (such as smog), toxic waste in the soil, poisoned food, high levels of radiation, and noise. They also include acid rain, cigarette smoke, alcohol, and drugs. Loud music at a rock concert or issuing from the next dorm room can qualify as pollution, since it may have harmful effects that we would not want under normal conditions. I have given only a broad definition of pollution in terms of what is normally harmful. Some people may have higher resistance to smog or radiation, so that what is harmful to one person may not be so to another. So there is an element of relativity in pollution. Nonetheless, the concept seems sufficiently clear to work with.

Three factors determine the severity of a pollutant: its chemical nature (how harmful it is to various types of living organisms); its concentration (the amount per volume of air, water, soil, or body weight); and its persistence (how long it remains in the air, water, soil, or body).

We divide a pollutant's persistence into three types: degradable, slowly degradable, and nondegradable. Degradable pollutants, such as human and animal sewage and soil, are usually broken down completely or reduced to acceptable levels by natural chemical processes. Slowly degradable pollutants, such as DDT, plastics, aluminum cans, and chlorofluorocarbons (CFCs) often take decades to degrade to acceptable levels. Nondegradable pollutants, such as lead and mercury, are not broken down by natural processes.

We know relatively little about the short- and long-range harmful potential of most of the more than 70,000 synthetic chemicals in commercial use on people and the environment. The Environmental Protection Agency (EPA) estimates that 80% of cancers are caused by pollution. We know that half of our air pollution is caused by the internal combustion engine of motor vehicles and that coal-burning stationary power plants produce unacceptable amounts of sulfur dioxide (SO_2). The World Health Organization (WHO) estimates that about 1 billion urban people (about one-fifth of humanity) are being exposed to health hazards from air pollution and that emphysema, an incurable lung disease, is rampant in our cities. Studies tell us that smog (a mixture of smoke and fog) is hazardous to our health and that it has caused thousands of deaths in such cities as London, New York, and Los Angeles.

In the United States 80% of our freshwater aquifers are in danger, so that a large percentage (perhaps more than 30%) of our population is drinking contaminated water. By 1991 the EPA had listed 1,211 hazardous waste sites for cleanup, with the estimated cost of cleanup per site being $26 million. Acid rain is killing our forests and lakes.

Technology, which enhances the quality of our lives, also causes an exponential growth of dangerous pollution. Can we have both high technology and high quality of life? Or must we make radical changes in our value system, reassessing our "need" for bigger and more powerful machines?

Because of its importance in the environmental debate, the four sections of this chapter deal with air pollution. In Chapter 14 we discuss water pollution, pesticides, and hazardous waste. We have already considered the relation of pollution to population in Chapter 11. In this introduction, we have discussed general considerations of the topic: How serious is pollution on balance? Next we treat air pollution, especially carbon dioxide (CO_2), sulfur dioxide (SO_2), and acid rain. Then we focus on one of the long-term spin-offs of pollution, the greenhouse effect, or global warming. How serious is it? At the end of the chapter, we discuss the problem of a depleting ozone layer.

AIR POLLUTION

The air we breathe may be described as a witches' brew of dust, dirt, poisonous gases, and a host of unseen, but foul-smelling pollutants. These pollutants hide the sun, dirty our clothes, and make us ill. An unseen enemy is often the most dangerous.

Most of the Earth's air is located in the *troposphere,* which extends about 11 miles (17 kilometers) above sea level. The air in the troposphere consists mainly of two gases: nitrogen (78%) and oxygen (21%); the other 1% consists of argon, carbon dioxide, traces of other chemicals, and small amounts of water vapor. Above the troposphere lies the stratosphere, extending 11–30 miles (17–48 kilometers) above sea level. It contains less air, but contains most of the ozone that acts as a global sunscreen, blocking out most of the sun's harmful ultraviolet (UV) rays. The Earth's biochemical cycles may be altered by human input. Burning fossil fuels and Earth's forests adds about 25% of the carbon dioxide to the atmosphere, causing increased global warming (see the next section).

Sources of stationary fuel consumption, such as electric power plants, paper mills, iron mills, smelters, and petroleum refineries, burn fossil fuels such as coal and oil and deposit large amounts of sulfur dioxide into the atmosphere, causing industrial smog over our cities and acid rain, which is polluting our lakes and rivers and killing trees all over the world. When the carbon and sulfur in fossil fuels are burned, the carbon is converted to carbon dioxide (CO_2) and carbon monoxide (2CO) and the sulfur into sulfur dioxide (SO_2). Unburnt carbon is released into the atmosphere as particulate matter—soot. The problem of SO_2 emissions has lessened in the United States, but in the developing nations, especially China and India, it is getting worse. Beijing, the capital of China, is one of the most polluted cities of the world.

In the United States motor vehicles produce more air pollution than any other human activity. In Los Angeles, Denver, Phoenix, Cleveland, New York, and other cities, polluted air contributes to emphysema, asthma attacks, and lung cancer. But other cities are even worse: in São Paulo, Buenos Aires, Bangkok, Rome, and Mexico City, motor vehicles account for about 85% of all air pollution. The World Health Organization (WHO) estimates that more than 1.1 billion people live in urban areas where the air is unhealthy.

The damage done by air pollution in the form of smog in a region depends on several factors, especially the local climate, topography, the types of industry, and the population density. In areas with high precipitation, the rain and snow tend to cleanse the air. Wind tends to blow pollutants away from local areas, but often carries them to others—downwind—so that lakes and forests in New York's Adirondacks and Canada's Southeast are being harmed by upwind pollution produced by power plants in Illinois, Indiana, Ohio, and Pennsylvania. The EPA estimates that at current rates, by the year 2040, 43% of all Adirondack lakes will be devoid of life if sulfur dioxide and nitrogen oxide emissions are not reduced below levels mandated by the 1990 Clear Air Act.

A source of special concern is the phenomenon known as *thermal inversion*. When the sun warms the air near Earth's surface, the heated air expands and rises, carrying pollutants into the higher regions of the atmosphere. But in cities like Los Angeles and Mexico City, which are located in valleys surrounded by mountains, the cool polluted air gets trapped, sometimes for several days. For years the causes of Los Angeles's severe smog were in doubt. The city had passed strict laws banning outdoor burning, and industrial plants were required to install electrostatic precipitators to reduce particulate emissions; but air pollution remained at high levels. In 1951 California Institute of Technology scientist A. J. Haagen-Smit discovered the major causes of pollution: the automobile and oil refineries. The city authorities took firm steps to reduce harmful emissions from oil refineries, but the automobile and petroleum industries were too powerful. In spite of having the world's most rigorous air-pollution-control program, the Los Angeles area, with its 14 million people and 23 million motor vehicles, is still the most air-polluted city in the United States. It suffers from the effects of thermal inversions about half of the normal year, sometimes longer.

Progress has been made in reducing air pollution. Lead-free gasoline and catalytic converters in motor vehicles have greatly reduced harmful emissions in North America and Europe since the 1970s; but more cars are now on the road, and CO_2 emissions are at their highest peak ever. The Clean Air Act of 1990, which was fought by coal companies as being economically damaging, has proven to cost only about 10% of the estimate originally given by the industry when it opposed the new standards set by the Act. Sulfur dioxide emissions dropped 30% between 1970 and 1993 and are expected to fall further by 2000. But acid rain is still wreaking havoc with the environment in the Northeast. As I write this, Senator D. Patrick Moynihan of New York has introduced a bill (HR 2365) titled the *Acid Deposition Control Act,* which would require power plants to reduce SO_2 emissions a further 50% and nitrogen oxide (NO_x) emissions 70% over the levels prescribed by the 1990 Clean Air Act.

Perhaps the least acknowledged form of pollution is found indoors. EPA studies show that in the United States levels of 11 common pollutants are on average 2–5 times higher in homes and commercial buildings than outdoors. Sometimes they are 70 times higher. In automobiles—especially in traffic jams—levels may be up to 18 times higher than levels outdoors. Most people spend more than 75% of their time indoors, where they are safe from harmful ultraviolet rays and automobile exhaust, only to breathe stale, harmful air. The move to ban smoking from public buildings is improving matters, but for households where cigarettes, cigars, and pipes are smoked, people, especially children, are at risk. The EPA has correlated pollutants found in buildings to dizziness, headaches, coughing, sneezing, nausea, burning eyes, chronic fatigue, and flulike symptoms; the phenomenon has been designated *sick building syndrome.* According to the EPA 17% of the 4 million commercial buildings in the U.S. are "sick"—including

EPA headquarters. Indoor air pollution costs an estimated $100 billion per year in absenteeism, reduced productivity, and health costs.[1]

Besides smoke, other causes of indoor pollution are methylene chloride in paint strippers, which cause nerve disorders (including seizures) and diabetes; nitrogen oxides from unvented gas stoves and kerosene heaters, which irritate lungs and contribute to colds and headaches; carbon monoxide from faulty furnaces and unvented gas stoves; para-dichlorobenzene found in air fresheners and mothball crystals, which contribute to cancer; and styrene in carpets and plastics, which cause kidney and liver damage. Then there are asbestos and radon 222, which contribute to lung cancer. Formaldehyde in foam insulation, furniture upholstery, and particleboard, as well as in the adhesives in carpeting and wallpaper, is particularly damaging. Around 20 million Americans suffer from chronic breathing problems, dizzness, rashes, headaches, sore throat, sinus and eye irritation, and nausea that may be caused by low levels of formaldehyde, which the EPA has determined is carcinogenic.

Some have blamed capitalism for the terrible state we are in. In 1984 the City of Bhopal, India, was the site of a tragedy whose effects are still being played out. An explosion in a Union Carbide pesticide factory caused enormous amounts of industrial pollutants to be released into the air; some 3,000 people were killed, 20,000 were permanently disabled, and thousands more were left with severely damaged health. An editorial in the *Wall Street Journal* appeared to condone the tragedy, as the price we must pay for our free enterprise system: "It is worthwhile to remember that the Union Carbide insecticide plant and the people surrounding it were where they were for compelling reasons. India's agriculture has been thriving, bringing a better life to millions of rural people, and partly because of the use of modern agricultural technology that includes applications of insect killers."[2] George Bradford, editor of the anarchist paper *Fifth Estate,* lashed out at what he saw as bourgeois arrogance, which assumed that everyone needs technology for a "Better Way of Life." Bradford condemned the whole economic and social philosophy that permitted and continues to be responsible for this and many other threats to humanity. In the third world, businesses (sometimes owned by international companies based in industrialized countries) cut costs by allowing lax safety standards. Calling these large corporations "corporate vampires," Bradford accuses them of turning industrial civilization into "one vast, stinking extermination camp." Chemicals that are banned in the United States and Europe are now produced overseas, but even in the United States and Europe our industrial culture continues to endanger our lives. Modern lifestyles, dependent on the use of dangerous industrial processes, reek with harmful pollution. We must throw off "this Modern Way of Life," argues Bradford, for it only constitutes a "terrible burden" that threatens to crush us all.[3]

On the other hand, Julian Simon, whose ideas we met in Chapter 11, takes a position diametrically opposed to that of Bradford and similar to that of the editorial in the *Wall Street Journal*. Simon, the consummate cornucopian, argues that there is good evidence that pollution of air and water are actually decreasing rather than increasing. He argues against the doomsdayers who romanticize the glorious past, quoting a report on the pollution in London in 1890 that relates how horse-drawn carriages spattered pedestrians with mud and dung and fouled the air with a sickening odor. Citing life expectancy as the number one indicator of environmental quality, he produces statistics to support his thesis. Compared with the past, when most people died of infectious diseases like pneumonia, gastroenteritis, or tuberculosis, people today are dying of diseases related to old age, such as cancer and heart attacks. But, claims Simon, there is no evidence that the environment is any more carcinogenic than it was in the past. Our cities are, in fact, cleaner, and so is our water. All of our pollution problems have a technical solution. All it takes is money. "Purification requires the will to devote the necessary part of a nation's present output and energy to do the job. Many kinds of pollution have lessened in many places—for example, filth in the streets of the U.S., buffalo dung in the streams of the Midwest, organic impurities in our foods, soot in the air, and substances that killed fish in the rivers." Lake Erie, once said to be a dead lake, now is open to fishing. We have cause to be "cheerful." Of course, pollution is by definition bad and we need to keep it as low as possible, but we cannot get rid of it without getting rid of the beneficial uses of resources that it accompanies. But "exaggerated warnings can be counter-productive and dangerous."[4]

Simon is correct to remind us that we cannot live in a pollution-free world and that some pollution is the by-product of beneficial technology. Statistics vary from year to year on just how much air quality has improved. Most health authorities disagree with Simon on whether pollution causes cancer; and the CDC, which monitors changes in the rate of disease in the United States, recently reported that cases of asthma had doubled over the past several years. Urban air quality, especially ozone and CO_2, is still a serious problem. While legislation like the Clean Air Act of the 1970s led to lower levels of air pollution, since then the number of vehicles has grown from 120 million to over 195 million in 1996, offsetting some of the progress. The National Resources Defense Council estimates that every year about 64,000 people die prematurely from cardiopulmonary causes tied to particulate air pollution. In the most polluted cities, lives are shortened by an average of 1 to 2 years. Los Angeles tops the list of early deaths (5,873), followed by New York (4,024), Chicago (3,479), Philadelphia (2,599), and Detroit (2,123).[5] The point is to reduce pollution to much lower levels than are now measured. Simon is correct to point out that industry and technology have brought us many benefits in health, convenience, and life expectancy, but our problems remain immense. Bradford's indignation, however unwelcome, should give us pause. He argues that people in other countries deserve the same safety standards that we expect. We may add that since ultimately we are all affected by the quality of the atmosphere, long-term self-interest should persuade us to require that corporations maintain high standards everywhere. Contractualists and cornucopians will dispute Bradford's claim that all people have a natural desert to optimal environmental conditions, but they should be moved by the self-interest considerations.

What more can be done to lower pollution levels? The Clean Air Act of 1990 that required industry to take measures to limit SO_2, NO_x, and other pollution emissions, was a significant step against air pollution, but emissions must be cut further. Coal-burning plants should be phased out, as other forms of energy replace coal (see Chapter 15). Stricter emission standards for fine particulates are needed. The American Lung Association estimated in 1994 that 91 million Americans were at risk of suffering health problems because of dangerous levels of particulate-matter pollution—such as coal dust. Fuel-efficiency standards for cars and trucks must improve and be globally applied. We need to cut our dependency on oil; electric cars have been introduced on the market and should be promoted. Bicycle use should be enhanced and public transportation made economic and accessible to reduce the need for automobile use. Train transport should replace trucks wherever feasible. We have to strive to reduce air pollution worldwide, especially since the developing nations are using industrialized nations as a model of the good life. We have to learn to live more simply, showing that the good life is only possible in the long run with policies and practices that take the environment seriously. We will discuss the matter of sustainable living further in the following chapters.

THE GREENHOUSE EFFECT

On July 23, 1988, Dr. James Hansen, Director of the National Aeronautics and Space Administration (NASA) Goddard Institute for Space Studies, told the U.S. Senate Committee on Energy and Natural Resources, "It's time to stop waffling so much and say that this evidence is pretty strong that the greenhouse effect is here." Taking place in Washington, D.C., during the hottest summer on record, Hansen's testimony found a receptive audience. Newspapers issued headlines announcing the greenhouse effect and gave Hansen's testimony pride of place. Many scientists agreed with Hansen's assessment, though generally were reluctant to give it the depth of conviction that Hansen expressed. Other scientists demurred, arguing that the evidence for global warming was ambiguous.

What is the greenhouse effect? It is the theory first set forth by the Swedish chemist Svante Arrhenius in 1896 that the atmospheric gases that hover over the earth keep our planet warm in a manner analogous to the glass windowpanes of a greenhouse. In a greenhouse the sun's rays (energy in the form of light) are allowed in through the glass, but the heat generated is trapped by the glass. The same phenomenon occurs when you keep your car windows closed on a sunny day. The heat is trapped inside, so that it is warmer in the car than outside.

Likewise, the sun's energy reaches the Earth in the form of light, infrared radiation, and small amounts of ultraviolet radiation. Earth's surface absorbs much of this solar energy, some of it used by plants for photosynthesis, and transforms it to heat energy, which rises back into the *troposphere.* But water vapor (mostly in the form of clouds), CO_2, methane (CH_4), ozone (O_3), chlorofluorocarbons (CFCs), and nitrous oxides (NO_x) prevent some of the heat energy from escaping just as do the glass panes of the greenhouse. They absorb the heat and warm the Earth. Without this heat-trapping blanket, the Earth's surface would cool to about 0°F (–18°C) instead of maintaining an average temperature of 57°F (14°C). Most of our planet would be frozen like Mars.

The problem, then, is not the greenhouse process, but its *increased* activity. It's too much of a good thing. For the past 8,000 years the Earth's average temperature has never been warmer than 1°C (1.8°F) above the norm, and the last time it was 2°C (3.6°F) warmer was 125,000 years ago. Hansen presented evidence that the Earth has begun to get warmer; and by current trends the polar ice caps and glaciers will gradually melt, causing the ocean's level to rise several feet. With this melting, millions of people living on islands and along coastlines would be displaced as their land became flooded by the rising oceans. While people in northern Canada, northern Russia, and Greenland might rejoice over warmer weather, most of the temperate zones would become much warmer. Forest fires would be more frequent. Perhaps the widespread Florida forest fires in 1998 are a token of what is to come. Droughts could destroy many ecosystems, and the heat waves could make the long summers very uncomfortable. Air conditioner use would increase, demanding more energy and creating more pollution, which in turn would ironically create a greater greenhouse effect. Weather patterns would change, with droughts occurring in historically fertile areas and thunderstorms occurring in desert areas, negatively affecting agriculture and causing global starvation. Increased sea temperatures would give rise to coral bleaching, resulting in the destruction of coral reefs all around the world. Since Hansen's report, climate models project that the surface temperature of the Earth will rise by 2° to 6°F (the average being 3.5°F) with global sea levels rising by 6 to 37 inches between 1990 and 2100, the average being around 19 or 20 inches (48 centimeters), and continuing to rise thereafter. This would cause severe flooding along the coastal regions of the Earth, where about one-third of the world's people live. The agricultural lowlands and deltas in parts of Bangladesh, India, and China would be devastated. Many of the Caribbean islands would be deluged, the Netherlands covered with water, and the coastal areas of the United States, including Los Angeles, San Francisco, Long Island, the Carolina Isles, and the coasts of Georgia and Florida, would disappear within 25 or 50 years. Ocean salt water would contaminate coastal aquifers, and rising sea levels could flood sewage and sanitation systems, spreading infectious diseases. Such climate change would create favorable conditions for insects, with further negative effects on agriculture and human health, including the spread of tropical diseases such as malaria. Hurricanes, typhoons, tornadoes, and inland flooding would increase in power and frequency. Norman Myers has warned that by 2050 global warming could produce between 50 and 150 million refugees, causing social instability.

Ironically, the greenhouse effect could result in a devastating "mini" Ice Age in Europe. If the North Atlantic ice caps melt, they would send an incursion of fresh cold water into the "conveyor belt" of warm salt water that is responsible for Europe's temperate climate, thus wreaking havoc.

What is the evidence for the increased greenhouse effect? First, measurements since the 1950s show that the levels of greenhouse gases in the atmosphere are gradually increasing. Second, since 1860 the average global temperature has steadily risen about 1°F (0.6°C). Eleven of the years between 1980 and 1997 were the hottest in the 119-year records of temperature measurements. The 1990s have been the hottest decade since record keeping began, 1997 being the hottest year of all (Figure 13-1). But it may be surpassed by 1998. Government scientists reported on June 8 that each of the first five months of 1998 broke record global temperatures, 1.76°F above the 1960–1990 benchmark average of 61.7°F.[6] The increased frequency of El Niños in the 1990s is believed to be connected to this increase in temperature. Third, measurements of air trapped in ice-core samples show that the levels of greenhouse gases have been increasing since preindustrial times: between 1860 and 1997 the concentrations of CO_2 in the atmosphere grew from 280 ppm (parts per million) to 363.6 ppm, higher than at any time in the past 150,000 years. Concentrations of CO_2 are now about 30% higher than they were before the industrial revolution, and the amount is projected to reach 560 ppm over the next 60 years. Fourth, the Larsen B ice shelf, stable for many centuries, is beginning to fall apart and disperse into the Antarctic Peninsula. In 1995, 1,100 square miles of shelf abruptly went to pieces. In February 1998, another 75 square miles started melting and crumbling into the sea. The area around the peninsula has warmed about 4.5°F since the 1940s. Fifth, a new study in *Nature* (April 23, 1998), by a team of climate scientists led by Michael E. Mann, showed evidence from tree rings,

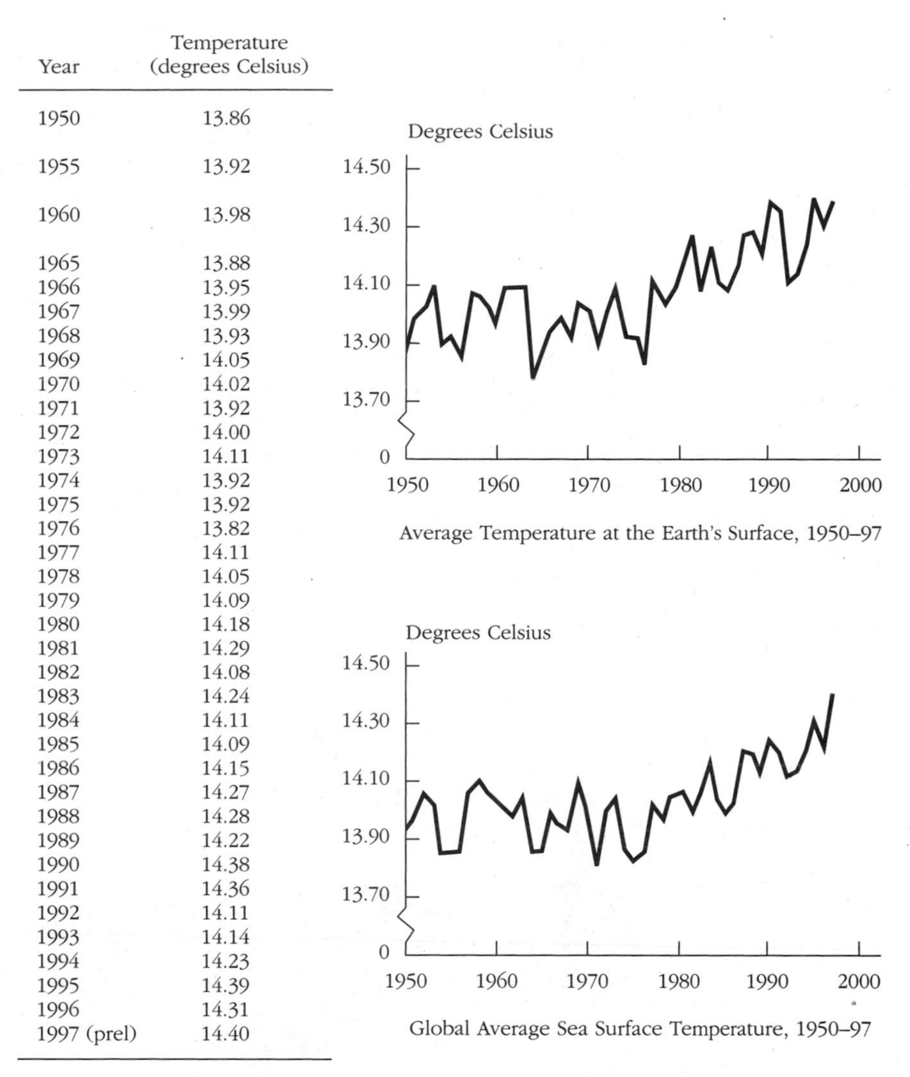

Global Average Temperature 1950–97

Year	Temperature (degrees Celsius)
1950	13.86
1955	13.92
1960	13.98
1965	13.88
1966	13.95
1967	13.99
1968	13.93
1969	14.05
1970	14.02
1971	13.92
1972	14.00
1973	14.11
1974	13.92
1975	13.92
1976	13.82
1977	14.11
1978	14.05
1979	14.09
1980	14.18
1981	14.29
1982	14.08
1983	14.24
1984	14.11
1985	14.09
1986	14.15
1987	14.27
1988	14.28
1989	14.22
1990	14.38
1991	14.36
1992	14.11
1993	14.14
1994	14.23
1995	14.39
1996	14.31
1997 (prel)	14.40

Figure 13-1 Global temperature changes.

ice, and lake sediment layers going back nearly 600 years that gives the strongest support to date of a rapidly rising climate. "This anomalous warming starts in the nineteenth century and accelerates right through the twentieth century. It is not related to earlier natural variations in the record but seems attributable to greenhouse gases," reports Mann.

While natural processes, especially volcanic eruptions like that of Mount Pinatubo in the Philippines in 1991, account for some of the increased greenhouse gases in the atmosphere, a great deal of it is caused by

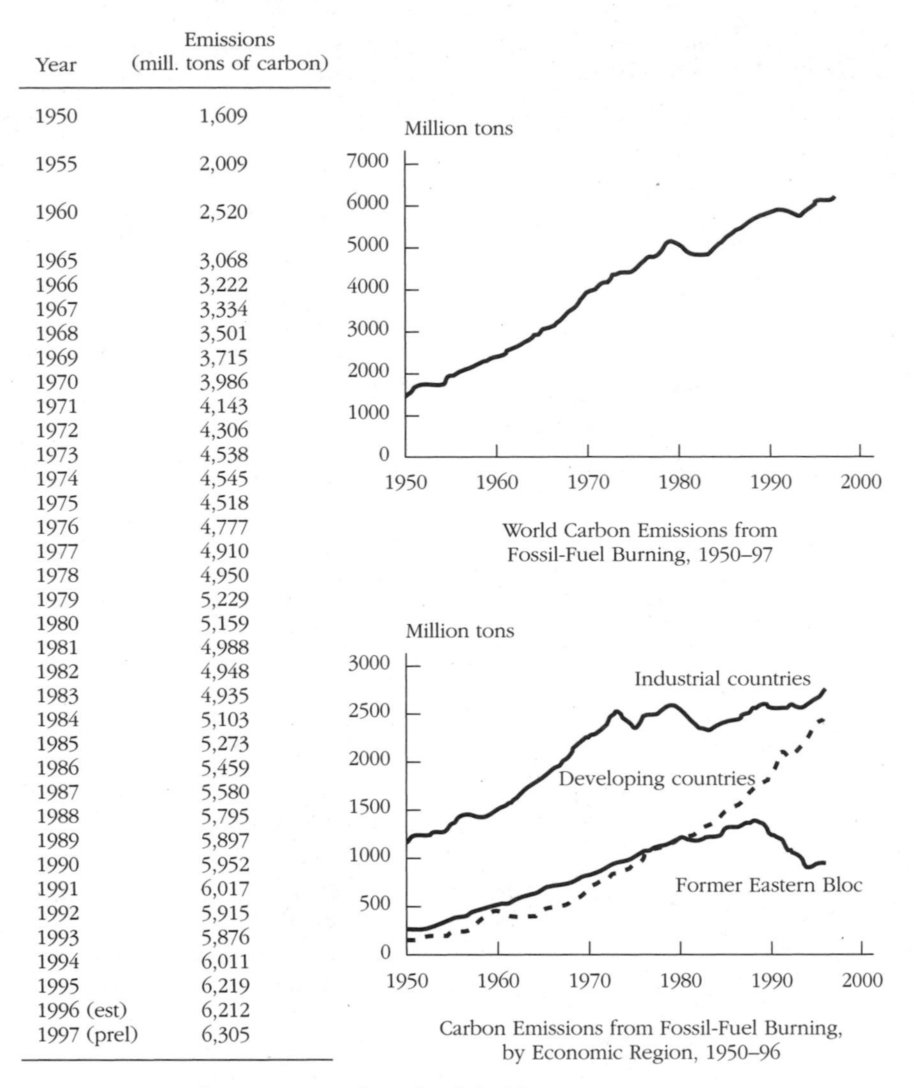

World Carbon Emissions from Fossil-Fuel Burning, 1950–97

Year	Emissions (mill. tons of carbon)
1950	1,609
1955	2,009
1960	2,520
1965	3,068
1966	3,222
1967	3,334
1968	3,501
1969	3,715
1970	3,986
1971	4,143
1972	4,306
1973	4,538
1974	4,545
1975	4,518
1976	4,777
1977	4,910
1978	4,950
1979	5,229
1980	5,159
1981	4,988
1982	4,948
1983	4,935
1984	5,103
1985	5,273
1986	5,459
1987	5,580
1988	5,795
1989	5,897
1990	5,952
1991	6,017
1992	5,915
1993	5,876
1994	6,011
1995	6,219
1996 (est)	6,212
1997 (prel)	6,305

Figure 13-2 Carbon emissions from fossil-fuel burning.

human activities. In 1997 annual global emissions of carbon from the burning of fossil fuels rose 107 million tons to a new high of 6.3 billion tons (Figure 13-2). The richer industrialized countries produce about 70% of the CO_2 emissions—mostly from burning fossil fuels. The United States emits about 23%, China 12%, Russia 9.5%, and Japan 5%. Carbon output in the United States expanded 8.8% between 1990 and 1996, with a 3.5% increase in 1996. Japan's carbon output increased 12.5% over the same six years.

Eastern Europe and the former Soviet Union account for 21% of the emissions since 1950.[7]

While scientists debate whether the evidence points to a significant trend in global warming, a growing consensus recommends that we act prudently, lowering greenhouse gases (CO_2, methane, ozone, and others) wherever possible. The debate is over the seriousness of the threat. Stephen Schneider, head of the Interdisciplinary Climate Systems at the National Center for Atmospheric Research in Boulder, Colorado, and Christopher Flavin and Molly O'Meara of World Watch Institute argue that the situation is grave and we must act now to reduce levels of greenhouse gases.[8] On the other hand, Dixy Lee Ray and Louis Guzzo, in their book *Trashing the Planet,* contend that there is very little evidence for a devastating greenhouse effect and that imponderables such as the El Niño current, volcanic activity, and/or solar patterns are much more significant than human pollution.[9] They agree that there are reasons to support significant lowerings of CO_2 and other pollutants by weaning ourselves from fossil fuels, but they feel we shouldn't get hysterical about the sky falling or the sea rising without good evidence. Scientists ought to exercise judicious restraint in coming to their conclusions. Ray and Guzzo are also skeptical about the depletion of the ozone layer. They argue that the evidence does not support the doomsday conclusions.

Assuming that we accept the evidence of the majority of climatologists, what can and should be done? First of all, we must reduce our use of fossil fuels and turn to alternatives already available, like solar power, public transportation, and electric cars. In December 1997, in Kyoto, Japan, more than 150 nations agreed to a protocol under which industrialized nations would reduce 1990 levels of carbon dioxide and other heat-trapping gases by 5% by about 2010. Second, we must stop the burning of the rain forests and at the same time plant more trees (see Chapter 16). Trees breathe CO_2, removing it from the atmosphere.[10] Planting more trees in our cities, towns, and countryside and throughout the globe would help offset the deleterious effects of CO_2 on the environment. Some scientists have called for a conversion to nuclear power because it produces only a small fraction of CO_2 per unit of electricity compared to coal. Others are wary of the hazards of nuclear power production. We will consider nuclear power in Chapter 15.

DEPLETION OF THE OZONE LAYER

Ozone (O_3) consists of three oxygen atoms bonded together. Most of the ozone in the Earth's atmosphere is concentrated in the ozone layer, located in the *stratosphere* about 15 to 25 miles above the Earth's surface. Ozone is also found near the ground and is a major component of smog. In the atmosphere we breathe it is a pollutant, but in the stratosphere it is a blessing. If all the ozone in the stratosphere were packed together to the density of air at sea level, it would make a layer about ⅛ inch thick. The importance of ozone is twofold: It traps reflected solar heat, contributing to the greenhouse effect, and it protects us from the sun's harmful ultraviolet light or radiation. Ultraviolet (UV) radiation is dangerous to the health, causing skin cancer, especially deadly melanoma, and cataracts. It also damages land plants and the sea's phytoplankton. Since phytoplankton are at the bottom of the ocean's food chain, their depletion could have a hazardous effect on many species of fish that depend on it for food. Phytoplankton also absorb large amounts of carbon dioxide, so their depletion could exacerbate the greenhouse effect. Ultraviolet rays may cause animals to go blind and affect their immune systems. It can kill the eggs of frogs, toads, and other amphibians and fish. There is evidence that the ozone layer is being depleted. Satellites monitoring ozone levels over Antarctica have reported an ozone hole that forms every spring over the South Pole, measured as large as 1.2 million square miles in size. It is getting larger every year. Satellite reports also show that the ozone layer over North America is thinning.

What is destroying ozone? Chlorine released by volcanoes may be one natural cause. The primary chemicals are chlorofluorocarbons (CFCs), which consist of chlorine and fluorine atoms bonded to a carbon atom. CFCs do not appear naturally in the atmosphere. They are entirely anthropogenic. They were produced for use as coolants in refrigerators and air conditioners and as propellants in aerosol spray cans; they have been used in the production of polystyrene foam. By 1993, over 20 million tons of CFCs had escaped into the Earth's atmosphere. CFCs drift upward into the ozone layer, where they meet ozone molecules, which they transform or catalyze into oxygen molecules:

$$\text{CFC} + O_3 \rightarrow \text{CFC} + O_2 + O$$

Thus the ozone that protects the Earth is reduced.

According to the U.N. Environment Program, additional ultraviolet radiation reaching the Earth is already responsible for the increased rates of melanoma deaths (from 2,340 in 1962 to 6,570 in 1992) and could cause as many as 300,000 cancer deaths worldwide each year. The EPA estimates that ozone thinning will account for 12 million new cases of skin cancer in the 1980s and 1990s. The United States is the largest producer of CFCs, and uses 30% of the world's CFCs. Car air conditioners account for 20% of all CFC use in the United States. One atom of chlorine, from one molecule of CFC, can destroy 100,000 molecules of ozone.

Halon, used in fire extinguishers, moth crystals, and solvents, and methyl bromide, used in pesticides, are even more lethal to the ozone layer and account for about 25% of ozone depletion.

What can be done? We should stop producing ozone-depleting chemicals immediately, since it will take at least a couple of decades for the ozone layer to heal itself. Substitutes for CFCs (containing less chlorine—the source of danger), such as *hydrofluorocarbons* (HFCs), have been found and should be utilized as soon as possible. We should not be making halon fire extinguishers. The United States and Scandinavian countries have banned the use of aerosol spray cans that use CFCs, and they and other industrial nations, having signed agreements in 1987 (Montreal) and 1992 (Copenhagen), are switching to substitutes for CFCs. A $240 million fund was established to assist developing countries to phase out their CFC use. However, current trends show that the United States is continuing to increase CFC output and is likely greatly to exceed its agreed-upon target. So far, populous countries like China and India have refused to sign the ozone agreements.

Meanwhile, it's wise to avoid sun exposure, especially between 10 A.M. and 3 P.M. When you do go out into the sun, wear sunscreen with a sun protection factor of 20 or more for UV-A and UV-B, sunglasses, and hats. And, most of all, work for a worldwide ban on CFCs and other ozone-depleting chemicals.

We have seen the far-reaching implications of global air pollution and discussed how the problem is related to some of the other concerns described in this book. Our quest for the good life has been interpreted as demanding high resource consumption in a manner that creates enormous air pollution, including an enhanced greenhouse effect and depletion of the ozone layer—which in turn have a negative effect on the good life. What should be done about this problem? Again, you will have to weigh the evidence and decide which arguments and statistics contain the most merit. Then go on to decide what you can do to decrease air pollution.

Review Questions

1. Is the indictment of big business by radical ecologists (such as George Bradford) justified? What are the implications of Bradford's indictment? What sort of world do you think that he would want us to live in? Is Bradford a "Luddite"? (Luddites were people in England in the early nineteenth century who went around destroying machines because they believed that the industrial revolution was evil.) How might someone in the business community respond to Bradford?
2. How serious do you think the problem of air pollution is? What can be done about it?
3. How serious do you think the problem of the enhanced greenhouse effect is? What should be done about it?

4. Automobile exhaust is a major cause of carbon dioxide, which enhances the greenhouse effect. Should you drive a car? If so, under what conditions? We will discuss automobile transport in Chapter 15.
5. How serious do you think the problem of ozone-layer depletion is? What should be done about it?
6. Is there a link between population growth and the kinds of pollution we have been discussing in this chapter? Explain what the possible relation is.

Suggested Readings

Barth, M., and J. Titus, eds. *Greenhouse Effect and Sea Level Rise: A Challenge for This Generation*. New York: Van Nostrand Reinhold, 1984.

Easterbook, Gregg. *A Moment on the Earth*. New York: Viking, 1995.

Ehrlich, Paul, and Anne Ehrlich. *Betrayal of Science and Reason*. Washington, D.C.: Island Press, 1996.

Flavin, Christopher. "Slowing Global Warming," in *State of the World 1990*. Washington, D.C.: Worldwatch Institute, 1990.

McKibben, Bill. *The End of Nature*. New York: Random House, 1989.

Ray, Dixy Lee, and Louis Guzzo. *Trashing the Planet*. Washington, D.C.: Regnery Gateway, 1990.

Schneider, Stephen. *Global Warming*. San Francisco: Sierra Club, 1989.

Simon, Julian. *The Ultimate Resource 2*. Princeton: Princeton University Press, 1996. Especially chapters 15–17.

Notes

1. I am indebted to G. Tyler Miller, *Living in the Environment,* 11th ed. (Belmont, Calif.: Wadsworth, 1998), pp. 474–75, for much of the data in the last two paragraphs.
2. *Wall Street Journal,* 13 December 1984.
3. George Bradford, "We All Live in Bhopal," in *Fifth Estate* 19:4 (Winter 1985), p. 319. Reprinted in *Environmental Ethics,* ed. L. Pojman (Belmont, Calif.: Wadsworth, 1998).
4. Julian Simon, *The Ultimate Resource 2* (Princeton: Princeton University Press, 1996), especially chaps. 15–17. The quotations are from the selection reprinted in *Environmental Ethics,* ed. L. Pojman, pp. 390, 392. Gregg Easterbook, in *A Moment on the Earth* (New York: Viking, 1995), says similar things. See especially chap. 12.
5. *Breath-Taking: Premature Mortality Due to Particulate Air Pollution in 239 American Cities,* a May 1996 report by the Natural Resources Defense Council. Available at http://www.nrdc.org/find/aibresum.html
6. William M. Stevens, "Global Temperature at a High for the First 5 Months of 1998," *New York Times*, 8 June 1998, pp. 1, 16.
7. Lester Brown, Michael Renner, and Christopher Flavin, eds., *Vital Signs 1998* (New York: Norton, 1998), p. 66.
8. Stephen Schneider, *Global Warming* (San Francisco: Sierra Club, 1989); Christopher Flavin, "The Heat Is On: The Greenhouse Effect," in *The Worldwatch Reader*, ed. Lester Brown (New York: Norton, 1991), reprinted in *Environmental Ethics,* ed. L. Pojman; Molly O'Meara, "The Risks of Disrupting Climate," *Worldwatch* 10:6 (December 1997).
9. Dixy Lee Ray and Louis Guzzo, *Trashing the Planet* (Washington, D.C.: Regnery Gateway, 1990).

10. A reviewer noted that oxygen procurement is important for plants as well as animals. "It is a common misconception that oxygen procurement is a problem faced only by animals, and that gas exchange in green plants consists exclusively of intake of carbon dioxide and release of oxygen. Certainly, this is the exchange that takes place in association with photosynthesis; but the carbohydrate products of photosynthesis are of little value to the plant if they cannot be respired to provide usable metabolic energy. *Hence plants, like animals, are constantly taking in oxygen and releasing carbon dioxide as they carry out the process of cellular respiration.* When a green plant is exposed to bright light, both photosynthesis and respiratory gas exchange are usually taking place; since the rate of photosynthesis then greatly exceeds the rate of respiration, the *net* effect is one of uptake of carbon dioxide and release of oxygen. The reverse is true, of course, when the green plant is in the dark or when it has no leaves in winter. Respiratory gas exchange, then, is as necessary for plants as it is for animals." [my ital.] (W. T. Keeton and J. L. Gould, *Biological Science,* 4th ed. New York: Norton, 1986, pp. 272f.)

CHAPTER FOURTEEN

Water Pollution, Pesticides, and Hazardous Waste

All is water.

—THALES

In 1993, 400,000 Milwaukee residents became sick with fever and severe diarrhea, and more than 100 died. The cause was traced to *Cryptosporidium,* a protozoan parasite about one one-hundredth the size of a speck of dust. The Centers for Disease Control and Prevention (CDC) in Atlanta report that this disease, mostly found in developing countries, is the major cause of traveler's diarrhea. But 2% of the diarrhea cases in U.S. hospitals are caused by this parasite. "The bad news is there's no treatment for it, but the good news is it runs its course in about 14 days in healthy people," said Dr. Dennis Juranek of the CDC. The CDC report states, "*Cryptosporidium* lives in the intestines of animals and humans and is spread by contact with anything from diapers to water tainted by farm runoff. Filters are the only protection from the parasite, which cannot be killed by chlorine or other chemicals." But home filters aren't fine enough to screen out the pest. Boiling water is recommended.[1]

In this chapter we continue our discussion of pollution. First, we discuss water as a resource and as sources of water pollution. Second, we concentrate on pesticide use, which may be beneficial to agriculture but poses a threat to human and animal life. What can we say about the moral status of our present use of pesticides? Third, we consider the problem of toxic waste and what is being done to clean it up. Finally, we look at the ethical implications of pollution.

WATER

When the well's dry, we know the worth of water.

—BENJAMIN FRANKLIN

When Benjamin Franklin uttered those words two and a half centuries ago, American wells were overflowing with fresh, clean water. They no longer are. Much of our water is neither fresh, clean, nor overflowing. Our cisterns are leaky, our water polluted, and, what's more, we're using it up at an alarming rate. If we don't appreciate the worth of water now, as our world's and nation's wells continue to dry up or issue polluted, disease-ridden content, we soon will.

The first Western philosopher, the engineer and astronomer Thales (625–545 B.C.), living on the Greek isle of Miletus, asked what was the nature of ultimate reality. His answer—*water.* He noticed that water is necessary for the production and sustenance of life, that it is everywhere—it certainly must have seemed so to a Greek islander. Look past the coastline and you'll find a sea of water; dig under the ground and you're bound to find water. It rises as mist from the sea and falls down to Earth as rain. He noted that it exists as a liquid over a wide range of temperature change—from 0°C to 100°C—and that, as a liquid, it can carry away objects and dissolve many compounds. Water cleanses and comforts. Heat it and it becomes a gas like air; freeze it and it becomes solid, like a rock, expanding and bursting pipes, bowls, and even rocks. So Thales concluded that the Earth is just a hard flat cork—like ice—floating on a sea of water. He revered water and saw it as ultimate reality. Humans and animals are, in fact, largely constituted of water. Human bodies are 80% water. Water is our lifeblood. A normal human being can live for a month without food, but less than a week without water. It is also the lifeblood of nature.

Religions deem water a symbol of cleansing. Christians are baptized in water, symbolizing the washing away of their sins. Hindus, Muslims, and Orthodox Jews emphasize ritual cleansing before acts of worship and eating. The Kogi Indians of Colombia say three things are holy and needed at the beginning of life: mother, night, and water. These rituals point to the primordial power of water.

Earth has lots of water. The oceans alone make up 71% of the Earth's surface. The problem is that 97.5% is salt water, and only about 2.5% freshwater. And of that 2.5%, two-thirds is in the form of ice.

The United States uses 339 billion gallons of groundwater a day. Although four trillion gallons of water fall on us daily in the form of precipitation, much of that disappears in evaporation and runoff, and our rivers and aquifers are being dangerously dirtied and depleted. We in the United States use 1,300 gallons per person a day, three times as much as the average European and several times more than the average person in most developing countries.

Agriculture and industry demand lots of water. It takes about 100,000 gallons (380,000 liters) of water to make an automobile, 800 gallons (3,000 liters) to produce a pound (454 grams) of grain-fed beef in an animal factory, and 26 gallons (100 liters) to produce 2 pounds (1 kilogram) of paper. Worldwide, domestic purposes account for 8% of water used; industrial purposes account for 23%; and agriculture 69%. In the United States, agriculture accounts for 41%, industry 11%, and public use 10%. Power-plant cooling uses the remaining 38%.

Water in the form of rain is spread unevenly over the globe. The rain forests of Brazil and Central Africa, the American Northeast, and much of Canada get abundant supplies, while the American Southwest, Mexico, Central Asia, and the Near East get relatively little. Similarly, while the Eastern seaboard has an abundant supply of rivers and lakes, the Southwest, mostly desert, must harvest water from aquifers and transfer water, especially from the Colorado River, via tunnels, aqueducts, and underground pipes. Nevada, Arizona, and southern California are deserts that would be virtually uninhabitable were it not for huge water transfers. Las Vegas, Nevada, couldn't exist without the help of the Hoover Dam. Yet, as the population of that region of the country continues to grow, the water supply is increasingly under stress.

Sometimes we get too much water too fast, as when flash floods crash down upon a population, wreaking havoc—as occurred in Bangladesh in 1988, when disastrous floods covered two-thirds of the small nation (the size of Wisconsin but with a population of 120 million, almost half that of the United States). Two million homes were leveled, 30 million people were homeless, and 2,000 drowned.

A water crisis faces the politically volatile Near East. The Arab nations and Israel have growing populations, but very limited water supplies. Israel diverts much of the water from the Sea of Galilee and the Jordan River for its own use, while surrounding countries, Syria and Jordan, complain that they are not getting their share. King Hussein of Jordan has stated that the only thing he'd go to war for is water.

About 150 of the world's 214 major river systems are shared by two or more countries, which can produce disputes on water rights. One such dispute is that among Ethiopia, Sudan, and Egypt over access to water from the Nile River. Ethiopia controls the headwaters that contribute 86% of the Nile's water, thus controlling its downstream flow into Sudan and Egypt. Famine-ridden Ethiopia and Sudan want to use a greater share for their own purposes. If they divert large quantities, Egypt, which is mostly desert, could become desperately in need of water. The population of the region is growing at a high rate. Egypt's population is expected to increase from the 1995 figure of 60 million to about 100 million around 2026—that's 1 million additional people every 9 months. Similarly, Turkey controls the headwaters of the Tigris and Euphrates Rivers, which flow into Syria and Iraq. Turkey is in the process of diverting more of these waters for its own growing population. Where will the growing populations of Syria and Iraq get their water?

A series of dams and reservoirs attempt to trap water and make it available for drinking and recreation in the Southwest, from Texas to Colorado, westward to California. Two problems with reservoirs are that they often take up precious bottom land, the most fertile soil land in the area, and that they expose a large surface area to evaporation—especially in the hot summer months characteristic of the Southwest. On the whole, it would be better if we left the water in the ground and used wells to tap into it at a slower pace. As it is, we are depleting our aquifers, especially the gigantic Ogallala Aquifer, the largest known aquifer in the world, which spans several Western states from Wyoming to Texas and New Mexico. We are pumping out the Ogallala Aquifer at eight to ten times its replacement rate. Throughout the United States groundwater is being withdrawn at four times the replacement rate.

Several techniques are being introduced to slow down the groundwater depletion rate, including new forms of irrigation, but much more needs to be done. Many states and municipalities do not even charge for water use. The desert city of Las Vegas gives out this precious commodity gratis. In the desert cities of the Southwest, including Los Angeles and San Diego, affluent people plant water-guzzling grass rather than utilizing native flora, and they water their lawns and golf courses, questionable uses of this scarce resource. Desalination, turning salt water into freshwater, is being practiced in the Near East, North Africa, and parts of Florida; but it is very costly, the product is usually brine-laden, and, at present, sufficient quantities cannot be produced to meet the increased demands.

In the United States tremendous water waste occurs in flushing toilets and watering lawns, as well as through leaky pipes. Some municipalities like Los Angeles are requiring toilets that use only 1.6 gallons a flush. Rebates are given for turning in water-wasteful toilets. Perhaps we should promote the Swedish Clivus Multrum, a waterless toilet that uses bacteria to break down human and food wastes; the result is a dry, odorless, solid fertilizer.

WATER POLLUTION

A World Bank study in 1995 estimated that polluted water causes about 80% of the diseases in developing countries and the deaths of about 10 million people annually. Typhus, cholera, bacterial and amoebic dysentery, enteritis, hepatitis, giardiasis, and schistosomiasis are among the leading diseases carried by contaminated water. Chemicals (such as lead) and pesticides (such as DDT, dieldrin, and PBC), which are serious health hazards, are also carried by water.

Water pollution can be traced to two types of sources, called *point sources* and *nonpoint sources*. Point-source pollution refers to waste dumped at a specific discharge point—for example, from factories or sewage plants. Nonpoint-source pollution cannot be traced to a single source. Examples include runoff of fertilizers, pesticides, and livestock manure. Factories often find it cheaper to pollute rivers, lakes, and the ocean, risking small fines, than paying for waste treatment or removal.

When fertilizers, containing such nutrients as nitrogen and phosphorus, get into rivers and lakes, a process known as *cultural eutrophication* takes place. **Eutrophication** happens naturally through erosion and runoff of nitrogen and phosphorus compounds into the aquatic system, but this natural process is exacerbated by **cultural eutrophication,** the artificial enrichment from human activities—synthetic fertilizers and manure from agriculture as well as the discharge of sewage. The excessive nutrients can cause gargantuan growth of algae. The algae die and sink to the bottom of the lake or stream. Oxygen-consuming **bacteria** decompose dead algae, depleting oxygen content from the water. Fish, which depend on oxygen, die, and bacteria decompose the dead fish, further depleting oxygen. If this process is not stopped, the natural chemical cycling system of the lake can be destroyed and the lake becomes devoid of life. About 33% of the medium-sized lakes and about 85% of the large lakes near large population centers in the United States suffer from cultural eutrophication.

Of special concern is ocean and sea pollution. Over half the people in the world live within 60 miles (100 kilometers) of the coast, while nine of the ten largest cities in the world are coastal cities. One would think that the oceans could absorb massive amounts of pollutants, and so they have, but there may be a limit even here. Companies sometimes dump wastes, including sewage sludge, into the ocean because they see it as cheaper than land disposal. But the pollutants drift back toward the coast. The coasts of Bangladesh, India, Pakistan, Indonesia, the Philippines, and Malaysia are severely polluted. California is also experiencing degradation of its coastal habitats: San Francisco Bay, the largest estuary in the western part of the United States, has shrunk by 60% in the past century due to land reclamation and can no longer support commercial fishing. Off the Palos Verdes peninsula near Los Angeles, a city sewage plant contributed to the progressive elimination of a 7.8-square-kilometer kelp (seaweed) forest as it increased its discharge twofold between 1928 and 1966. Sludge laced with toxins and heavy metals covered 95% of the peninsula floor.

Some years ago my family was vacationing on the Mediterranean coast off Spain. While swimming we all noticed a brown substance sticking to our arms and legs. It turned out to be half-dissolved feces! We then discovered that a sewage system from the city was dumping its effluence 1,000 feet from the public beach. The next day we were all sick. Other vacationers have related similar incidents occurring along different parts of the Mediterranean coast. About 85% of the sewage from large cities along the Mediterranean Sea is discharged into the sea—untreated. These areas have a coastal summertime population of 200 million.

Michael Specter reported in the *New York Times* (May 3, 1992) that nearly half of Europe's harbor seals were destroyed by viral infections in the 1980s. Ulcerative mycosis has destroyed fleets of fish such as the

Atlantic menhaden. "The sea's most delicate and commanding coral reefs, from the Florida Keys to Australia's Great Barrier Reef, have begun to die; some slowly as logging silt chokes off sunlight; others suddenly as divers squirt in cyanide to stun fish for aquariums."

G. Tyler Miller points out that in the United States about 35% of all city sewage ends up in coastal waters. Most of our harbors are badly polluted from this sewage and from industrial waste. Scuba divers talk of swimming through clouds of partially dissolved feces and of bay and harbor bottoms covered with toxic sediments known as *black mayonnaise.* "They see lobsters and crabs covered with mysterious burn holes, and fish with cancerous sores and rotting fins."[2]

Miller also notes that California's famous Santa Monica Bay is the filming site for the popular TV show *Baywatch,* "which gives the illusion of a clean California beach lifestyle. What viewers don't know is that the bay is so polluted that the actors get extra pay each time they enter the water and are chemically cleaned afterwards."[3]

Our sewage treatment centers are overstressed. Perhaps new technology will help some, but as the population grows, technology may not be able to keep up with our needs for clean water.

Worldwide, drinking water is often unsafe to drink. Many rivers in eastern Europe, Latin America, Africa, and Asia are severely polluted, even though they continue to supply large populations with drinking water. Aquifers are increasingly contaminated by pesticides and fertilizer runoff. In China most of the large cities get their drinking water from polluted sources. In Russia half of all tap water is unfit to drink.

While the United States' Safe Drinking Water Act of 1986 (sometimes called the Clean Water Act) represents a major step forward in the fight for safe water, contaminated drinking water is still a serious problem. The Milwaukee experience was the most prominent example of the problem, but between 1993 and 1995 over 53 million Americans received tap water laced with high levels of lead, pesticides, and chlorine by-products. Recently, the *Los Angeles Times* reported on a study that stated that nationally, 1,172 water systems, serving 11.6 million people, reported contamination with fecal coliform, which causes gastrointestinal illnesses. "Lead, which causes permanent loss in mental capacity in children, appeared to be the second most common water contaminant nationally. Violations of federal standards for lead were reported in 2,551 systems nationwide, serving 10.3 million people."[4]

However, it is difficult to determine acceptable levels of chemicals in water. How extensively should we treat water? People sometimes forget that natural water has its own impurities, and that the treatment itself—say, with chlorine—can also be a health hazard. Jonathan Tolman has pointed out that the water standard of Milwaukee met Environmental Protection

Agency (EPA) standards in 1993, and that a particularly virulent strain of *Cryptosporidium* caused the outbreak there. (A meat-packing company had dumped contaminated animal waste into a public sewer.) Far higher amounts of that parasite have been discovered in the Ohio, Missouri, and Mississippi Rivers without outbreaks occurring. The lesson seems to be that, even with our best efforts to maintain clean water, we cannot, at least at this point in history, eliminate all risk in drinking water.[5]

In addition to buying filters, which sometimes help, but, as in the case of the *Cryptosporidium* parasite, may not, many people are opting for bottled water. But one must exercise caution here. More than one-fourth of the 7,000 domestic brands of bottled water sold by the $2.7-billion-a-year bottled water industry comes from the same sources that supply tap water, which is regulated much more strictly than bottled water. You should check with the International Bottled Water Association (IBWA), the testing agency for bottled water, to determine whether the brand you are drinking meets its requirements.

PESTICIDES: IS *SILENT SPRING* A REALITY?

In 1962 Rachel Carson published *Silent Spring,* in which she documented the effects of DDT (dichlorodiphenyltrichloroethane) and other pesticides on human health.[6] She begins her work by depicting the halcyon loveliness of a prosperous American town, which is suddenly charged by a strange blight.

> Some evil spell had settled on the community: mysterious maladies swept the flocks of chickens; the cattle and sheep sickened and died. Everywhere was a shadow of death. The farmers spoke of much illness among their families. There had been several sudden and unexpected deaths, not only among adults but even among children, who would be stricken suddenly while at play and die within a few hours. There was a strange stillness. The birds, for example—where had they gone? Many people spoke of them, puzzled and disturbed. The feeding stations in the backyards were deserted. The few birds seen anywhere were moribund; they trembled violently and could not fly. It was a spring without voices. On the mornings that had once throbbed with the dawn chorus of robins, catbirds, doves, jays, wrens, and scores of other bird voices there was no sound; only silence lay over the fields and woods and marshes.

What had happened? An onslaught of synthetic chemicals had bombarded the landscape and invaded the streams and rivers. Nature hadn't time to adjust. "The chemicals to which life is asked to make its adjustment are no longer merely the calcium and silica and copper and all the rest of the minerals washed out of the rocks and carried in rivers to the sea; they are the synthetic creations of man's inventive mind, brewed in his laboratories, and having no counterparts in nature. . . . The figure is staggering and its implications are not easily grasped—500 new chemicals to which the bodies of men and animals are required somehow to adapt each year, chemicals totally outside the limits of biologic experience." These "elixirs of death" are causing widespread cancer and genetic mutations, as well as wreaking havoc on birds, fish, and wildlife:

> These sprays, dusts, and aerosols have the power to kill every insect, the "good" and the "bad," to still the song of the birds and the leaping of fish in the streams, to coat the leaves with a deadly film, and to linger on in the soil—all this though the intended target may be only a few weeds or insects. Can anyone believe it is possible to lay down such a barrage of poisons on the surface of the earth without making it unfit for all life? They should not be called "insecticides," but "biocides."

Carson then explains that pesticides have the paradoxical consequence of producing a greater insect problem than the one they are combating. In a "triumphant vindication of Darwin's principle of survival of the fittest," new strains of pests evolve that are resistant to the pesticides; and so ever stronger doses of the poison must be applied, which in turn lead to ever more resistant strains. In a current example of this, new strains of malaria-carrying mosquitoes have developed in many parts of the world, including Central Africa; these strains resist the traditional insecticides and threaten the lives and health of many thousands and millions of people. Carson's book startled the nation and launched a series of blistering attacks on her work, but eventually led to a government ban on the use of DDT in 1972.

Carson's doomsday message has been challenged. Michael Fumento, in *Science Under Siege: Balancing Technology and the Environment* (1993), argues that, while there are dangers in pesticide use, we've gone too far in the other direction.[7] He argues that most of the studies showing chemicals like Alar or dioxin to be cancer-causing were conducted under questionable circumstances. For example, the EPA has banned Alar, a pesticide sprayed on apples to enhance their quality and life span. The decision was an economic tragedy for the apple growers of the state of Washington; furthermore, Fumento argues, the process that led to the banning illustrates a dangerous antiscientific, antirational trend in which scare tactics are substituted for reason and evidence. He argues that well-intentioned environmentalists often overreact and so do more harm than good. These experiments involved giving mice daily doses of Alar that were 266,000 times higher than the amount that was ingested every day by the children said to be at risk. Fumento's thesis: We should not condemn a proven pesticide on insufficient evidence, for by so doing we can unjustly harm farmers and unwisely harm our own interests in pest-free food.

The case for the wise use of pesticides has also been set forth by Dixy Lee Ray and Louis Guzzo in their book *Trashing the Planet*.[8] Accusing Carson of "lyrical hysteria," they seek to show that the overall effects of pesticides like DDT have been beneficial. DDT was a welcome substitute for such toxic insecticides as arsenic, mercury, and lead. The result of using DDT to kill body lice in World War II was that "no Allied soldier was stricken with typhus fever (carried by lice) for the first time in the history of warfare. In World War I, by contrast, more soldiers died from typhus than from bullets." DDT was credited with bringing the number of cases of malaria in Sri Lanka down from 2.8 million in 1948 to 17 in 1963. When DDT was outlawed, the number of cases started to climb again. Ray and Guzzo admit that one can abuse a good thing, but argue that DDT and other pesticides have been given an undeserved condemnation:

> The attack on DDT rested on three main allegations: that DDT caused the death of many birds and could lead to the extinction of some bird populations; that DDT was so stable that it could never be eliminated from the environment; and that DDT might cause cancer in humans. None of these charges has ever been substantiated.

Their argument against these charges rests on the inability to trace a clear causal path from the use of DDT to the death of birds or cancer in humans. Environmentalists argued that DDT caused young birds to die before hatching—because their eggshells were too thin to protect them. Ray and Guzzo argue that the thin eggshell phenomenon predates DDT and can be caused by diets low in calcium or vitamin D, fright, and diseases. "Experiments designed to show a toxic effect from eating DDT failed, even though the experimenters fed their birds (pheasants and quail) from 6,000 to 20,000 times more DDT than the 0.3 parts per million residue of DDT found in food. Quail fed 200 parts per million in all their food throughout the reproductive period nevertheless hatched 80% of their chicks, compared with an 84% hatch in the control group. No shell-thinning was reported." Similarly, Ray and Guzzo argue that DDT breaks down harmlessly in the natural environment. One study fed volunteer humans up to 35 milligrams of DDT per day for periods of 21 to 27 months, with no ill effects then or in the nearly 30 years since. Most of the DDT is excreted, with some small residue, up to 12 parts per million, stored in human fat. No harm has ever been detected from these trivial amounts." Similarly, they argue that 93% of all DDT is broken down in water within 38 days. According to their findings, DDT is not carcinogenic. Ray and Guzzo conclude that there is no evidence that the use of PCBs, dieldrin, dioxin, Alar, or DDT causes cancer. Without these insecticides and herbicides, we would not have our present abundance of fresh fruits and vegetables. Pesticides increase food supplies and lower food costs.

Ray and Guzzo don't discuss crop-duster spraying, but they probably would caution against it. Under current practices, low-flying airplanes spray about 25% of all pesticides used on crops in the United States, but only about 1% of these pesticides actually reaches the target. The rest is carried away by air or gets into the water, food, nontarget organisms, and soil.

Most environmentalists dispute Ray and Guzzo's findings that DDT does not cause the thinning of eggshells in birds. They cite evidence that high DDT levels in birds can cause a calcium deficiency in their eggshells, so that the shells are much thinner than normal. These shells break easily and so fail to protect the chicks inside the egg. The near disappearance of the peregrine falcon was attributed to DDT. In 1975 only about 120 of these birds were left in the United States. After the banning of DDT, the number grew to 1,750 in 1994. Such pollution was responsible for the drastic decline of the California brown pelican, osprey, prairie falcon, sparrow hawk, and bald eagles in the 1960s, though all have been restored after the banning of DDT.

Ray and Guzzo's denial that pesticides such as DDT are health hazards and, particularly, carcinogens, has been challenged by many scientists. The World Health Organization and the U.N. Environmental Program estimate that each year about 25 million agricultural workers in developing countries are seriously poisoned by pesticides, resulting in about 220,000 deaths. Farm workers in these countries typically apply the pesticide by hand and the regulations are lax. In the United States it is estimated that about 300,000 farm workers become ill from pesticide use each year. A 1987 EPA report concluded that pesticides were the third most serious environmental health threat in the United States, and a 1993 study of Missouri children revealed a statistically significant correlation between childhood brain cancer and use of various pesticides in the home, including flea-and-tick collars, No-Pest strips, and chemicals used to control roaches, ants, mosquitoes, and termites. A 1995 study found that children whose yards were treated with pesticides were four times more likely to suffer from cancers of muscle and connective tissue than were children whose yards weren't treated. But because there is a lag time—sometimes as many as 15 or 20 years—between the cause of cancer and its manifestation, it is difficult to prove conclusively that a given agent is the identifiable cause of any particular instance of cancer.[9]

A major danger of pesticides is **biological magnification,** the increased concentration of pollutants in living organisms. Many synthetic chemicals, such as DDT, dieldrin, and PCBs, are soluble in fat, but insoluble in water, and are slowly degradable by natural processes. That is, when an organism takes in DDT or PCB, the chemical becomes concentrated in the fatty tissue of the animal and is degraded very slowly. When a large fish consumes plankton or small fish containing these pesticides, the chemicals become more concentrated still (Figure 14-1). Seemingly very minute levels of these chemicals in an organism can cause serious harm or death.

On November 18, 1963, Robert LaFleur of Louisiana's Division of Water Pollution Control made a phone call to the Public Health Service (PHS) in

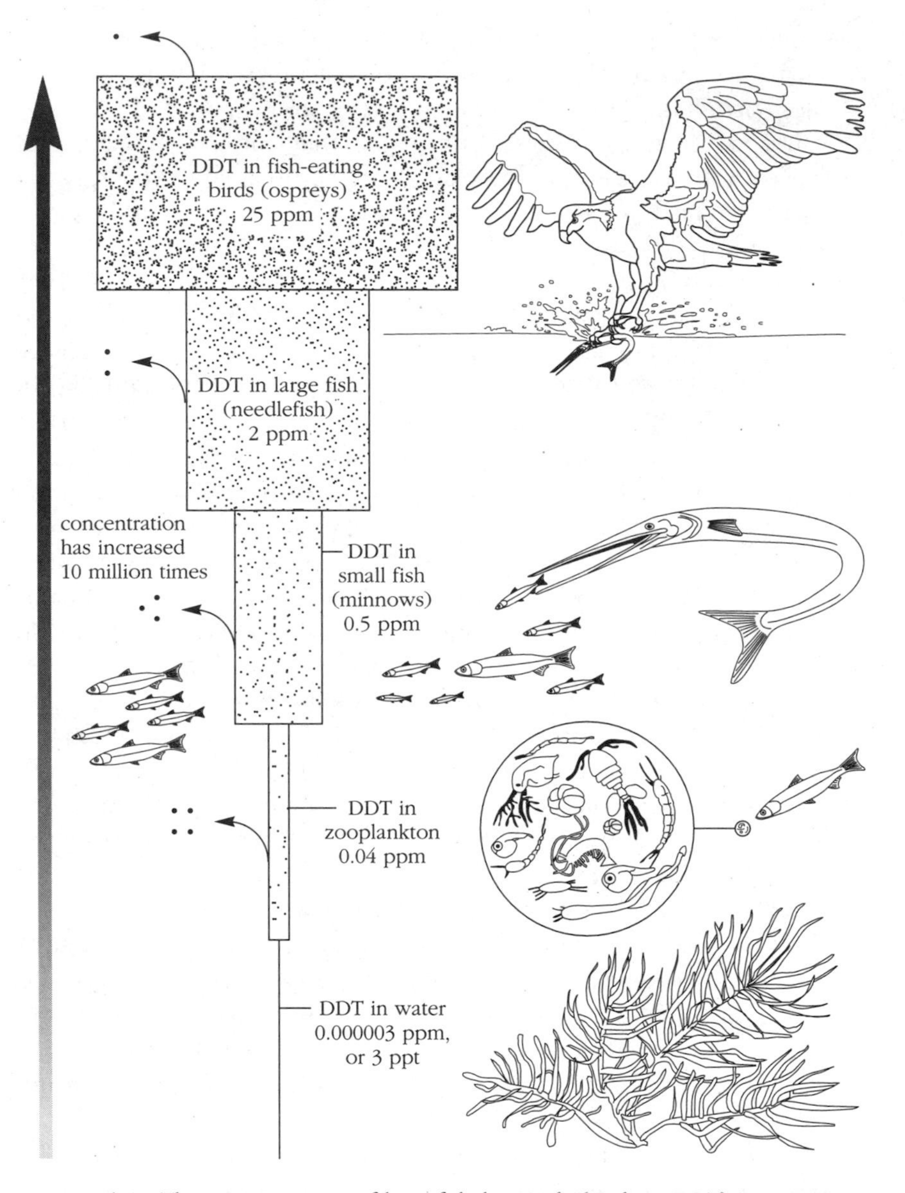

Figure 14-1 The concentration of harmful chemicals (such as DDT) in organisms may be magnified as much as 10 million times in a food chain. Dots represent DDT, and arrows show small losses of DDT through respiration and excretion. Adapted from G. Tyler Miller, *Living in the Environment,* 3rd ed. (Belmont, Calif.: Wadsworth).

Washington. The call was noteworthy for two reasons: (1) It is unusual for a state agency to seek federal help in dealing with ecological matters. States do not normally welcome federal interference in their affairs. (2) The purpose of the call was to report the largest fish kill ever witnessed in the continental United States. The state of Louisiana had no other option but to call the federal government. It was confronted with a widespread and dangerous problem that it could not tackle alone. An estimated 5 million fish were floating dead on the lower Mississippi, a river that provides drinking water for more than 1 million people and supports a large portion of the U.S. fishing industry.

One such kill might have been shrugged off as a natural disaster, but fish kills were becoming an annual event in Louisiana along the Mississippi and Atchafalaya Rivers. The kills had begun in the 1950s; at least thirty large kills were reported during the summer of 1960, when dead fish were clogging the intake of the Franklin power plant, and fish were dying in the Bayou Teche, a stream used as a source of drinking water for the town of Franklin.

In November 1960 the kills reached epidemic proportions. Although a large number of carp, threadfin shad, and freshwater drum were found dead, catfish (a major food source for the poor of the region) composed 95% of the kill. State officials were puzzled by this disaster and duly investigated the matter. Temperature changes in the water, its alkalinity, and its dissolved oxygen content were analyzed and found normal. Only the fish were abnormal. Catfish were bleeding about the mouth and many were bleeding about the fins—a situation caused by distention of the swim bladder and digestive tract. The digestive tract was empty of food and contained only gas and some bilelike frothy material. Yet the river contained abundant food. What could be wrong? Dying fish were swimming at the surface, often inverted and very lethargic, and were easily caught by hand. The causes of this strange phenomenon remained a mystery for three more years.

Then in 1963 the Great Fish Kill occurred. In addition to large numbers of dead freshwater carp and catfish were the brackish water fish—mullet, sea trout, and marine catfish. It was at this point that LaFleur made his phone call to the federal government.

Scientists from the Division of Water Supply and Pollution Control came to Louisiana, where they found both the fishing industry and hunting activities closed down; ducks, which ate the fish, were dying too. The Mississippi River was a spectacle of floating corpses, a fish graveyard.

One of the scientists, Donald Mount, was struck by the similarities exhibited by all of the dying fish: convulsions, loss of equilibrium, some hemorrhaging, surface swimming, and empty digestive tracts. Weeks of examination passed with no clues to the causes. Finally, a breakthrough occurred when the river was dredged. Chloroform extract residues were found in the mud, suggesting the invasion of a foreign toxic substance. Scientists before Mount had hypothesized that the cause was a foreign substance, but neither they nor Mount knew what it was. Mount and his team took advantage of recently developed equipment that measured chemical residues in traces smaller than 1 part per billion (1 ppb). Utilizing this technology, Mount made his fish autopsies, discovering in the fishes' fatty tissue small amounts of dieldrin and DDT. They also found two unidentified substances, which they referred to as X and Y. After a long period of frustrating failure to identify the substances, they were both discovered to be varieties of one highly toxic chemical, a drug 15 to 30 times more poisonous than DDT—endrin.

Finding the source of this dread chemical would be no easy task. More than 100,000 industries resided along the Mississippi River, and more than 100 of these manufactured pesticides. Doing incredible detective work, Mount and his colleagues assembled clues that led 500 miles up the river to a site in Memphis, Tennessee. The source was a pesticide factory owned by the Velsicol Chemical Corporation of Chicago, which manufactured endrin. The company was uncooperative, denying any wrongdoing. Eventually, the principal investigator, Alfred Grzenda, made an independent search and wrote his findings: "All of the [dump] sites are subject to flooding. The plant manager denied that any solids from the endrin plant were being hauled to the Hollywood Dump, but we noted drums labeled 'isodrin scraps' at the dump on April 15. Isodrin is one of the compounds used in the manufacture of endrin." Samples of water taken from the sewers and waterways around the Velsicol plant revealed quantities of endrin exceeding all previous reports of concentrations of chlorinated hydrocarbon insecticides in water.

Secretary of the Interior Stewart Udall condemned this violation, and Senator Abraham Ribicoff of Connecticut called for hearings, but business interests attempted to play down the gravity of the tragedy. Senator Everett Dirksen of Illinois accused the PHS of making wild accusations against Velsicol and of "unjustly crucifying" this good company. Velsicol was never brought to court, was never subpoenaed, nor were its plants investigated until 1965 when Velsicol changed management and took a more responsible policy under threat of court action.

Mount commented that it was understandable that the business community failed to realize the magnitude of the problem of endrin toxicity. "Perhaps it is difficult to understand that any substance in water in such minute concentrations as 0.1 ppb could be acutely toxic to fish. However, one must consider that in just two hours the blood of a catfish can attain an endrin concentration of 1,000 or more times greater than that of the water in which the fish swims; understanding then comes more readily."[10]

Fish kills occur regularly. In Michigan in 1968, 700,000 salmon died in hatcheries due to DDT contamination. In Canada a similar incident occurred in 1969. In Maine traces of DDT were found in dead lobsters as far as 100 miles off its coast.

Entomologist David Pimentel has assessed the progress and problems of pesticide use since *Silent Spring* was written.[11] On the one hand, much progress has been made, so that humans and wildlife are less directly affected by the poisons in pesticides. On the other hand, pesticide-resistant insects have replaced their less damaging ancestors. Furthermore, pesticides have destroyed some of the natural enemies of certain pests, so that more crops are now lost to insects than when *Silent Spring* was written. Because of better overall agricultural techniques and fertilizers, the total picture is positive, but there is a long way to go in reducing the use of pesticides. The fact that only about 1% of insecticide sprayed on a field hits the intended target is problematic. It is like using saturation bombing to wipe out mosquitoes. The use of natural predators, such as spiders, is being introduced in many places, including China, with good results. Spiders are "smart bombs" that hit specific targets and, for the most part, are harmless to humans.

HAZARDOUS WASTE

The magnificent Niagara Falls on the New York–Canadian border is one of the scenic wonders of our nation. Shockingly, the city of Niagara is one of the most polluted cities in our nation. The waterfalls provide inexpensive electricity, and inexpensive electricity invites industry. Along the shorelines of the Niagara River the spiraled pipes of distilleries issue forth effluents and fill the air with the odor of chlorine and sulfides.

Nearby is Love Canal, named after its builder, William Love. In the 1960s a young cement technician and his wife, Tim and Karen Schroeder, bought a home along this canal and began to raise a family. They built a fiberglass swimming pool enclosed by a redwood fence. One October morning in 1974 Karen noticed that the pool had risen two feet above ground. Although perplexed by this phenomenon, they decided to wait until the next summer to pull it up and replace it with a sturdier cement pool. But when they pulled up the pool that summer, they beheld a large gaping hole filled with a rancid yellow and blue liquid. The chemical liquid gradually rose until it covered the entire yard. It attacked the redwood posts with such a caustic bite that after some days the fence collapsed. When the waters finally receded later that summer, the gardens and shrubs were withered and scorched.

How did these liquid chemicals get into the Schroeders' backyard? Between 1947 and 1952 Hooker Chemical Company dumped 22,000 cubic tons of chemical wastes into an old canal and covered it with dirt. In 1953 Hooker sold the canal area to the school board for $1 with the stipulation that Hooker be absolved of all future responsibility for the conditions of the land. A school and hundreds of homes were built along the canal.

Already in 1959 people had begun noticing that a strange black sludge leached through their basement walls and gave off malodorous fumes. But Niagara reeks with such odors, so the inhabitants took them for granted.

On November 21, 1968, Karen's third child (Sheri) was born with a congenital heart problem, bone blockage of the nose, partial deafness, deformed ears, and a cleft palate. She had developmental delays and an enlarged liver. When her teeth came in, they appeared in double rows.

In 1976 the city of Niagara supported Hooker when the New York State Department of Environmental Concern discovered dangerous hydrocarbons (PCB) in basements. One part per million (1 ppm) is considered a serious environmental concern. Investigators found more than 1,000 ppm in these basements. Later when Karen's parents' basement was tested, the air pollution level was 1,000 times above the acceptable level.

During this period Tim had broken out in a rash, suffered fatigue, and found it difficult to stay awake during the day. The Schroeders' daughter Lauri's hair began falling out, and Karen was suffering throbbing pains in her head. They suspected that the surrounding chemical pollution was the culprit, but being people of limited means, they had nowhere to go. Later benzene was found in the soil in the area; it is a carcinogen that also causes headaches, fatigue, loss of weight, and dizziness.

Other families began to share their stories. It was discovered that the cancer rate around Love Canal was 35% above the national average and that 31% of the babies were born with birth defects, typically hearing loss and retardation. Between 30% and 40% of the pregnant women in any given year had miscarriages. A young boy died of unexplained kidney failure.

Led by a courageous young mother, Lois Gibbs, the community around the Hooker site organized and contacted legislators and public health authorities. As a result of their actions, in 1978 New York State Public Health Authorities launched a full investigation into the matter. Over 80 dangerous chemicals, including 14 chlorinated hydrocarbons, were discovered. Later another 400 chemicals were found. PCB levels were found at 100,000 ppm. Tetradioxin, used to make Agent Orange, whose safety level is 1 part per trillion, was identified, and Hooker finally admitted that it dumped 200 tons of it into the canal (this is 1,000 times the amount used to defoliate the Vietnamese jungles). New York State spent over $250 million cleaning up the site and another $67 million relocating more than 700 families from the area. It sued Hooker and Occidental Petroleum, which had bought up Hooker, for $2 billion. In 1985 Occidental Petroleum settled out of court and paid many of the families settlements of between $2,000 and $400,000. The state is still seeking $250 million in punitive damages.

A hazardous waste is any discarded chemical that can adversely affect one's health or the environment. The EPA characterizes hazardous wastes by four qualities:

1. *Ignitability* or *flammability* (waste oils, used organic solvents, and PCBs)
2. *Corrosiveness* to materials such as metals or human tissue (strong acids, strong bases)
3. *Reactivity*—being unstable enough to explode or release toxic fumes (cyanide solvents)
4. *Toxicity*—toxic if handled in ways that release the chemical into the environment (DDT, dioxins, PCBs, arsenic, mercury, and lead)

Radioactive wastes, mining wastes, and oil- and gas-drilling wastes are not on this list. Many environmentalists are attempting to get Congress to include these categories.

Over 275 million tons of hazardous waste are produced in the United States each year. The EPA lists 34,000 hazardous waste sites in the United States, with 1,200 sites on a National Priority List for cleanup. The General Accounting Office estimates that the number of sites is between 103,000 and 425,000. The largest number of priority sites are in New Jersey, Pennsylvania, California, Michigan, and New York. It costs about $26 million to clean up a site, so the total costs of cleanup could exceed $500 billion.

In 1976 Congress passed the Resource Conservation and Recovery Act (RCRA), which called on the EPA to create "cradle to grave" oversight of hazardous waste. In 1980 the Comprehensive Environmental Response, Compensation, and Liability Act, known as the Superfund program, was passed by Congress. This law provided a $1.6 billion fund to be financed by the federal and state governments and by taxes on chemical and petrochemical companies; the fund would be used to clean up hazardous-waste sites. In 1986 a new Superfund program, the Superfund Amendment and Reauthorization Act (SARA), was passed and provided $9 billion to the cleanup program.

THE ETHICAL IMPLICATIONS OF POLLUTION

Everyone is against pollution. By definition it is bad stuff, harmful. We don't want it in our backyards. Unfortunately, it is not possible to avoid pollution. To live is to pollute. Nature itself creates pollution. For example, volcanic eruptions are the leading cause of CO_2 emissions, which enhance the greenhouse effect. Should we concentrate on controlling volcanic eruptions? Those environmentalists who would prohibit interfering with nature would probably object to this; but a commonsense ethic would seem to approve intervention in this and other natural processes for the good of life on Earth.

Humans and animals, who must eat, excrete wastes back into the environment. Some of that waste contains competitive life, such as bacteria and parasites, who threaten our health and lives. Some of that waste will naturally get into the water supply and be imbibed by other humans and animals. To make a steel cabinet is to create steel filings; to fly an airplane or drive a car is to need fossil fuel; to protect one's crop is to kill pests—perhaps with a petrochemical pesticide. But it isn't as though humans create all the pollution. Impurities are already in nature. Background radiation is contained in rock formations; bacteria and unsafe minerals are in rivers and streams. It is impossible to live without risk of pollution. All we can do is decide which pollutants are more threatening to us than others and concentrate on reducing their threat. William Baxter puts the point this way:

> Low levels of pollution contribute to human satisfaction but so do food and shelter and education and music. To attain ever lower levels of pollution, we must pay the cost of having less of these other things. I contrast that view of the cost of pollution control with the more popular statement that pollution control will "cost" very large numbers of dollars. The popular statement is true in some sense, false in others; sorting out the true and false senses is of some importance. The first step in that sorting process is to achieve a clear understanding of the difference between dollars and resources. Resources are the wealth of our nation; dollars are merely claim checks upon those resources. Resources are of vital importance; dollars are comparatively trivial.[12]

The point is that we can't treat pollution in a vacuum. We are stuck with the difficult process of making trade-offs between goods and levels of pollution. In a democracy our only hope is to educate the populace sufficiently so that they value the right things. How much are we willing to pay for clean air and water? An electric car is now beginning to be marketed that produces far less pollution than the gas-driven automobile with its internal combustion engine. Unfortunately, it is not nearly as efficient and is more costly to purchase. Should we change over from gas-dependent vehicles to the more costly electric ones? Perhaps the issue is complicated by the fact that the car user doesn't pay directly for the full cost of using a car. If we had to pay for the effects of the pollution that cars and trucks cause, we would probably have to pay three or four times the present price of gasoline. Instead we pay with poorer health, higher medical insurance, and the like. Perhaps we also have to factor in the price of our military expedition to Kuwait during the 1991 Gulf War, which was primarily fought over oil.

We cannot go back to nature, to living in hunter-gatherer societies: few can even maintain themselves in simple agricultural societies. We have recreated ourselves as a *civilized* society; and we need clothes, electricity, refrigeration, sanitation systems, trains, insulated homes, books, computers, bicycles—and perhaps cars and airplanes. With such needs we cannot avoid using resources that produce pollution. We need food and need to protect our food from competing pests, so some form of pesticide—even natural predators like spiders—will be needed. Still, we can live more simply—I will argue in Chapter 18 that we can and should live in a manner that minimizes resource use, waste, and pollution and promotes sustainable ecological living. As our technology and population grow and our ability to pollute does too, these decisions will occupy us for a long time.

Review Questions

1. What is the value of water? Should we pay more for it? How serious is the problem of water pollution? How can we preserve clean water? How might you use less water?
2. Go over Rachel Carson's position on pesticides. Why should "insecticides" be called "biocides"? Is Carson persuasive? What are the main reasons for and against the use of pesticides?
3. Consider Fumento's and Rae and Guzzo's arguments for pesticides. How strong are their considerations?
4. Discuss the problem of toxic waste. How serious is this problem and what can be done about it?
5. Do you agree with Baxter that pollution is just the opposite side of the coin of resource use? Do you also agree that on the principle "waste is a bad thing," we cannot help but bring about evil?
6. Discuss ways of reducing pollution while maintaining a high quality of life. What should our priorities be? What are the implications for population growth? for technology? for the ways we lead our daily lives?

Suggested Readings

Baxter, William. *People or Penguins: The Case for Optimal Pollution*. New York: Columbia University Press, 1974.

Carson, Rachel. *Silent Spring*. Boston: Houghton Mifflin, 1962.

Fumento, Michael. *Science Under Siege: Balancing Technology and the Environment*. New York: Morrow, 1993.

Miller, G. Tyler. *Living in the Environment*, 10th ed. Belmont, Calif.: Wadsworth, 1998.

National Geographic. *Special Edition: Water.* November 1993.

Ray, Dixy Lee, and Louis Guzzo. *Trashing the Planet*. Washington, D.C.: Regnery Gateway, 1990.

Notes

1. Based on a report of the Centers for Disease Control and Prevention in Atlanta to the wire service, 10 April 1993.
2. See G. Tyler Miller, *Living in the Environment*, 10th ed. (Belmont, Calif.: Wadsworth, 1998), p. 525, for further discussion of the points made in the preceding paragraphs.
3. Ibid.
4. Melissa Healy, "Studies Point Up Contamination of Drinking Water," *Los Angeles Times*, 2 June 1995. Reprinted in *The Environment: Opposing Viewpoints*, ed. David Bender and Bruno Leone (San Diego, Calif.: Greenhaven Press, 1996).

5. Jonathan Tolman, "EPA Regulates Against Unreal Risks to Drinking Water," *Human Behavior,* 4 February 1994. Reprinted in *The Environment: Opposing Viewpoints,* ed. David Bender and Bruno Leone.
6. Rachel Carson, *Silent Spring* (Boston: Houghton Mifflin, 1962).
7. Michael Fumento, *Science Under Siege: Balancing Technology and the Environment* (New York: Morrow, 1993). Reprinted in *Environmental Ethics,* ed. L. Pojman (Belmont, Calif.: Wadsworth, 1998).
8. Dixy Lee Ray and Louis Guzzo, *Trashing the Planet* (Washington, D.C.: Regnery Gateway, 1990).
9. These studies are discussed in Miller, p. 625.
10. The Mississippi Fish Kill incident is set forth in Frank Graham, Jr., *Since Silent Spring* (Boston: Houghton Mifflin, 1970), pp. 94–108. My discussion is based on this report.
11. David Pimentel, "Is Silent Spring Behind Us?" in *Silent Spring Revisited*, ed. G. Marco, R. Hollingworth, and W. Durham (Washington, D.C.: American Chemical Society, 1987). Reprinted in *Environmental Ethics,* ed. L. Pojman.
12. William Baxter, *People or Penguins: The Case for Optimal Pollution* (New York: Columbia University Press, 1974).

CHAPTER FIFTEEN

Energy: The Ethics of Power

A city of one million people plunges into darkness as the cables supplying its electric lifeblood unexpectedly fail. Workers rush in emergency diesel generators to keep hospitals, traffic lights, and other vital services going. As officials struggle to restore power, high-rise residents fear using elevators during the sporadic periods when power is on, lest they suddenly become trapped. Food spoils. Water becomes a problem. Downtown businesses struggle to survive. "If you didn't laugh you'd cry," one resident tells a local reporter.[1]

This major shutdown of electrical power happened in Auckland, New Zealand, in February 1998, crippling that international city for weeks. To utility experts, the Auckland disaster illustrates how "the lights can suddenly go out anywhere in the world, causing human suffering and financial hardship." Businesses in the United States lose about $26 billion every year from power disturbances. As we become more dependent on electric energy (in the United States consumer use has increased 35% in the past 10 years, while transmission capacity is up only 18%), more and more outages are expected. Malfunction, overload, and ecoterrorism are likely to plague our system to greater degrees.

In this chapter we will examine questions connected with the use of energy. What sorts of energy sources are morally acceptable or optimal from an environmental perspective? Can we move toward more energy-efficient technologies? What should we be doing about nuclear energy? Do the benefits of nuclear power outweigh the dangers of nuclear accidents and contamination from nuclear wastes? First we discuss energy and its sources, followed by a more detailed examination of the potential benefits and problems associated with nuclear power. Finally, we will discuss the matter of energy as it relates to sustainable transportation.

ENERGY

Energy enables us to do things. When suitably organized, it is power. Energy cannot be created or destroyed. It can only be changed from one form to another. The first law of energy, or *first law of thermodynamics,* states that we cannot get energy for nothing, but that it takes energy to get energy. The second law of energy, or *second law of thermodynamics,* states that in any conversion of energy from one form to another, high-quality or useful energy is always degraded to lower-quality, less useful heat energy that cannot be recycled back to high-quality energy. That is, we can't break even with regard to quality of energy.

Earth's main source of energy is the sun, which provides 99% of our energy. Without the sun the Earth's average temperature would be −400°F (−240°C), too cold for life. Solar energy recycles chemicals and produces several forms of *renewable* energy, such as wind, water power (hydropower), and biomass. The other 1% of energy is called *commercial energy*. Most of it comes from burning mineral resources extracted from Earth's crust; these resources include coal, natural gas, oil, geothermal, and nuclear energy. The United States is the world's largest consumer of energy. With about 4.5% of the world's population, it uses 25% of its commercial energy. About 93% of this energy comes from nonrenewable sources: oil (39%), natural gas (24%), coal (22%), and nuclear (8%). About 7% comes from renewable sources, such as hydroelectric (3.4%), geothermal (3%), and solar and wind (0.6%).

Primitive societies use human and animal power and were and still are fueled by wood, charcoal, and dung. Industrial societies need abundantly more energy and more sophisticated energy-producing mechanisms. It takes a lot of energy to heat or air-condition a home, provide electricity for refrigerators, lighting, computers, television, cars, and airplanes. It takes a tremendous amount of electricity to run a hospital or factory. Some people say that they want to reject modern technology, but even anarchistic, anti-technological organizations like *Fifth Estate* in Detroit use electric lightbulbs and computers.

In evaluating energy sources we need to compare the relative costs of obtaining and producing various types of energy. (This does not include the costs of pollution caused by the various types.) It takes energy to produce energy. Oil and gas must be located, pumped from the ground, transported to refineries, processed, and then transported to users. The usable quantity of energy is called its *net energy*—the total useful energy available from the resource over its lifetime, minus the amount of energy used or wasted in finding, processing, and transporting it to users. For example, suppose it takes 90 units of energy to produce 100 units of available energy, resulting in a net of 10 units. The net energy ratio (NER) is 100/90, or 1.1. This represents a small net gain. When the ratio is less than one, we have a net loss.

Surface-mined coal has a high NER of 28.2 and underground-mined coal an NER of 25.8. Natural gas has an NER of 4.9 and oil 4.7. Passive solar for space heating in homes has an NER of 5.8. The NER of gasoline (refined crude oil) for transportation is 4.1. According to energy expert Amory Lovins, 84% of all commercial energy used in the United States is wasted, much of this unnecessarily. He puts the figure at $300 billion per year—an average of $570,000 per minute. Worldwide, the figure grows to over

$1 trillion. Lovins argues that the single most effective way to improve our economic, military, and environmental security is to reduce this wasted energy. How? Change our lifestyles: walk and bike for short trips, use mass transit, dress more warmly instead of putting up the thermostat, and use less energy-intensive lighting.[2] Our present use of high-quality electric energy for heating homes or household water is like using a chainsaw to cut butter or a sledgehammer to kill a fly.

The two least efficient energy mechanisms are the incandescent lightbulb (which wastes 95% of the energy input) and vehicles with internal combustion engines (wasting about 90% of the energy input). Why don't we improve our energy efficiency? The main reason is that there is an abundance of low-cost, underpriced fossil fuel. Since the pollution costs are not factored into the price of these fossil fuels, it looks like a good deal. See Chapter 17 for a discussion of this economic problem—that of factoring in pollution costs as part of the *full costs* of energy consumption. Currently, the United States has only about 2.3% of the oil reserves, but uses 30% of the oil extracted. We import 52% of the oil we use, and the percentage is expected to rise over the next 10 years. Most of the oil comes from the Mideast, where the political situation is volatile. In 1990, when Iraq invaded Kuwait, the United States and other Western nations fought the Gulf War largely to protect our oil supply. Christopher Flavin and Nicholas Lenssen of Worldwatch remarked, "Not only is the world addicted to cheap oil, but the largest liquor store is in a very dangerous neighborhood."[3] Oil prices are expected to rise sharply as reserves are depleted in the coming decades.

Transportation accounts for about 66% of all oil used in the United States. Americans have 35% of the world's cars and drive as many miles each year as the rest of the world combined. At any one time, 75% of vehicles on the road have only a single passenger. Better fuel efficiency and developing electric cars would reduce our consumption somewhat. But at present electric cars can run only about 100 miles before needing recharging—and are dependent on electricity-generating fossil-fuel-burning power plants. If technology can devise solar cells or wind turbines to recharge the batteries of electric cars, this would be a splendid ecological advancement. We should also be turning from trucks and planes to trains and barges to transport freight. In the United States, the Northeast and Midwest have a wonderful waterway in the Great Lakes and Hudson, Mississippi, Missouri, and St. Lawrence Rivers to transport goods. Trains are far more energy-efficient than trucks and cause a fraction of the pollution. We should also tax gasoline more, as Europeans do, to cover pollution costs. In the United States, 80% of the federal gasoline tax goes to highway construction and maintenance, with only 20% to mass transit. This distribution should be reversed. Trolley cars and light rail should be promoted. A light-rail system can carry about 400 people compared with the 40 passengers on a typical bus.

Modern mass transit is both cleaner and more efficient than motor vehicles with their internal combustion engines. Ironically enough, in the early part of this century all major U.S. cities had efficient electrical streetcar systems. Trolleys and streetcars were used by 20 million riders in 1920, and Los Angeles had the largest and one of the best systems in the country. In the 1940s a holding company called National City Lines, formed by General Motors, Firestone Tire, Standard Oil of California, and Phillips Petroleum, bought up the privately owned streetcar systems in 100 major cities. The old systems were systematically dismantled and replaced by buses and the promotion of cars. Gradually, the bus companies were allowed to fail in many of these cities, creating an increased demand for cars. The courts found the companies guilty of conspiracy to eliminate about 90 percent of the nation's light-rail systems and fined the corporate executive officers $1 each and each company $5,000. But by that time General Motors alone had made $25 million in additional bus and car sales.[4]

About one-third of the energy in industrial societies is used for heating, cooling, and lighting buildings, and most of it is wasted. The 110-story, twin-towered World Trade Center in Manhattan is an example of energy waste. It uses as much electricity as a city of 100,000 people for only about 53,000 employees. By contrast, Atlanta's 13-story Georgia Power Company building uses 60% less energy than conventional office buildings of the same size. G. Tyler Miller describes the building:

> The large surface of the building faces south to capture solar energy. Each floor extends out over the one below it, blocking out the higher summer sun to reduce air conditioning costs but allowing warming by the lower winter sun. Energy-efficient lights focus on desks rather than illuminating entire rooms. The Georgian Power model and other existing cost-effective commercial building technologies could reduce energy use by 75% in U.S. buildings, cut carbon dioxide emissions in half, and save more than $130 billion per year in energy bills by 2010.[5]

One practical energy improvement we can all make is to change from incandescent bulbs to fluorescent bulbs. Replacing 25 incandescent bulbs with fluorescent bulbs can save from $700 to $1,500 over their lifetime (which is much longer than that of incandescent bulbs).

Solar and wind energy are the hope of the future. Passive solar heating works by using the building itself to trap heat and letting radiation and convection currents distribute it. Heat is stored in internal masonry or dense material. The system is 90% energy-efficient (compared with oil-burning furnaces at 53%) and pollution-free. Solar collectors can efficiently be used for water heating. Active solar heating relies on water or air collectors to trap the heat. It is more complicated and thus more expensive than passive solar, as the collectors and heat transfer mediums need to be replaced frequently. Solar thermal power plants are presently being tested in California, Japan, and France; they could provide an inexpensive, clean source of energy in the future—replacing the pollution-heavy coal-burning power plants. Solar **photovoltaic cells,** which transfer solar energy through silicon panels into electrical energy, are expected to become economical in the twenty-first century. It is hoped that sometime in the next 50 years or so such solar cells could provide as much as 25% of the world's energy and 35% of energy in the United States. But we are a long way from that now.

Presently, the market costs of solar and wind energy are higher than those of traditional coal-burning energy processes, so that in the immediate future these sustainable energy forms will have a difficult time competing with traditional processes. But an experiment in Arizona suggests that Americans are willing to pay more for clean energy. A Phoenix-based group of people called Solar Partners recently agreed to pay $3 per month more for the energy from a 100-watt unit of a solar power plant. Even before the plant was opened on April 17, 1998, Solar Partners had signed up enough new members to announce that it will build two more solar power plants by the end of 1998. Members of Solar Partners said that they were paying extra costs for their children and grandchildren. Six marketers of green power recently announced that they have formed a trade association called Renewable Energy Alliance to "ensure that the restructuring of our nation's power markets ultimately leads to increased use of renewable energy sources and a cleaner environment." In late April the green power company AllEnergy announced that it will begin to produce electricity using methane gas from landfills as part of its green power.[6]

Improvement in the development and use of energy is a good example of how technology can be used for good. If we can develop affordable solar-paneled homes, factories, and office buildings, we could rid ourselves of much of the harmful pollution currently associated with coal-burning power plants. However, to reiterate, widespread solar energy is still decades, if not generations, away. In the meantime, we have to make difficult decisions on how to deal with the energy problem.[7]

What about nuclear energy? The United States presently gets about 23% of its electricity from nuclear power plants. Other countries, such as France, get far more. Should we support greater reliance on this form of energy? This is probably the most controversial and most difficult energy issue.

SHOULD WE SUPPORT NUCLEAR ENERGY?

On April 26, 1986, one of the nuclear reactors at the Chernobyl nuclear power plant north of Kiev in the former Soviet Union experienced a steam explosion. Two blasts blew the 1,000-ton roof off the reactor building, set the graphite core on fire, and flung 7,000 kilograms of radioactive debris into the atmosphere. Extensive fallout occurred in an area roughly 40 miles by 40 miles northwest of the reactor and in an area of similar size to the northeast. Winds carried these contaminated materials over parts of the Soviet Union and Europe, as far as 1,250 miles from the plant. The accident occurred when the engineers were performing an experiment to ascertain how long a generator could continue to function after it had been disconnected from the main power supply during a power outage. A series of human errors occurred in the process and combined to cause the Chernobyl disaster.

Fifty people died from the radiation poisoning, most of them firefighters and workers who tried to put out the more than 30 fires caused by the explosions. However, Greenpeace Ukraine has estimated that by 1995 as many as 32,000 deaths had been indirectly caused by the disaster. In a 20-mile radius 125,000 people were evacuated. An estimated 24,000 evacuees received serious doses of radiation. Although the Vienna report estimates that about 200 cases of hereditary disorders will occur, no birth defects were reported among the 1,950 children born to pregnant women evacuated from the Chernobyl area. More than 7,000 cases of thyroid cancer have been reported in the area exposed, but no cases of other types of cancer, including leukemia, have been reported. Estimates vary widely (ranging from 1,000 to 10,000) on how many cancer deaths will eventually result from the accident. No one really knows.[8] The total economic cost of the accident has been estimated at over $100 billion. The famous Ukrainian gymnast Olga Korbut, who was in a nearby city when the explosion occurred, gives this poignant report:

> I was in Minsk when Chernobyl happened, and they didn't tell us for three or four days. We were all outdoors, because it was close to the May 1 celebration, and we were planting gardens and enjoying the spring. . . . It has been five years, but people are still very frightened and very angry. Our food and water supply is contaminated from radiation.
>
> When I went into the schools in Byelorussia, I learned that the first-graders have never been in the forest, because the trees were so contaminated. . . . When children want to see what nature used to be like, they go into a little courtyard inside the building, and the teacher says, "This is a bird and this is a tree," and they are plastic. Isn't that sad?[9]

American proponents of nuclear power quickly pointed out that such an explosion would not have occurred in U.S.-built reactors, which have strong containment domes (Chernobyl's had no containment). But the public at large reacted in shock to Chernobyl. Opinion polls taken all over the United States and Europe right after the explosion at Chernobyl showed an overwhelming rejection of nuclear power as a solution to our energy crisis. In the minds of many public policy makers, nuclear power was dead.

But many scientists argue that, tragic as Chernobyl was, as an industrial accident it was relatively minor. The U.S. National Academy of Sciences estimates that nuclear plants cause 6,000 premature deaths annually compared to between 65,000 and 200,000 premature deaths caused by coal-burning power plants.[10] Proponents of nuclear power ask us to compare the harm caused by nuclear power with the more than 500 people killed shortly before Chernobyl when a Japanese airplane crashed or the 3,700 who were killed as a result of the chemical explosion at Bhopal. If we look at the situation rationally, they argue, we will realize that we must disengage ourselves from the harmful effects of fossil fuels, from the harmful ash, sulfur, and nitrogen oxides as well as from the carbon dioxide of coal- and oil-burning plants, which are responsible for acid rain, greenhouse gases, cancer, and emphysema. We must instead turn to nuclear power, they feel. They point out that a normally operating nuclear power plant emits no air pollution and typically less radiation than a coal-burning power plant. Nuclear power, producing only 2% of the carbon dioxide that coal-burning power plants produce and no methane, offers safe, clean, efficient, affordable energy.

These nuclear-energy advocates contend that it is mainly ignorance that promotes the fear of what they consider a safe form of energy. An understanding of the facts should dissipate most of our worries. What are the facts?

By *radiation* we typically mean the high-energy type of electromagnetic radiation, including X rays and *gamma rays*. They are more energetic than other forms of energy, such as light and radio waves, in that they can pass through human tissue, causing serious damage. Another form of energetic radiation consists of particles such as *alpha particles* and *beta particles*, which are ejected from the nuclei of radioactive atoms during radioactive decay. While these particles are not normally dangerous (they can be stopped by a sheet of paper or cardboard), they can harm humans and animals if ingested or inhaled in sufficient quantity—the dose of radiation is measured in terms of *rems*. The time it takes for a radioactive atom to decay to half its power is called its *half-life*. Half-lives vary from a few milliseconds to billions of years.

Radiation, called *natural background radiation,* or NBR, is part of our natural environment. Its sources are cosmic rays from outer space; radon, a gas produced by radium in rocks; and even water and food. NBR varies depending on the soil type and altitude. We are all exposed to low doses of radiation, which appear to be relatively innocuous. The average NBR is about 3/10 rem per year, so in a 70-year lifetime, a person will have received about 21 rems. Cancer has not been observed in humans for doses under 20 rems. It is doses above 50 rems, received acutely or chronically, that are harmful. NBR does not seem to be harmful. The dose from NBR in

Colorado is three times that in Louisiana, mostly due to radon in granite rocks, but the overall cancer death rate in Colorado is actually less than that in Louisiana. The National Research Council has stated that "No increase in the frequency of cancer has been documented in populations residing in areas of high natural background radiation."[11]

John Jagger, a former Oak Ridge National Laboratory radiation biologist, argues that nuclear energy is safe and that we need it. Citing an array of data, he argues that it is both cleaner and safer than coal or oil. He agrees that we should use solar and wind energy wherever possible, but these will never be sufficient to supply all our energy needs. He details safe ways of disposing of nuclear waste. Jagger argues that in spite of the loud protests of the antinuclear faction, no one has ever been killed in the United States from a nuclear accident, and that Chernobyl, while a tragic accident, was a rare combination of mistakes in which everything went wrong that could go wrong. Judging by other accidents, such as airplane accidents, it was a minor industrial accident. Nuclear power advocates say that the opposition to nuclear power is based mostly on ignorance and fear of the unknown.[12] The media have exacerbated the problem, tending toward hysteria. Witness the headlines of the Philadelphia *Evening Bulletin* in 1979 after the nuclear plant meltdown of 40% of the reactor core at Three Mile Island: "[Radiation] Is Spilling All Over the U.S." and "Nuclear Grave Is Haunting Kentucky." It turned out that the only important radiation released was a very small amount of radioiodine with a half-life of eight days.

Nuclear Waste

To set forth the argument against nuclear power, environmental philosophers Richard and Val Routley use the metaphor of a train crowded with passengers and freight, which makes a stop—where someone consigns a package of highly toxic and explosive gas.[13] This gas is packed in a thin container that may well break, releasing the gas, if an accident should happen, if the train should derail or have a collision, or if some passenger should interfere with the freight, trying to steal some of it. All of these things have occurred on previous journeys. "If the container should break, the resulting disaster would probably kill at least some of the people on the train in adjacent carriages, while others could be maimed or poisoned or sooner or later incur serious diseases." We would no doubt condemn the consigner for this dangerous act. What could the consigner possibly say to justify his act?

> He might say that it is *not certain* that the gas will escape, or that the world needs his product and it is his duty to supply it, or that in any case he is not responsible for the train or the people on it. These sorts of excuses, however, would normally be seen as ludicrous when set in this context. Unfortunately,

similar excuses are often not so seen when the consigner, again a (responsible) businessman, puts his workers' health or other people's welfare at risk. . . .

The matter of nuclear waste has many moral features which resemble the train case. How fitting the analogy is will become apparent as the argument progresses. There is no known proven safe way to package the highly toxic wastes generated by the nuclear plants that will be spread around the world as large-scale nuclear development goes ahead. The waste problem will be much more serious than that generated by the 50 or so reactors in use at present, with each one of the 2,000 or so reactors envisaged by the end of the century producing, on average, annual wastes containing 1,000 times the radioactivity of the Hiroshima bomb. Much of this waste is extremely toxic. For example, a millionth of a gram of plutonium is enough to induce lung cancer. A leak of even a part of the waste material could involve much loss of life, widespread disease and genetic damage, and contamination of immense areas of land. Wastes will include the reactors themselves, which will have to be abandoned after their expected lifetimes of perhaps 40 years, and which, some have estimated, may require 1½ million years to reach safe levels of radioactivity. . . .

The risks imposed on the future by proceeding with nuclear development are, then, significant. Perhaps 40,000 generations of future people could be forced to bear significant risks resulting from the provision of the (extravagant) energy use of only a small proportion of the people of 10 generations.[14]

Jagger and Alvin Weinberg reject arguments like the Routleys' as bordering on hysteria. They claim there is no insurmountable technical problem in disposing of nuclear waste. First of all, the Routleys and other antinuclear advocates neglect the fact that depleted fuel rods can be recycled, as Britain, France, India, and Japan do. Reprocessing waste would lower the half-life of the uranium and plutonium, ameliorate the problem of storage, and lower the cost of nuclear energy (probably to where it is less expensive than coal-burning technologies). Second, the transportation problem is not as serious as the Routleys make out. The containers of nuclear material are designed to withstand being crashed into solid walls at 80 mph, hit by railroad locomotives at similar speeds, engulfed in gasoline fires for 30 minutes, and sunk in water for 8 hours. To date the safety record on transportation of radioactive material is 100%. Third, the opponents of nuclear power confuse various types of radiation. Some types of nuclear waste, such as uranium-238, which has a half-life of 4.5 billion years, have such low radioactivity that a person can handle them for half an hour without receiving a significant dose of radiation. High-level radiation waste, which is produced mainly by nuclear power plants, consists mainly of depleted fuel rods, which contain strontium-90 and cesium-137, with half-lives of 30 years or less, so they are not a long-term problem. Jagger compares the problem to burning wood. "If you burn wood slowly over a long period of time, it will release heat at a low rate; but if you burn it quickly, the heat will be much more intense. The same applies to radioactive decay. For example, a miner can hold a chunk of uranium ore in his hand without being dangerously exposed to radiation. Thus, much of the popular concern about radioisotopes of very long half-life is a misplaced concern."[15] The main problem is with radioisotopes with intermediate half-lives, such as plutonium-239 (24,000 years) and americium-243 (8,000 years). Although this is dangerous material, it can be safely stored by being buried in very deep rock formations, such as the Canadian Shield, centered in Ontario Province and extending into Minnesota and Wisconsin, or in the Yucca Mountain, about 90 miles northwest of Las Vegas, Nevada, where the terrain is thought to be stable.[16] (See Box 15-1.) If, for some unanticipated reason, the area became unstable, the waste could be safely removed. The total amount of nuclear waste produced in the United States to date is less than would fill up a small (three-story) building.[17] Even this amount could be greatly reduced through reprocessing.

Recently Charles Hollister and Steven Nadis have argued that radioactive waste could be safely buried under the seabed. They maintain that "Many studies by marine scientists have identified broad zones in the Atlantic and Pacific that have remained geologically inert for tens of millions of years" and whose clay-rich muds would entomb radioactive materials.[18]

Finally, nuclear-energy proponents rebut the charge that the proliferation of nuclear reactors could enable more countries to build nuclear weapons. One sometimes hears the argument that we ought to suspend nuclear power because the processed uranium could be stolen and used to build nuclear bombs. The proponents of nuclear energy respond that altogether different technologies are required for nuclear reactors than for nuclear weapons. For example, the uranium-235 used in nuclear reactors is purified to about 3% of the total uranium present, whereas a bomb requires purification to about 90%.

But antinuclear environmentalists disagree with this assessment. In an article, "Nuclear Waste: The Problem That Won't Go Away," Worldwatch ecologist Nicholas Lenssen argues that our ignorance of the dangers of low-level nuclear radiation, our knowledge of the lethal effects of moderate doses, and our inability to guarantee that disposed nuclear waste will not harm future generations present a strong presumption against using nuclear power. We have a difficult enough problem in trying safely to dispose of the waste already produced without creating more.[19] Opponents also point to the former Soviet Union's dumping of large quantities of radioactive material in the environment. Mayak, a plutonium-production facility in southern Russia, had a nuclear-waste storage tank explode in 1957, contaminating 1,000 square miles. It is estimated that, altogether, Mayak has spewed 2½ times the amount of radiation as Chernobyl. Around the shores of Novaya Zemlya, an island off the coast of Russia, 18 defunct nuclear-powered submarines were dumped, along with 17,000 containers of radioactive waste. In

BOX 15-1

Nuclear Waste Disposal

As of 1995 there were 430 commercial nuclear reactors in 32 countries, producing 6% of the world's commercial energy and 17% of its electricity. In the United States there were 109 commercial nuclear power plants, generating about 23% of its electricity. No new power plants have been ordered since 1973, and the Department of Energy expects a decline in capacity of nuclear-power generation. Presently the issue of where to store nuclear wastes is preoccupying the industry and lawmakers. On May 12, 1998, after 24 years of preparation and litigation, the world's first deep, underground nuclear-waste storage site won a federal license to begin burying plutonium, beginning in June 1998. The Waste Isolation Pilot Plant, carved out of salt beds 2,150 feet below the desert near Carlsbad, New Mexico, would serve as the disposal site for hundreds of thousands of barrels of radioactive wastes from 55 years of buildup. The repository is to keep the plutonium safely sealed away from the environment for 10,000 years and beyond. The site is a salt formation that the Energy Department says has been stable for 250 million years. The Carlsbad plant cost $2 billion to build. It is the smaller of the Energy Department's two waste projects; the other is at Yucca Mountain in Nevada, where far more radioactive material is to be buried. As I write this book, opponents argue that the Carlsbad site is not safe because extensive oil and gas reserves in the area attract drilling. The site may accidentally be drilled; 10,000 years from now, its purpose may have been forgotten.

1993 Greenpeace caught Russians dumping large amounts of radioactive waste from nuclear submarines into the Sea of Japan.

Environmentalist Christopher Flavin, also of Worldwatch Institute, argues that, while we must move away from the use of fossil fuels, nuclear power is not the answer. Accidents like Chernobyl are too dangerous, the cost of constructing safe and efficient nuclear reactors too high, and the political opposition to nuclear power too great for us to consider reviving nuclear power in the foreseeable future. Even if there are no technological problems with designing safe reactors and disposing of nuclear waste, we have the human problem—and one horrendous example in Chernobyl—of how human fallibility upsets our idealized projections. Murphy's Law warns us, "If something can go wrong, it probably will." We should turn our attention to renewable energy forms, such as solar, wind, and biomass. Here is how Flavin sums up his position:

> The Faustian bargain of nuclear energy has been lost. It is high time to leave the path pursued in the use of nuclear energy in the past, to develop new alternative and clean sources of energy supply and, during the transition period, devote all efforts to ensure maximum safety. This is the price to pay to enable life to continue on this planet.[20]

A "Faustian bargain" refers to the legendary intellectual Faust, who sold his soul to the devil for magical knowledge and power. Is this an accurate metaphor for our attraction to nuclear energy?

Should we avail ourselves of what ideally could be inexpensive, clean, abundant energy? Should we make a Faustian bargain, betting on the ability of government and technology to secure the safety of the nuclear-power industry, or should we reject the offer of what could turn out to be a nuclear nightmare? Would it be better if we simplified our lifestyles, reducing our dependence on expensive fossil fuels and questionable energy production? To that question we turn.

SUSTAINABLE TRANSPORTATION: PEDAL POWER

Imagine that we invented a mighty Convenience Machine that would make our lives wonderfully more enjoyable and enable us to reach more of our goals. Unfortunately, using the machine would cost about 43,000 lives per year, more casualties than eight years of the war in Vietnam. Would you use the machine? Should we allow it to be sold on the market?

When I have posed this question to audiences, there is virtually universal agreement that we should not, for no amount of comfort equals the value of a single life.

The question then becomes, Why don't those who treat human life as sacred stop driving cars? The Convenience Machine in our thought experiment is the motor vehicle. Each year about 43,000 people lose their lives in automobile accidents in the United States, and another 30,000 lose their lives to diseases brought on by fuel emissions. Worldwide, about 250,000 people are killed in motor vehicle accidents each year, more than those killed by the atom bombs dropped on Hiroshima and Nagasaki in 1945. Since the introduction of the automobile in the United States 100 years ago, about 3 million people have been killed, more than in all the battlefields in all our wars.

The automobile provides us with enormous freedom and power. It makes us mobile and able to cover long distances in relatively small amounts of time. In moderation, the automobile is or could be a good thing, but we have become overdependent upon it and have misused it. We have turned what could be a wonderful servant into a tyrannical idol—the religion of car worship with its new urban cathedral, the multistoried park-

Table 15-1 Health Effects of Pollutants from Automobiles

Pollutant	*Health Effect*
Carbon monoxide	Interferes with blood's ability to absorb oxygen; impairs perception and thinking; slows reflexes; causes drowsiness and so can cause unconsciousness and death; if inhaled by pregnant women, may threaten growth and mental development of fetus.
Lead	Affects circulatory, reproductive, nervous, and kidney systems; suspected of causing hyperactivity and lowered learning ability in children; hazardous even after exposure ends.
Nitrogen oxides	Can increase susceptibility to viral infections such as influenza. Can also irritate the lungs and cause bronchitis and pneumonia.
Ozone	Irritates mucous membranes of respiratory system; causes coughing, choking, and impaired lung function; reduces resistance to colds and pneumonia; can aggravate chronic heart disease, asthma, bronchitis, and emphysema.
Toxic emissions	Suspected of causing cancer, reproductive problems, and birth defects. Benzene is a known carcinogen.

ing lot, reeks with the incense of greed, waste, and deleterious exhaust fumes. Here are some facts.

Our economy is overdependent on the motor vehicle. One-sixth of every dollar and one-sixth of every nonfarm job is related to the motor vehicle. It accounts for 20% of our GNP and an annual average of $300 to $500 billion in government subsidies, more than $1,000 per person above the direct costs to car users. It accounts for 63% of our oil use ($43 billion are spent to import oil each year, one-fourth of our national trade deficit).

The motor vehicle in the United States is the largest source of urban smog, accounting for 50% of our air pollution: 67% of our carbon dioxide emissions, 50% of our methane, 40% of our nitrogen oxides (all told, 50% of greenhouse gases, thus contributing to the greenhouse effect), and 25% of chlorofluorocarbons (CFCs), contributing to the breakdown of the ozone layer. Urban residents in cities throughout the world incur eye, nose, and throat irritation, asthma, headaches, heart attacks, cancer, and emphysema due to car-produced smog (see Table 15-1). Fuel emissions cause about 30,000 deaths annually in the United States. In 1992 the Natural Resources

Defense Council declared that the car was "the worst environmental health threat in many U.S. cities."[21]

Furthermore, cars and trucks clog up our highways and streets, causing traffic jams, which in turn greatly increase stress and produce hypertension, ulcers, and nervous disorders; sometimes they even lead to violence and murder. These traffic jams cause employees to arrive late at work. Government agencies estimate that more than $100 billion are lost each year due to traffic jams. Each day 5 million vehicles crowd the Los Angeles freeways, contributing to the 100,000 hours per day that are wasted in traffic jams in that city. A total of one-third of the city's land is given over to automobile-related roads, parking lots, and gas stations (one-half of downtown Los Angeles). Estimates are that the average American commuter spends 2 years of his or her life in traffic jams. In 1907 the average speed of a horse-drawn carriage through the streets of Manhattan was 11.5 mph. Today a vehicle averaging the power of some 300 horses does the same mile at a pace of 5 mph. "A car makes you wish the whole world was paved with concrete," a Chevy Camaro advertisement broadcasts. Indeed, it must seem so! At present 38.4 million acres nationwide are given to roads and parking lots (that's the size of the state of Georgia). As Jane Holtz Kay points out, "In built-up areas we devote more land to our cars than to our homes. In the wilderness we lay 370,000 miles of road on just the Forest Service's 300,000 square miles, more than a mile of road per square mile of wooden wilderness."[22]

As I mentioned at the beginning of the chapter, we've actually regressed with regard to mass transit. In the early part of this century the United States had an effective public transportation system in virtually all major cities. Trolleys, streetcars, and trains, once the pride of the nation, were beaten by the internal combustion engine—and so was the quality of our air. We also walk less, so that one-third of all Americans carry extra pounds. According to the Institute of Medicine, our total girth has risen 8% in less than a generation.[23]

We need to build more mass transit systems throughout the world, but for short trips, another mode of transportation is called for: the bicycle. It is an inexpensive alternative to automobile transport, 30 or 40 times cheaper (when gas and maintenance are included); my two bikes, which are more than 10 years old and with which I commute to work daily, each cost about $100, with a combined total of about $500 in maintenance over a 10-year period. The materials used to construct one middle-sized American car can construct 100 bicycles. In China and India the bicycle is becoming the preferred mode of transportation; and in some cities in the Netherlands between 40% and 50% of all trips are made by bike. Twenty percent of the half-million adults in old Toronto bike to work, school, or store. In the United States the use of bicycles as a means of transportation has increased in the past few years, from less than 1% to 2%.

Table 15-2 Energy Intensity of Selected Transport Modes in the United States, 1984

Mode	*Calories/Passenger Mile*
Bicycling	35
Walking	100
Transit rail	885
Transit bus	920
Automobile, single occupant	1,860

Sources: President's Council on Physical Fitness and Sports, Washington, D.C., private communication, 23 June 1988; Mary C. Holcomb and others, *Transportation Energy Data Book: Edition 9* (Oakridge, Tenn.: Oak Ridge National Laboratory, 1987).

The bicycle is the most efficient mode of transportation, using less energy per mile than any other mode, including walking. The energy available from an ear of corn will get you 3½ miles, and there's less noise and no distilling or refining problem! The bicycle uses 35 calories per passenger mile, walking 100, rail transport 885, and the automobile 1,860 (53 times more than the bicycle).[24] See Table 15-2.

Pedal power creates no air pollution, cuts down on stress, and provides a good source of exercise; employees arrive at work more refreshed and healthier than do those imprisoned in traffic jams. In urban areas bicycles move at the same average speed as cars and sometimes faster. In fact, police in Seattle frequently use bikes to catch car thieves.

Of course, biking can be dangerous too, especially when bicyclists use the same roads as powerful cars and trucks (90% of cycling accidents involve collisions with a car or truck). The greatest danger to cyclists is drivers in parked cars opening their doors without looking for oncoming bicyclists. I was hit by a door suddenly opening near the University of California at Berkeley. Fortunately, my helmet cushioned my head from the hard cement, saving me from a brain concussion. Brian Wong had a worse fate. Riding along University Avenue in Toronto, a driver in a parked car opened his door without looking. Wong's bike hit the door, and he catapulted into traffic, where he was run over by a bus and killed. The driver of the car was given a $105 ticket. Bicycle advocates were unsuccessful in getting the driver charged with criminal negligence. The police refuse to do more to such negligent drivers than issue tickets.

Still, bikes are far safer than cars. Outnumbering cars throughout the world 2 to 1, only 2% of traffic fatalities involve bicycles. But the cyclist should take precautions by wearing a helmet and using a rear-view mirror attached to the handlebar or helmet.

Cities like Davis and Palo Alto, California, have extensive bike paths, and St. Paul, Minnesota, is currently in the process of building a 17-mile bicycle freeway. The Netherlands has more than 9,000 miles of bicycle paths, and Japan provides not only bicycle paths but also large bicycle garages at train stations, thus combining public transportation with bicycle use. By shifting to nonmotorized transportation, countries could save millions of dollars in fuel costs. A Worldwatch researcher estimates, "If 10% of Americans who commute by car switched to bike-and-ride, more than $1.3 billion could be shaved off the U.S. oil import bill."[25] A 1983 study of American commuters showed that turning to the bicycle could save each commuter 150 gallons of gas per year.

In 1969 a group of University of Minnesota students dragged an automobile engine into downtown Minneapolis. "As a small crowd gathered, the students dropped the motor into a grave, covered it with dirt, and solemnly declared an end to the tyranny of the internal combustion engine over the lungs and lives of civilized people. As traffic rushed past on nearby streets, a young minister read the eulogy:

> Ashes to ashes, dust to dust,
> For the sake of mankind, iron to rust.[26]

Perhaps the university students' actions were overly dramatic and premature. The internal combustion engine is not likely to die a natural death in the near future. Let's hope, however, that a less-polluting, electrical car is on the horizon. But for reasons of the economy, health, and land use, the bicycle, combined with increased public transportation, should replace the automobile as the preferred mode of transportation in the United States and throughout the world. As author and cyclist James McGurn writes, "The bicycle is the vehicle of a new mentality. It quietly challenges a system of values that condones dependency, wastage, inequality of mobility, and daily carnage. . . . There is every reason why cycling should be helped to enjoy another Golden Age."[27]

The time has come to dethrone the automobile—our destructive Convenience Machine—as the American vehicular idol and to replace it with ecologically sounder modes of transportation, with the person-powered two-wheeled vehicle as the centerpiece.

Review Questions

1. Go over the arguments for safe nuclear energy. Do the arguments convince you that radiation is not the evil that it is sometimes made out to be? that we live with radiation all the time? that the data suggesting that small amounts are injurious are suspect?
2. Consider Flavin's contention that we should resist the Faustian bargain:

> The Faustian bargain of nuclear energy has been lost. It is high time to leave the path pursued in the use of nuclear energy in the past, to develop new al-

ternative and clean sources of energy supply and, during the transition period, devote all efforts to ensure maximum safety. This is the price to pay to enable life to continue on this planet.

Recall that a "Faustian bargain" refers to the legendary intellectual Faust who sold his soul to the devil in order to gain magical knowledge and power. Is this an accurate metaphor for our attraction to nuclear energy? Explain why or why not.

3. Review the major arguments against nuclear power. Which is the strongest and which the weakest? How do antinuclear environmentalists use Murphy's Law to their favor? Can you think of a response by proponents of nuclear energy?
4. According to those who support nuclear energy, scientists do know of permanent and safe ways to dispose of nuclear waste; only political opposition, fired by an almost hysterical public reaction to the idea, prevents progress in this area. Do you agree with their assessment?
5. Consider Jagger's claim that solutions to the waste problem are not close at hand only because of political opposition. Furthermore, there is no harm in letting the wastes stay where they are—at the power plants themselves—for another decade or so, since their radiation is decaying *steadily* with time. Does this help in assessing the merits of renewing a commitment to nuclear energy as the best hope in providing clean energy for our world? Go over the arguments again and explain your thoughts on this difficult and vital subject.
6. How feasible is the switch from automobiles to bicycles? Do you see any problems with the analysis in this chapter?
7. Would switching to electric cars and more efficient mass transit be more realistic ways to make progress in transportation?

Suggested Readings

Fremlin, J. H. *Power Production: What Are the Risks?* New York: Adam Hilger, 1989.

Jagger, John. *The Nuclear Lion.* New York: Plenum, 1991.

Kaku, M., and J. Trainer, eds. *Nuclear Power: Both Sides.* New York: Norton, 1983.

Kay, Jane Holtz. *Asphalt Nation: How the Automobile Took Over America and How We Can Take It Back.* New York: Crown, 1997.

Lester, R. "Is Nuclear Industry Worth Saving?" *Technology Review* (October 1982).

Marshall, E. "Lessons of Chernobyl." *Science* 233 (26 September 1986), pp. 1375–76.

Miller, G. Tyler. *Living in the Environment*, 10th ed. Belmont, Calif.: Wadsworth, 1998, chap. 16.

Pojman, Louis, ed. *Environmental Ethics*, 2nd ed. Belmont, Calif.: Wadsworth, 1998. Contains articles by Jagger, Flavin, and Lenssen on nuclear energy.

Routley, Richard, and Val Routley. "Nuclear Power: Some Ethical and Social Dimensions." In *And Justice for All.* Edited by Tom Regan and Donald VanDeVeer. Totowa, N.J.: Rowman and Littlefield, 1982.

Shrader-Frechette, K. S. *Nuclear Power and Public Policy.* Boston: Reidel, 1980.

Weinberg, Alvin. *Continuing the Nuclear Dialogue.* LaGrange, Ill.: American Nuclear Society, 1985.

Notes

1. "Will the Lights Go Out More Often with Deregulation?" *New York Times,* 1 June 1998, p. 1.
2. Amory Lovins, "Technology Is the Answer (But What Is the Question?)," in G. Tyler Miller, *Living in the Environment,* 10th ed. (Belmont, Calif.: Wadsworth, 1998), p. 426.
3. Christopher Flavin and Nicholas Lenssen, "Reshaping the Power Industry," in *State of the World 1994,* ed. Lester Brown (New York: Norton, 1994).
4. For a good discussion of how the automobile took over the United States, see Jane Holtz Kay, *Asphalt Nation: How the Automobile Took Over America and How We Can Take It Back* (New York: Crown, 1997).
5. Miller, *Living in the Environment,* p. 403.
6. "Green Power Off to Encouraging Start," *New York Times,* 1 June 1998, p. EN 15.
7. Perhaps we should say, not the "energy problem," but the "consumption problem," since we simply use enormously more energy than is necessary for a good life.
8. The most authoritative study is the report of the International Chernobyl Conference, sponsored by the European Commission, the International Atomic Energy Agency (IAEA), and the World Health Organization (WHO), held in Vienna in April 1996, published as *One Decade After Chernobyl—Summing Up the Consequences of the Accident* (Vienna: IAEA, 1996). My data are based on this study.
9. Olga Korbut, quoted in G. Tyler Miller, *Living in the Environment,* p. 431.
10. Cited in G. Tyler Miller, *Living in the Environment,* p. 447. The National Safety Council reports the following list of accidental deaths due to injury in the United States.

	1992	***1991***	***1990***
Motor vehicles	40,982	43,536	46,814
Railway	223	215	220
Electric generating plants, including transmission	139	132	160
Lightning	53	75	89
Gas distributed by pipeline	21	20	33
Radiation	0	0	0

11. *Health Effects of Exposure to Low Levels of Ionizing Radiation.* (Washington, D.C.: National Academy Press, 1990).
12. Jagger, John. *The Nuclear Lion.* (New York: Plenum, 1991), and "Nuclear Power Is Safe and We Need It," in *Environmental Ethics,* 2nd ed., ed L. Pojman (Belmont, Calif.: Wadsworth, 1998), pp. 414–26.
13. These philosophers have subsequently changed their names to Richard Sylvan and Val Plumwood.
14. Richard Routley and Val Routley, "The Nuclear Train into the Future," in *And Justice for All: New Introductory Essays in Ethics and Public Policy,* ed. Tom Regan and Donald VanDeVeer (Totowa, N.J.: Rowman and Littlefield, 1982), pp. 116–18.
15. Jagger, *The Nuclear Lion,* p. 473. Against the Routleys and other doomsdayers, he quotes the 1995 Department of Energy fact sheet:

    Various kinds of radioactive materials have been moved around our country for decades. Shipments of spent nuclear fuel regularly go to or from nuclear power plants, government research facilities, industrial complexes, and other facilities. After more than 25 years—and more than 2,500 shipments of spent fuel—there has not been a single death or injury because of the radioactive nature of the cargo.

    Jagger points out that "No other industrial technology, and certainly no other electric power technology, even approaches [the] safety record [of nuclear power]." (Letter to author, 13 May 1998).

16. A major report by the Energy Department on the suitability of Yucca Mountain is being prepared even as I write this chapter. Congress was scheduled to take up the issue in 1998. Opponents of using Yucca Mountain claim that no guarantee exists that Yucca Mountain will be completely safe from earthquakes, water damage, or a well-planned terrorist attack for the 10,000 years or so that the nuclear wastes will be buried there. Proponents argue that opponents are motivated by fear, not facts, and that Yucca Mountain is a safe storage site. The mounting wastes must be placed somewhere, and Yucca Mountain is needed, at least as a temporary site. The wastes will be contained in such a way as to permit transference should that be required. President Clinton indicated he would veto any bill establishing a temporary site in Nevada.
17. I have read statements by opponents of nuclear energy that we have an enormous amount of nuclear waste—over 80,000 tons of it. This is misleading. Because uranium and plutonium are 20 times as dense as water, 80,000 tons of nuclear waste occupies a volume equivalent to a cube 50 feet on each side. All of the high-level nuclear waste from all of the nuclear reactors in the United States for all the 35 years in which we have had nuclear power would fill a small house. This is not the impression one gets from the bald statement of "80,000-odd tons."
18. "Burial of Radioactive Waste Under the Seabed," *Scientific American,* January 1998.
19. Nicholas Lenssen, "Nuclear Waste: The Problem That Won't Go Away," in *State of the World,* ed. Lester R. Brown (Washington, D.C.: Norton, 1992), reprinted in *Environmental Ethics,* ed. L. Pojman, pp. 427–34.
20. Christopher Flavin, "Nuclear Power's Burdened Future," *Bulletin of Atomic Scientists* (July 1987), p. 26.
21. Jane Holtz Kay, *Asphalt Nation,* p. 81.
22. Ibid., p. 83.
23. Ibid., p. 109.
24. A calorie is the amount of heat (energy) required at a pressure of 1 atmosphere to raise the temperature of 1 gram of water 1 degree centigrade.
25. Marcia Lowe, "Pedaling into the Future," in *Worldwatch Reader,* ed. Lester Brown (Washington, D.C.: Norton, 1991). I have used Lowe's research in this article.
26. Ed Ayers, "Breaking Away," *Worldwatch* 6:1 (January-February 1993). Ayers points out that the sprocket chain-driven bicycle and the gas-fueled combustion-engine car were both invented in the same year, 1885, but the history of the two vehicles has been one wherein the bigger gadget has pushed the smaller off the road. Might makes right—or, at least, right of way.
27. Quoted in Marcia Lowe's *The Bicycle: Vehicle for a Small Planet* (Washington, D.C.: Worldwatch Institute, 1989), p. 46.

CHAPTER SIXTEEN

Preservation of Wilderness and Species

In wildness is the preservation of the World.
—HENRY DAVID THOREAU, "Walking"

A wilderness, in contrast with those areas where man and his works dominate the landscape, is hereby recognized as an area where the Earth and its community of life are untrammeled by man, where man himself is a visitor who does not remain.
Wilderness Act of 1964

THE WILDERNESS: PRESERVATION OR CONSERVATION?

The Restoration Thesis

Suppose a rich deposit of copper were discovered in your favorite state wilderness area. A mining company offered to pay a large sum of money for the right to mine the mineral for one year. In order to extract the copper ore efficiently, it would have to destroy a large part of the forest, constructing roads and edifices. There would be some drainage into the nearby river and lots of noise during the weekdays. But at the end of the year, the company would remove the roads and edifices, fill the mine with debris, and plant new trees, so that in a short time the forest would be restored to its former beauty. If you're worried about the length of time, then suppose that fertilizers with enhanced growth hormones would speed up the growth prospects. Within a few years, hardly anyone would know the difference. The question is, Would it be morally right to lease the area to the mining company? Would it make a difference to your answer that the money the state received from the company would be used to build better schools, hire better teachers at higher salaries, and generally promote the living conditions of the poor? What do you think?

This process of destroying and then restoring part of a natural environment for human benefit is known as the *restoration thesis*.[1] Anthropocentrist conservationists (discussed in Chapter 9) accept the restoration thesis as good stewardship of the wilderness. They would agree to the offer. The prospects of promoting the human good would outweigh the value the wilderness had as a habitat for other living beings and the pleasure it provided to nature lovers. Besides, only part of the park would be closed down and then for only a year—or a few years, if we consider the time to restore it. Most preservationists, including deep ecologists, biocentrists, and restoration ecologists, would no doubt reject the offer as sacrificing something of inestimable intrinsic worth.[2] Nature—the wild—is irreplaceable. It has deep intrinsic value.[3] The new-growth forest would be discontinuous with the old. It would be the product of human control and management, so not really a *wild*-erness. Even though it might eventually look exactly like the old forest, its origins would be different. It would lack its continuity with ages, perhaps thousands of years, of natural development. What we value in the wilderness is its *wildness,* its being outside human control, outside our technological dominance.

In his article "What's Wrong with Plastic Trees?" Martin Krieger depicts the wilderness as a state of mind, as an "attitude toward a collection of trees, other plants, animals, and the land on which they all exist. The idea that a wilderness exists as a product of an intellectual movement is important. A wilderness is not discovered in the sense of some man from a civilization looked upon a piece of territory for the first time. It is the meaning that we attach to such a piece of territory that converts it to a wilderness."[4] He argues that it is humans and culture, especially urban culture, that endow the wilderness with special value. People tend to romanticize the wild from their comfortable fully furnished, centrally heated, electricity-abundant homes in communities with grocery and clothing stores. They wistfully long for undisturbed nature, which they really desire only as temporary escapes from city life. Those who have to live in the wild know its inconveniences and would often just as soon trade it for some creaturely comforts.

Even if our "restored wilderness" were only restored to a plastic simulacrum, that might well be good enough—even better. Perhaps we could use the latest supertechnology to create giant Jurassic Parks, filled with spectacular plastic mountains, trees, and flowers, with artificial rivers and lakes in abundance. "What's wrong with plastic trees?" Krieger asks. "My guess is that there is very little wrong with them. Much more can be done with plastic trees and the like to give most people the feeling that they are experiencing nature. We will have to realize that the way in which we experience nature is conditioned by our society—which more and more is seen to be receptive to responsible interventions."

Preservationist biocentrists object to the restoration thesis and to the kind of reasoning used by conservationists. A "restored" nature is not natural. At least two things are missing: (1) the natural process that produced the natural entity in the first place; (2) an appreciation for the wilderness as valuable in itself. Robert Elliot, in "Faking Nature," compares natural areas such as a wilderness to works of art. He argues that just as knowing that the Vermeer painting you own is a replica lowers its value, so likewise knowing that the experience you are having is only a replica of the natural original lowers the value of that experience. Fakes lack the value of the original entity. Here is the crucial passage:

> Imagine that I have a piece of sculpture in my garden which is too fragile to be moved at all. For some reason it would suit the local council to lay sewerage pipes just where the sculpture happens to be. The council engineer informs me of this and explains that my sculpture will have to go. However, I need not despair because he promises to replace it with an exactly similar artifact, one which, he assures me, not even the very best experts could tell was not the original. The example may be unlikely, but it does have some point. While I may concede that the replica would be better than nothing at all (and I may not even concede that), it is utterly improbable that I would accept it as full compensation for the original. Nor is my reluctance entirely explained by the monetary value of the original work. My reluctance springs from the fact that I value the original as an aesthetic object, as an object with a specific genesis and history.
>
> Alternatively, imagine I have been promised a Vermeer for my birthday. The day arrives and I am given a painting which looks like a Vermeer. I am understandably pleased. However, my pleasure does not last for long. I am told that the painting I am holding is not a Vermeer but instead an exact replica of one previously destroyed. Any attempt to allay my disappointment by insisting that there just is no difference between the replica and the original misses the mark completely. There is a difference and it is one which affects my perception, and consequent valuation, of the painting. The difference of course lies in the painting's genesis.
>
> I shall offer one last example which perhaps bears even more closely on the environmental issue. I am given a rather beautiful, delicately constructed, object. It is something I treasure and admire, something in which I find considerable aesthetic value. Everything is fine until I discover certain facts about its origin. I discover that it is carved out of the bone of someone killed especially for that purpose. This discovery affects me deeply and I cease to value the object in the way that I once did. I regard it as in some sense sullied, spoilt by the facts of its origin. The object itself has not changed but my perceptions of it have. I now know that it is not quite the kind of thing I thought it was, and that my prior valuation of it was mistaken. The discovery is like the discovery that a painting one believed to be an original is in fact a forgery. The discovery about the object's origin changes the valuation made of it, since it reveals that the object is not of the kind that I value.[5]

Elliot notes that John Muir regarded the Hetch Hetchy Valley as a place where he could have direct contact with nature. He valued it, not just because of its beauty but because it was "a part of the world that had not been shaped by human hand. . . . The news that it was a carefully contrived elaborate *ecological* artifact would have transformed that valuation immediately and radically. . . . We value the forest and river in part because they are representative of the world outside our dominion, because their existence is independent of us. . . . Origin is important as an integral part of the evaluation process."

Eugene Hargrove offers the following supportive data. Due to the effects of tourism, the famous ancient cave paintings at Lascaux in France were in danger of irreparable damage. So the authorities built a full-scale model of the cave nearby. Hargrove argues that, while the aesthetic experience of the replicated cave is not as valuable as the original, nevertheless, "the knowledge that the original still exists enhances the experiences afforded by the representation."[6] Hargrove then compares this situation with threatened damage to natural objects. An extremely beautiful passageway into Mammoth Cave, Kentucky, called Turner Avenue, was being damaged by tourism, so the authorities photographed it, closed it off from aesthetic experience, and allowed only the indirect aesthetic experience that comes from beholding the pictures. Hargrove argues, analogous to the first situation, that knowledge that the original still exists enhances the experience afforded by the representation.

Conservationists who advocate the restoration thesis argue that the comparison of natural objects with art misses a crucial point. A restored forest is not a fake forest. A Disneyland plastic forest might be a fake, but not the restored forest. A restored forest follows the same biological laws as the original—but humans have intervened to manage those laws, so it need not be any less natural than the original.

But the preservationist will still object that value is lost since the genesis of the wilderness or ecosystem has been changed. To that the conservationist replies, "What's so great about genesis?" We might not *now* have adequate knowledge to replicate fully the wonders and homeostasis of a natural system—"nature knows best"—but we can learn and perhaps even improve on nature. The value we place on an *original* in art is itself a cultural value, not a natural one. In appealing to the criterion of *originality* in nature against anthropocentric tampering, the preservationists are themselves guilty of using an anthropogenic value—does nature give a darn what its origins are? Does this constitute a valid criticism of the preservationist view?

Or consider this thought experiment: God creates an incredibly beautiful ecosystem, one that surpasses in stability and integrity the Grand Canyon, Yellowstone, and the Brazilian rain forest all put together. But it would not be *natural* in the sense of being a product of biological processes or *wild* in the sense of being untouched by intentional purposes. It would be like the Garden of Eden with the Grand Canyon thrown in. We'll call it *Super Grand Canyon II*. Suppose God has to maintain it by His power. Would it be any less valuable for not being *natural* or *wild*? If you say no, then imagine that by the year 2150 humans acquire the knowledge and power to construct *Super Grand Canyon III,* almost as glorious as *Super Grand Canyon II* but far more beautiful than the Grand Canyon. Would it be less valuable for not being *wild* or *natural* (that is, produced by natural causes, rather than human "artificial" causes)?

The Callicott–Rolston Debate

Conservationists believe in the "wise use" of nature for the human good. Human beings should be good stewards of natural resources, ensuring that there be enough and as good left over for posterity. (Why should we care about posterity; what has posterity ever done for us? See Chapter 5 for the relevant discussion.) Recently Baird Callicott has argued that no less than the apostle of environmentalism, Aldo Leopold (see Chapter 9), actually supported the conservationist program over against the preservationist's resistance to economic incursions.[7]

Leopold aimed to combine ecological knowledge with the promotion of the human good, seeking a relationship of harmony between the two entities. He favored **sustainable development**—that is, human economic activity that preserves ecological integrity and enhances its health. Callicott offers examples of how wise-use policy can enhance the **biotic community.** He cites studies of two similar desert oases 30 miles apart where Papago Indians live—one in Organ Pipe Cactus National Monument in the United States and the other the Quitovac, across the border in Mexico. The U.S. government designated the Organ Pipe area a bird sanctuary and stopped all cultivation there in 1957. The Mexican area is still being farmed. Thirty-two species of birds have been observed in the Organ Pipe protected area, but 65 species in the Mexican oasis. The ethnobotanist Nabhan, who is responsible for the study, remarked, "When the people live and work in a place, and plant their seed and water their trees, the birds go live with them. They like those places. There's plenty to eat and that's when we are friends with them."[8] Other examples of **symbiosis** between humans and nature include the Kayapo Indians of the Amazon rain forest who fish, hunt, gather, and cultivate agricultural plots (swiddens). "In sharp contrast to the displaced Euro-Brazilian peasants who are entering the region, the Kayapo, through a complex cycle of planting, manage to cultivate a forest clearing for nearly ten years, instead of merely three or four. But after a decade of cultivation, a Kayapo plot is not simply abandoned. Instead, the Kayapo manage the regeneration of the forest by planting useful native species—first, fast-growing short-lived early succession plants like banana, and later, long-lived canopy trees like Brazil nut trees and coconut and oil palm. Thus, their fallows become permanent resource patches from which they obtain fruit, nuts, medicines, thatch, and other materials in perpetuity."[9]

Callicott also points out that before the coming of the Europeans in the fifteenth century, the American Indians had slashed and burned the forests and generally impacted the environment. Nevertheless, they also maintained a symbiotic relationship with nature. They provide a model of *sustainable development,* living off nature, but doing so in a responsible way.

Callicott isn't suggesting that we return to this Indian model of culture. That's impossible at this point. But their culture does provide an example of how we might adapt our lives to nature's cycles and processes while benefiting from her. A "win-win" philosophy of conservation, *sustainable development,* is possible. "The human–nature relationship is an ongoing, evolving one. We can, I am confident, work out our own, postmodern, technologically sophisticated, scientifically informed, sustainable civilization, just as in times past the Minoans in the Mediterranean, the vernacular agriculturalists of Western Europe, and the Incas in the Andes worked out theirs."

Callicott points out that one-third of the lower 48 states is publicly owned land, managed by the Forest Service, the Bureau of Land Management, and the National Park Service. On much of this land, the western rangeland, cattle and sheep graze, wreaking havoc on the ecosystem (while the public subsidizes the "pathetic livestock industry" in the process). Better to end these grazing practices now before more damage is done to the land. We could use these rangelands in ways that preserve ecological integrity and still provide economic benefits, ripping out the roads, and restoring native wildlife populations, such as deer. Dry farming is also a possibility.

In most cases, the wise course is not to go to either extreme, *either* the extreme of complete laissez-faire capitalism with its ignorance of or indifference to the environment *or* complete untouched wildernesses, but to use technology wisely. "Can't we be good citizens of the biotic community, like the birds and the bees, drawing an honest living from nature and giving back as much as or more than we take?"

Callicott's conservationist proposal has been attacked by the preservationist Holmes Rolston III in "The Wilderness Idea Reaffirmed." Rolston argues that Callicott is wrong on several counts. First of all, Callicott's comparison of Indian culture with what modern Americans have done is ludicrous. The impact is of a different order of magnitude. The Indians had little agriculture, no motors (not even wheels), no domestic animals, and no beasts of burden. Even obtaining hot water was difficult. Your chances of harming an environment like the Americas when you live on the landscape with foot, muscle, and arrows is minimal. We, on the other hand, with our powerful technology, have accelerated the rate of change in the environment logarithmically:

> Ought implies can; the Indians could not, so they never thought much about the ought. We in the twentieth century can, and we must think about the ought. When we designate wilderness, we are not lapsing into some romantic atavism, reactionary and nostalgic, to escape culture. We are breaking through culture to discover, nonanthropocentrically, that fauna and flora can count in their own right (an idea that Indians also might have shared). We realize that ecosystems sometimes can be so respected that humans only visit and do not remain (an idea that the Indians did not need or achieve). A "can" has appeared that has generated a new "ought."[10]

Second, Rolston argues that it's misleading for Callicott to use the studies of Quitovac and Organ Pipe communities as strong evidence for his wise-use policy. This is an unusual case. The location is a desert, where water is a limiting factor. Birds in the desert will naturally flock to where feeders offer food, water, and shelter, as they do at Quitovac. But the same study shows that community is "not nearly as diverse in mammals." Deer, javelina, and even rodents are far more abundant at Organ Pipe than in Quitovac, where the domesticated dogs, horses, and cattle limit the presence of wild animals. But the important issue isn't sustainable development. We ought to have rural nature with a high bird count. The issue is whether we should preserve the wilderness for its own sake.[11]

What worries Rolston is Callicott's willingness to sacrifice the wilderness to human use. Rolston quotes passages from Callicott's work to show that he is essentially an anthropocentrist, who holds that all so-called intrinsic value in nature is "grounded in human feelings" and "projected" onto the natural object that "excites" the value. "Intrinsic value ultimately depends upon human valuers." Then Rolston exclaims:

> Talk about dichotomies! Only humans produce value; wild nature is valueless without humans. All it has without humans is the potential to be evaluated by humans, who, if and when they appear, may incline, sometimes, to value nature in noninstrumental ways. . . . Nature in itself (a wilderness, for example) is without value. There is no genesis of wild value by nature on its own. Such a philosophy can value nature only in association with human habitation. But that—not some elitist wilderness conservation for spiritual meditation—is the view that many of us want to reject as "aristocratic bias and class privilege."[12]

In sum, Rolston claims that Callicott's failure to appreciate the significance of the wilderness is the consequence of his anthropocentrism, his failure to realize that nature has intrinsic value. This undermines Callicott's interpretation of Leopold, who did value the wilderness for its own sake and sought to preserve both human sustainable development in rural culture *and* the wilderness. That is, affirming sustainable development should not mean denying wilderness:

> On Earth, man is not a visitor who does not remain: this is our home and we belong here. Leopold speaks of man as both "plain citizen" and as "king." Humans too have an ecology, and we are permitted interference with, and re-arrangement of nature's spontaneous course: otherwise there is no culture. When we do this there ought to be some rational showing that the alteration is enriching, that natural values are sacrificed for greater cultural ones. We ought to make such development sustainable. But there are and should be places on Earth where the nonhuman community of life is untrammeled by man, where we only visit and spontaneous nature remains. If Callicott has his way, revising wilderness, there soon will be less and less wilderness to visit at all.[13]

Who is right? Whether Callicott can be justly accused of being an anthropocentrist is unclear. He has defended the idea that humans, as part of nature, have primary duties to each other. The deeper issue is whether and to what extent nature has objective value. Is nature intrinsically or objectively valuable or only instrumentally (or potentially) so? It makes a difference in one's attitude toward the wilderness.

Both Callicott and Rolston accept the biocentric model of Leopold: "A thing is right when it tends to preserve the integrity, stability, and beauty of the biotic community; it is wrong when it tends otherwise" (see Chapters 7 and 9). We turn next to a theory that rejects that view in favor of the notion of wildness as the locus of intrinsic value.[14]

Ecocentrism Versus Wildness

In the paper "Refocusing Ecocentrism: De-emphasizing Stability and Defending Wildness," Ned Hettinger and William Throop challenged the Leopold–Callicott–Rolston "integrity-stability" thesis as sufficient to account for our value of the wild. They point out that empirical studies show that populations fluctuate irregularly and that the predator/prey model in which numbers of predators and prey oscillate predictably over time ignores many factors that affect population size. The also note that some ecologists suggest that "many interacting populations are chaotic systems, in the mathematical sense of "chaos."[15] That is, although the systems are fully deterministic, we cannot predict outcomes. Furthermore, "some stable ecosystems are not very diverse, such as east coast U.S. grass ecosystems where *Spartina alterniflora* grows in vast stands that are simple in species composition but quite stable." Michael Soulé claims that "the idea that species live in integrated communities is a myth."[16]

Callicott has responded to this critique of the stability-integrity thesis by modifying Leopold's famous dictum to read: "A thing is right when it tends to disturb the biotic community only at normal spatial and temporal scales. It is wrong when it tends otherwise."[17] But it is doubtful whether this is a satisfactory reply. As Hettinger and Throop point out, the response suggests that it is morally permissible to destroy other species, so long as we do so at a pace "comparable to normal extinction frequencies in evolutionary history." They point out that this also has the implausible consequence that restoration projects are impermissible if they proceed at rates faster than normal natural frequencies.

Hettinger and Throop do not reject the stability-integrity model altogether. They recognize that species in an ecosystem are interdependent, that keystone species are a significant factor in the survival of an ecosystem, and that increased diversity can often add to the stability of the system. But one should recognize its limitations, especially the fact that other values besides the vague "stability and integrity" are also important:

> We think that advocates of ecocentric ethics should shift the emphasis away from integrity and stability toward other intrinsically valuable features of the natural systems, such as diversity, complexity, creativity, beauty, fecundity, and wildness. . . . [W]e think that the value of wildness plays a central role in this nexus of values. Emphasizing wildness provides the most promising general strategy for defending ecocentric ethics.[18]

By *wild* they mean something not *humanized* in the relevant respect, not influenced, controlled, or altered by *humans*. "While one person walking through the woods does little to diminish its wildness, leaving garbage, culling deer, or clear-cutting do diminish wildness, although in different degrees." While hikers trekking through a wilderness probably do not alter it, clear-cutting and slash-and-burning techniques do. They point out, as evidence for their thesis, that the admiration we feel for a person's attractive features is likely to diminish when we learn that they were produced by plastic surgery; that people prefer the birth of a child without the use of drugs or a cesarean section; and that "picking raspberries discovered in a local ravine is preferable to procuring the store-bought commercial variety." This, of course, resembles Elliot's arguments, discussed earlier in this chapter. People also typically value the things not controlled by humans, such as "the weather, the seasons, the mountains, and the seas. This is one reason why the idea of humans as planetary managers is so objectionable to many. . . . Humans need to be able to feel small in comparison with something nonhuman which is of great value. Confronting the other helps humans to cultivate a proper sense of humility." They go on to argue that humans now appropriate between 20% and 40% of photosynthetic energy produced by terrestrial plants, that through our increased technology we have the power to affect wildness in profoundly destructive ways.

Further evidence for their view of the value of the wild is our need to "confront, honor, and celebrate the 'other.' In an increasingly secular society, 'nature' takes on the role of the other. Humans need to be able to feel small in comparison with something nonhuman which is of great value." Although Hettinger and Throop do not mention it, this feeling seems similar to Rudolf Otto's notion of religion as the confrontation with the Wholly Other, the supernatural and fascinating quality provoked by the idea of God. If what they are saying is accurate, is the wild, then, an object of religious experience? What kind of religion? How like or different from the Judeo-Christian experience of encountering a personal transcendent God? Is this related to the Gaia hypothesis (discussed in Chapter 9)? We begin to open a host of deep and difficult questions.

Wildness is a value-enhancing property. It also is a "root" value that is a significant source of other values, such as creativity, diversity, beauty, and fecundity. As Eugene Hargrove argues, "Our aesthetic admiration and appreciation for natural beauty is an appreciating of the achievement of complex form that is entirely unplanned. It is in fact because it is unplanned and independent of human involvement that the achievement is so amazing, wonderful and delightful."[19] Hettinger and Throop qualify their thesis by stating that *wildness* is not an absolute value, but can sometimes be overridden by other anthropocentric and nonanthropocentric reasons. For example, Antarctica might be more intrinsically valuable if it had more diversity, even if that were caused by decreasing its wildness.

In a paper given February 20, 1998, Throop argued that if we reflect on the idea of wildness we will likely come to believe that it is valuable: "We have a range of intuitions which cohere with the claim that we intrinsically value wildness. Wildness intensifies or adds to the original value."[20] He seems to be rejecting the Rolston thesis that nature has intrinsic, objective value, but rather argues that we are so constructed through the evolutionary process that we need wildness in the same way we need food, drink, sleep, and love.

Here Throop might have called on the foremost sociobiologist of our age, Edward O. Wilson, who points out that we are part of nature and, as such, have a deep tendency to identify ourselves with the rest of life. He calls this tendency to prize nature *biophilia*. The more we develop an understanding of our natural roots, the more we will build an ethic that reflects biophilia. It is worth quoting Wilson at length:

> The human heritage does not go back only for the conventionally recognized 8,000 years or so of recorded history, but for at least 2 million years, to the appearance of the first "true" human beings, the earliest species composing the genus *Homo*. Across thousands of generations, the emergence of culture must have been profoundly influenced by simultaneous events in genetic evolution, especially those occurring in the anatomy and physiology of the brain. Conversely, genetic evolution must have been guided forcefully by the kinds of selection rising within culture.

> Only in the last moment of human history has the delusion arisen that people can flourish apart from the rest of the living world. Preliterate societies were in intimate contact with a bewildering array of life forms. Their minds could only partly adapt to that challenge. But they struggled to understand the most relevant parts, aware that the right responses gave life and fulfillment, the wrong ones sickness, hunger, and death. The imprint of that effort cannot have been erased in a few generations of urban existence. I suggest that it is to be found among the particularities of human nature, among which are these.
>
> People acquire phobias, abrupt and intractable aversions, to the objects and circumstances that threaten humanity in natural environments: heights, closed spaces, open spaces, running water, wolves, spiders, snakes. They rarely form phobias to the recently invented contrivances that are far more dangerous, such as guns, knives, automobiles, and electric sockets.
>
> People are both repelled and fascinated by snakes, even when they have never seen one in nature. In most cultures the serpent is the dominant wild animal of mythical and religious symbolism. Manhattanites dream of them with the same frequency as Zulus. This response appears to be Darwinian in origin. Poisonous snakes have been an important cause of mortality almost everywhere, from Finland to Tasmania, Canada to Patagonia; an untutored alertness in their presence saves lives. We note a kindred response in many primates, including Old World monkeys and chimpanzees: the animals pull back, alert others, watch closely, and follow each potentially dangerous snake until it moves away. For human beings, in a larger metaphorical sense, the mythic, transformed serpent has come to possess both constructive and destructive powers: Ashtoreth of the Canaanites, the demons Fu-Hsi and Nu-kua of the Han Chinese, Mudamma and Manasa of Hindu India, the triple-headed giant Nehebkau of the ancient Egyptians, the serpent of Genesis conferring knowledge and death, and, among the Aztecs, Cihuacatl, goddess of childbirth and mother of the human race, the rain god Tlaloc, and Quetzalcoatl, the plumed serpent with a human head who reigned as lord of the morning and evening star.[21]

As further evidence to the universal sense of mystery connected with the snake, Wilson notes that the picture of two serpents entwined on the staff of Mercury is today the universal emblem of the medical profession. We have a sympathetic-antipathetic relationship to nature as Other. We find ourselves both powerfully drawn to her and frightfully repelled by her danger—nature both attracts and repulses us. Wilson's ideas of biophilia and biophobia seem corroborations of the complex powers of what Hettinger and Throop call the *wild*. Indeed, Wilson goes on to include wilderness as part of the primordial element of our psyche:

> To biophilia can be added the idea of wilderness, all the land and communities of plants and animals still unsullied by human occupation. Into wilderness people travel in search of new life and wonder, and from wilderness they return to the parts of the earth that have been humanized and made physically secure. Wilderness settles peace on the soul because it needs no help; it is beyond

> human contrivance. Wilderness is metaphor of unlimited opportunity, rising from the tribal memory of a time when humanity spread across the world, valley to valley, island to island, godstruck, firm in the belief that virgin land went on forever past the horizon.

Suppose that a need for the *wild* has been evolutionarily grounded in our nature? Can we change our nature to satisfy that need in other ways: through religion, art, culture, or other creative activities? Are we plastic enough to transform ourselves into more civilized and urbanized beings who have less need for the wild? Or is the wild a primordial need so deeply rooted in us that we reject or ignore it only at our peril? Are we oversocialized or is further socialization the way of our salvation? Is urbanity our cure from nature or our curse, the instrument of alienation from our mother, nature?

Daniel Botkin, in his book *Discordant Harmonies,* has argued against the preservationist thesis that nature always knows best and left undisturbed by humans is like a harmonious symphony.[22] He gives several examples of ecosystems deteriorating due to natural processes. Moose on Isle Royale in Michigan exceeded the carrying capacity of the island and suffered a tremendous dieback that almost wiped them out. Eventually, through human intervention, including the introduction of wolves by humans, the species was saved on the island. Glacial activity radically changed the ecosystems of North America and Europe during the Pleistocene Age. Rain forests along the west coast of New Zealand's South Island gradually disappeared, as chemical elements necessary for the plants leached downward into the soil, below the reach of the trees. In the late 1960s, during a drought, elephants destroyed a large section of Kenya's Tsavo National Park, leaving it desertlike. In this last case, Richard Laws, one of the world's foremost experts on the elephant, warned early on of the coming disaster and recommended that 3,000 elephants be shot to keep the population within the food supply. His opponents argued against him, trusting in nature's harmonies ("nature knows best"). The park warden, David Sheldrick, believed that nature used droughts to maintain or restore ecological balance. Over 6,000 elephants starved to death, and a large section of Tsavo was turned into a desert.[23] Botkin argues that sometimes human intervention is necessary to preserve an ecosystem and that those who are purists regarding ecological preservation and natural harmony are simply blind to the facts. Nature is discordant as well as harmonious. We have the power, especially through technology, to affect nature for good and evil, to offset some of the discord. (Botkin offers use of computer projections as an example of innovative technological power.) "The answer to the question about the human role in nature depends on time, culture, technologies, and people. There is no simple, universal (external to all people, cultures, times) answer. However, the answer to this question for our time is very influenced by the fact that we are changing nature at all levels—from the local to the global—that we have the power to mold nature into what we want it to be or to destroy it completely, and that we know we have that power."[24] Through the wise use of nature, we can improve her and benefit ourselves.

Botkin's voice, that of the new conservationist, is still a minority. Perhaps he overestimates the power of technology and our ability to improve on nature, but his work should be taken seriously.

WHY DO SPECIES MATTER?

Why should we be concerned with the preservation of **species**?* Why is **biodiversity** so important that we must make sacrifices to preserve and enhance it? Who cares if the snail darter, a fish of no known use to humans, perishes in the process of building a dam that may save the economy of the Tennessee Valley? Here is Edward O. Wilson's answer:

> Why should we care? What difference does it make if some species are extinguished, if even half of all the species on earth disappear? Let me count the ways. New sources of scientific information will be lost. Vast potential biological wealth will be destroyed. Still undeveloped medicines, crops, pharmaceuticals, timber, fibers, pulp, soil-restoring vegetation, petroleum substitutes, and other products and amenities will never come to light. It is fashionable in some quarters to wave aside the small and obscure, the bugs and weeds, forgetting that an obscure moth from Latin America saved Australia's pastureland from overgrowth by cactus, that the rosy periwinkle provided the cure for Hodgkin's disease and childhood lymphocytic leukemia, that the bark of the Pacific Yew offers hope for victims of ovarian and breast cancer, that a chemical from the saliva of leeches dissolves blood clots during surgery, and so on down a roster already grown long and illustrious despite the limited research addressed to it.
>
> In amnesiac revery it is also easy to overlook the services that ecosystems provide humanity. They enrich the soil and create the very air we breathe. Without these amenities, the remaining tenure of the human race would be nasty and brief. The life-sustaining matrix is built of green plants with legions of microorganisms and mostly small, obscure animals—in other words, weeds and bugs. Such organisms support the world with efficiency because they are so diverse, allowing them to divide labor and swarm over every square meter of earth's surface. They run the world precisely as we would wish it to be run, because humanity evolved within living communities and our bodily functions

---

* The usual definition of a *species* is a set of organisms that freely interbreed under natural conditions but not with members of other species.

> are finely adjusted to the idiosyncratic environment already created. Mother Earth, lately called Gaia, is no more than the commonality of organisms and the physical environment they maintain with each passing moment, an environment that will destabilize and turn lethal if the organisms are disturbed too much. A near infinity of other mother planets can be envisioned, each with its own fauna and flora, all producing physical environments uncongenial to human life. To disregard the diversity of life is to risk catapulting ourselves into an alien environment. We will have become like the pilot whales that inexplicably beach themselves on New England shores.[25]

Wilson argues that biodiversity is valuable for economic reasons, for pharmaceutical and nutritional benefits, and because of its role in the web of life, so crucial to our survival and flourishing. The good news is that in 1973 the U.S. Congress passed the Endangered Species Act, which has been enlarged to protect 750 endangered species. The bad news is that worldwide we are losing the battle. He estimates that each year we are losing at least 27,000 species.[26] Corroborating Wilson's arguments, the Worldwatch Institute 1992 report "Life Support: Conserving Biological Diversity" claimed that an unprecedented biological collapse has begun that will lead to mass extinctions worldwide. We have had an adverse effect on other species through overhunting, overfishing, increased pollution, and the destruction of natural habitats. The passenger pigeon, the grey dodo, the Stephen Island wren, great auk, Steller's sea cow, Schomburgk's deer, sea mink, Antarctic wolf, Carolina parakeet, and the American chestnut are some of the more famous examples of species we are responsible for extinguishing. The continuation of this process could have colossal implications for humanity.[27] On the macro-level, the California condor, the peregrine falcon, the Northern spotted owl, and several species of whales are presently endangered. On the micro-level, the very life-support system we all depend on for survival is at stake. If we are to survive, we must halt this loss of biodiversity, the ecosystems, species, and genes that together constitute life on Earth. The remaining wildlands and areas impacted and degraded by human activities must be protected. The need to save the remaining ancient forest ecosystems in the United States has never been more urgent. Here are some facts.

Biologists have identified over 1.5 million species, a small fraction of the estimated 10 to 100 million species on Earth (Figures 16-1 and 16-2). Insects make up over half of all species, and at least one-third of these are beetles.

When Jonathan Schell, in *The Fate of the Earth,* warned that after a nuclear holocaust insects would take over the world, he misstated the point. They already have! Beetles and ants are two of the most prolific, successful phyla in the world.[28] It is true that they are not in danger; species of mammals, birds, plants, and trees—and the future of human beings—are.

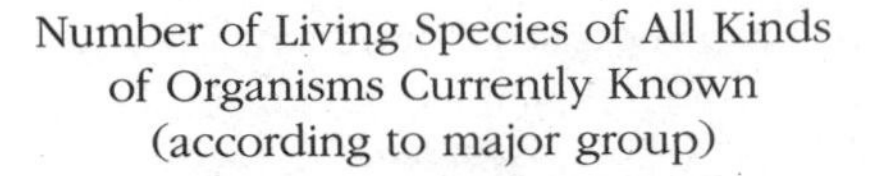

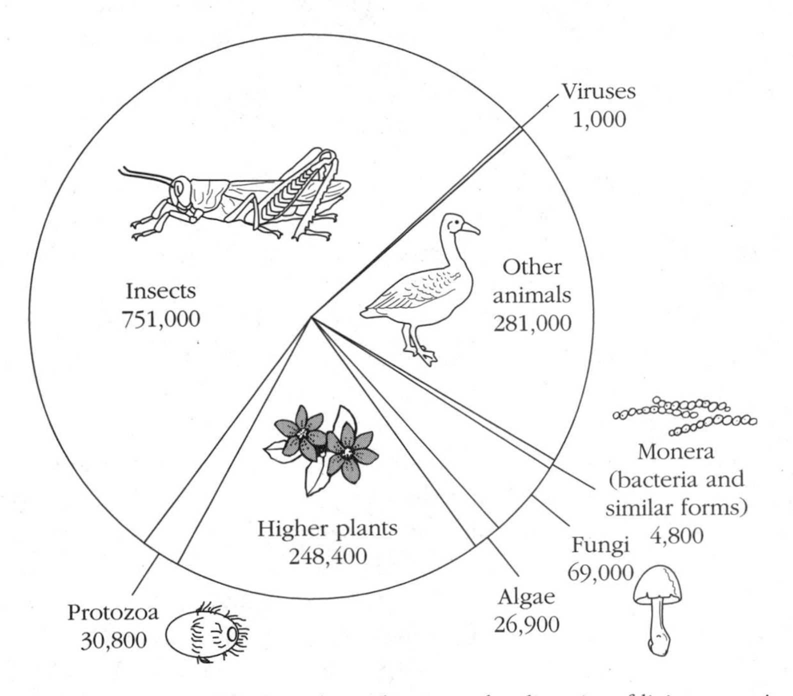

Figure 16-1 Insects and higher plants dominate the diversity of living organisms known to date, but vast arrays of species remain to be discovered in the bacteria, fungi, and other poorly studied groups. The grand total for all life falls somewhere between 10 and 100 million species.

About one-half of all species live in the tropical rain forests; 30% of the world's bird species are in the Amazon, and 16% in the Indonesian rain forests. There are 440 butterfly species in the United States, but one entomologist counted 429 butterfly species in only 12 hours in the Amazon rain forest. Ecologists have identified 275 kinds of trees in a single hectare (2.47 acres) of rain forest, 72 of them yielding fruits, vegetables, wild chocolates, and latex.

We live in an interconnected web of life, which has evolved over millions of years. We have no way of knowing how much benefit for humanity and animals exists in a single species. Except when they cause disease, we take microorganisms for granted, hardly recognize their existence; but if their work of turning detritus into nutrients were interrupted, we would soon perish. Donella Meadows explains the environmental services performed by diverse species:

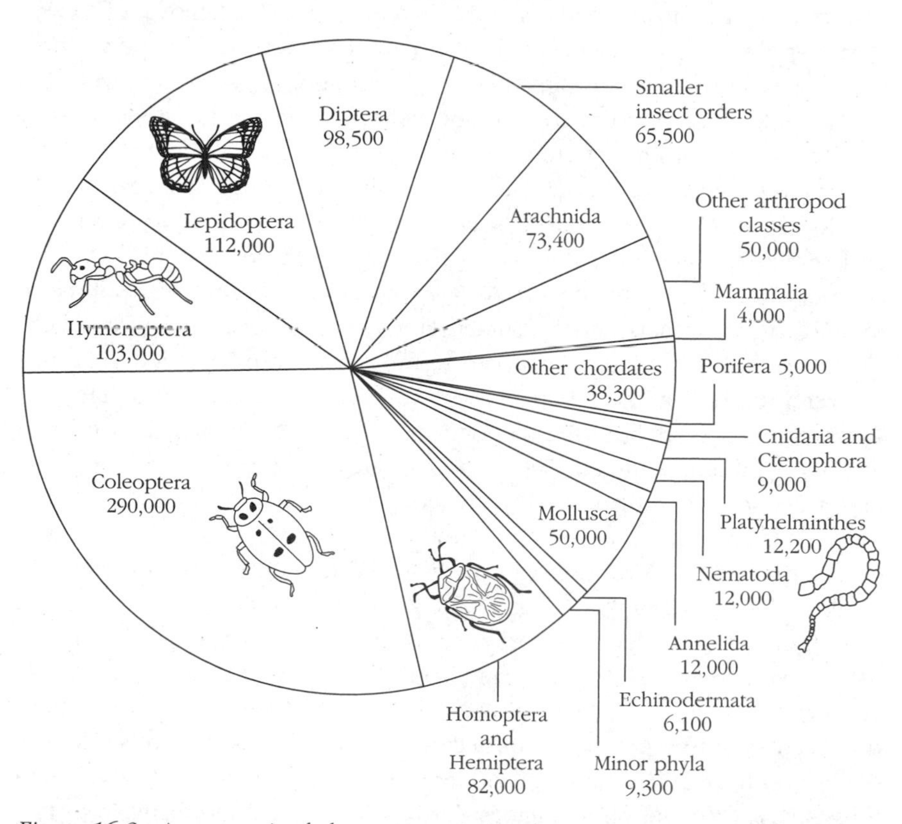

Figure 16-2 Among animals known to science, the insects are overwhelming in number. Because of this imbalance, most animal species live on the land; but most phyla (Echinodermata, etc.), the highest units of classification, are found in the sea.

> How would you like the job of pollinating trillions of apple blossoms some sunny afternoon in May? It's conceivable, maybe, that you could invent a machine to do it, but inconceivable that the machine could work as elegantly and cheaply as the honeybee, much less make honey on the side. Suppose you were assigned to turn every bit of dead organic matter, from fallen leaves to urban garbage, into nutrients that feed new life. Even if you knew how, what would it cost? A host of bacteria, molds, mites, and worms do it for free. If they ever stopped, all life would stop. We would not last long if green plants stopped turning our exhaled carbon dioxide back into oxygen. Plants would not last long if a few genera of soil bacteria stopped turning nitrogen from the air into nitrate fertilizer.[29]

Biodiversity contains the stored information and wisdom of nature. It is our natural library. Allowing a species to perish is like tearing a page out of a book of life, one which may prove important at some later date. Here are some recent discoveries of the benefits. An adult frog can eat its own weight in insects daily; diminishing frog populations in India have been linked to higher rates of malaria and crop damages. The removal of a **keystone species** causes drastic upheaval in an ecosystem. Other species previously excluded or kept in check by predation may now invade it, fundamentally altering the community. Reimplanting the keystone species may restore the community to its original state.

Such is the case of the sea otter, which lives in the Pacific Ocean off the Northwest coast from Alaska to southern California. Hunted by explorers and settlers for its fur, by the end of the nineteenth century this keystone species came close to extinction. As it disappeared, its prey, the sea urchin, multiplied in numbers and consumed large portions of kelp and other inshore seaweeds. The kelp, anchored to the sea bottom, constituted an underwater forest. But as the kelp was devoured, a seafloor desert replaced it. Conservationists restored the sea otter to its original habitat, and it once again preyed on the urchin. The sea forests grew back to their former glory, and a host of other algal species, crustaceans, squid, fishes, animal plankton, and even gray whales returned to the coastal area.

Pharmaceutical products and medicines are another critical benefit of biodiversity. A seemingly insignificant plant, the Madagascar periwinkle, was discovered by chance to produce the ingredient for drugs now being used to cure leukemia. In the Pacific Northwest, as noted earlier, the Pacific yew trees produce the chemical taxol, which has proven to have tremendous cancer-curing potential. For decades considered useless, this tree was eradicated from 90% of its original range by the clear-cutting of old-growth forests. Over 25% of all medicines derive directly and indirectly from forest plants; and as the yew tree demonstrates, the undiscovered potential of these plants is tremendous. From the saliva of the rain-forest leech has come the anticoagulant *hirudin,* which has proved important for treating hemorrhoids, rheumatism, thrombosis, and contusions, conditions where clotting blood is sometimes painful and dangerous.

Remarkable food sources remain undiscovered in rain-forest flora and fauna. About a dozen fruits currently dominate American and European markets. Wilson has noted that at least 3,000 other species are available in the tropics. A whole host of vitamin-rich tubers, vegetables, and other plant species are waiting in the wings. A winged bean of New Guinea has been called the "one species supermarket." The entire plant is edible, from its spinach-like leaves to young pods usable as green beans. Its young seeds are like peas; and its tubers, which may be boiled, fried, baked, or roasted, are richer in protein than potatoes. Mature winged bean seeds resemble soybeans and can be ground into flour or liquefied into a delicious beverage. The plant grows at a phenomenal pace, reaching 12 feet in a few weeks. It can also be used to raise soil fertility for other crops. The Amazon river turtle, an easily cultured food source, weighing 120 pounds, is a far better source of meat than cattle. It can be fed aquatic vegetation at minimal cost, producing 22,000 pounds of meat per acre—400 times the yield of cattle raised in nearby pastures.

Global Deforestation

Worldwide, only one-fourth of the original native forests remain. Half of the tropical forests have been destroyed, and in the United States, 95% of the original forest cover has been removed or converted to secondary forests, tree plantations, and farmlands. This loss of primary natural forests is the single greatest contributor to the loss of diversity. Preserving the tropical rain forests is particularly important because that is where the greatest amount of biodiversity occurs. A fifth or more of all the Earth's plant and animal species could be doomed to extinction within the next two decades unless radical measures are taken to save them. An area the size of a football field is being destroyed every second. Current satellite mapping studies suggest that the rate of deforestation continues to increase worldwide. Burning of the Brazilian rain forest is responsible for 336 million tons of CO_2 emissions—20% of the world's total—annually. The burning of the Indonesian rain forest accounts for 192 million tons annually, or 12% of the world's total CO_2 emissions.[30] This destruction is being caused by logging, the building of highways, slash-and-burn techniques designed to turn parts of the rain forest into grazing or cropland, and human settlements. Settlers or cattle ranchers cut the forest and burn it during the dry season and then use the ash to fertilize the crops. The land, however, is not suited for farming. It has no topsoil and is nutrient-poor, and it gives only a few years (as little as two or three) of use. The land is left a "wet desert," often completely ruined. Then the settlers or ranchers move on to a new area, slash and burn, and use the ash to fertilize the crops for a few more years. The tragedy is that sustainable economic use is possible. Harvesting latex, native meats, nuts, and fruit could be profitable, but short-term gain has driven the incursions. Violence has often accompanied the wreckage. In the 1980s the Amazon rubber tappers, led by Chico Mendes, attempted to halt the destruction of their forest. On December 22, 1988, Mendes was shot to death as he walked out of his house. Two cattle ranchers were indicted, convicted of the murder, and sent to prison—from which, a few years later, they escaped. The destruction of the rain forest continues unabated. Thousands of species are being lost every year. Wilson has pointed out that the tragedy is that poor nations possess most of the world's rain-forest riches and are under economic pressure to exploit them for short-term gain. The Brazilian environmentalist Jose Lutzenberger argues that stewardship of the rain forest should be an international, not just a national concern: "This talk of 'we can do with our land what we want' is not true. If you set your house on fire it will threaten the homes of your neighbors."[31]

Loss of Biodiversity in the United States

Only 5% of the original wilderness area of the United States is left, with a total of 10% of the land federally protected. However, many of these areas are badlands, desert, rock, and ice, and we have not saved much relative to other nations. The northern spotted owl, chosen by wildlife scientists as an indicator species of the overall forest health in the Pacific Northwest, is critically endangered, suggesting that the old-growth forest is also in decline. If logging continues at the same rate as during the 1980s, almost all of the old-growth habitat and the remaining healthy streams will be degraded in only 10 years.

Biological diversity depends on large continuous habitats. It is increasingly common for quite small areas or islands of old-growth forests (or similarly important natural areas) to be kept intact, surrounded by inroads of human habitation. The edges of these forest islands of old growth can deteriorate rapidly from exposure to winds, exotic species, and changes in the microclimate. In the Pacific Northwest, it has been observed that for each 25-acre clear-cut, an additional 35 acres of old-growth is degraded by these "edge effects." Due to the checkerboard cutting and fragmentation of roadless areas throughout the regions, 37% of all the old-growth now occurs in islands smaller than 400 acres in size, thus posing a huge threat to the habitats of many species.

Global Warming

The impact of global warming will probably make the world's current biological collapse look pale in comparison. Rapidly rising temperature will overwhelm many species' and ecosystems' ability to adapt. Widespread die-offs of forests and disruption of animal migrations will result if actions are not taken soon to slow global warming. See Chapter 13 for a further discussion of this problem.

Ancient forests act as an important carbon storehouse, and the old-growth forests of the Pacific Northwest contain the greatest biomass of any forests in the world, including tropical rain forests. The loss of these forests will further exacerbate the problem of global warming. Clear-cutting and then burning the remaining undergrowth contributes to greenhouse gas emissions in the Pacific Northwest. The conversion of forest lands account for 25% of all carbon dioxide emissions in the United States.

Wilson argues that respecting biodiversity is, first of all, prudent, but in such a way as to claim a place in a global ethic:

> The ethical imperative should therefore be, first of all, prudence. We should judge every scrap of biodiversity as priceless while we learn to use it and come to understand what it means to humanity. We should not knowingly allow any species or race to go extinct. And let us go beyond mere salvage to begin the restoration of natural environments, in order to enlarge wild populations and stanch the hemorrhaging of biological wealth. There can be no purpose more enspiriting than to begin the age of restoration, reweaving the wondrous diversity of life that still surrounds us.
>
> The evidence of swift environmental change calls for an ethic uncoupled from other systems of belief. Those committed by religion to believe that life was put on earth in one divine stroke will recognize that we are destroying the Creation, and those who perceive biodiversity to be the product of blind evolution will agree. Across the other great philosophical divide, it does not matter whether species have independent rights or, conversely, that moral reasoning is uniquely a human concern. Defenders of both premises seem destined to gravitate toward the same position on conservation.
>
> The stewardship of environment is a domain on the near side of metaphysics where all reflective persons can surely find common ground. For what, in the final analysis, is morality but the command of conscience seasoned by a rational examination of consequences? And what is a fundamental precept but one that serves all generations? An enduring environmental ethic will aim to preserve not only the health and freedom of our species, but access to the world in which the human spirit was born.[32]

Do Species Have Intrinsic Value and Rights?

It is one thing to accept the need for a global environmental ethic that includes biodiversity, based on our interdependence with the web of life. It is another to hold that species themselves have intrinsic worth, even rights. The Pittsburgh University philosopher Nicholas Rescher says, "When a species vanishes from nature, the world is thereby diminished. Species do not just have instrumental value. . . . They have a value in their own right—an intrinsic value."[33] And Callicott writes, "The preciousness of individual [animals] . . . is inversely proportional to the population of the species. . . . [T]he human population has become so disproportionate from the biological point of view that if one had to choose between a specimen of *Homo sapiens* and a specimen of a rare even if unattractive species, the choice would be moot."[34] And Richard Routley asks us to consider whether killing the last animal of its species is intrinsically worse than just killing the animal. He thinks there is something worse.[35]

Many people agree with Rescher, Callicott, and Routley: Species have intrinsic value and inherent rights to exist. But there are problems with this sort of thinking. First of all, supposing that you agree with these philosophers, why do you think species have intrinsic value? Note, they do not have the same qualities as individuals. Individuals, but not species, can be sentient, feel pain, be touched, fed, watered, nourished, harmed, and killed. Neither the dogwood species nor the dog species can suffer pain or experience joy. We can be said to have an obligation not to cause an individual animal unnecessary pain or suffering, but how can we have an obligation not to cause a species unnecessary pain or suffering? Suppose that individual animals have rights. How can species have them? Are species superorganisms?

In a seminal article "Why Do Species Matter?" Lilly-Marlene Russow argues that while members of a species may have rights and may have instrumental value for us, species themselves do not have intrinsic value or moral rights. If they did, then we would have an obligation to preserve them even if they had no place in an ecosystem and did us positive harm. She is skeptical about the whole notion of objects having intrinsic value:

> Unsurprisingly, the stumbling block [to attributing intrinsic value to species] is what this intrinsic value might be grounded in. Without an explanation of that, we have no nonarbitrary way of deciding whether subspecies as well as species have intrinsic value or how much intrinsic value a species might have. The last question is meant to bring out issues that will arise in cases of conflict of interests: is the intrinsic value of a species of mosquito sufficient to outweigh the benefits to be gained by eradicating the means of spreading a disease like encephalitis? Is the intrinsic value of the snail darter sufficient to outweigh the economic hardship that might be alleviated by the construction of a dam? In short, to say that something has intrinsic value does not tell us *how much* value it has, nor does it allow us to make the sorts of judgments that are often called for in considering the fate of an endangered species.[36]

She notes that if one believes that diversity itself is a value, one would have the obligation of creating a new species wherever possible, even harmful or bizarre ones that serve no purpose.

Russow, then, considers the comparison between species and natural wonders or works of art, suggesting that species might have some aesthetic value. The world would be a poorer place without bald eagles in the same way it would be poorer without the Grand Canyon or a great work of art. In these cases the experience of seeing these things is inherently worthwhile. Since in some cases diversity is a component of aesthetic appreciation, part of the previous intuition would be preserved. "There is also room for degrees of selectivity and concern with superficial changes: the variety of rat that is allowed to become extinct may have no special aesthetic value, and a bird is neither more nor less aesthetically pleasing when we change its name."[37]

But there are problems with this line of reasoning. While the lion and blue whale may have significant aesthetic qualities, some species seem to lack them. Furthermore, while things that have an aesthetic value may be compared and ranked, aesthetic value is not absolute and can be overridden for moral reasons:

> Someone who agrees to destroy a piece of Greek statuary for personal gain would be condemned as having done something immoral, but someone who is faced with a choice between saving his children and saving a "priceless" painting would be said to have skewed values if he chose to save the painting. Applying these observations to species, we can see that an appeal to aesthetic value would justify putting more effort into the preservation of one species than the preservation of another; indeed, just as we think that the doodling of a would-be artist may have no merit at all, we may think that the accidental and unfortunate mutation of a species is not worth preserving. Following the analogy, allowing a species to become extinct for mere economic gain might be seen as immoral, while the possibility remains open that other (human?) Good might outweigh the goods achieved by the preservation of a species.[38]

Still, a fundamental confusion exists, for we really do not admire species at all. They really do not have aesthetic or any other value. What we value is the existence of individuals with appropriate aesthetic qualities. Russow argues that changing the focus from species to individuals allows us to make sense of our feelings about endangered species in two ways:

> First, the fact that there are very few members of a species—the fact that we rarely encounter one—itself increases the value of those encounters. I can see turkey vultures almost every day, and I can eat apples almost every day, but seeing a bald eagle or eating wild strawberries are experiences that are much less common, more delightful just for their rarity and unexpectedness. Even snail darters, which, if we encountered them every day would be drab and uninteresting, become more interesting just because we don't—or may not—see them everyday. Second, part of our interest in an individual carries over to a desire that there be future opportunities to see these things again (just as when, upon finding a new and beautiful work of art, I will wish to go back and see it again). In the case of animals, unlike works of art, I know that this animal will not live forever, but that other animals like this one will have similar aesthetic value. Thus, because I value possible future encounters, I will also want to do what is needed to ensure the possibility of such encounters—i.e., make sure that enough presently existing individuals of this type will be able to reproduce and survive. This is rather like the duty that we have to support and contribute to museums, or to other efforts to preserve works of art.[39]

Russow believes that individual animals can have aesthetic value to a greater or lesser degree, that they are valued for their beauty, awesomeness, intriguing adaptations, rarity, and many other reasons. She concludes: "We have moral obligations to protect things of aesthetic value, and to ensure . . . their continued existence; thus we have a duty to protect individual animals (the duty may be weaker or stronger depending on the value of the individual) and to ensure that there will continue to be animals of this sort. . . . We value and protect animals because of their aesthetic value, not because they are members of a given species."[40]

I must leave a further analysis of the problem of the objective value of species to you to work through. Suffice it to say, even if species themselves don't have rights or intrinsic value, individual animals may well have such properties, and both fauna and flora have instrumental value, probably much more than we have realized. Given the whole web of life of which we are part and on which we depend for our health and survival, it behooves us to practice the Golden Rule with regard to nature—to do unto nature as we would nature do unto us. As Stephen Jay Gould has noted, nature holds all the cards and has immense power over us. We had better sign a contract with her while she is willing to make a deal. "If we treat her nicely, she will keep us going for a while. If we scratch her, she will bleed, kick us out, bandage up, and go about her business at her planetary scale. . . . She will uphold her end [of the bargain]; we must now go and do likewise."[41]

Review Questions

1. Consider the thought experiment (regarding copper mining) discussed at the opening of this chapter. Is the restoration thesis valid or invalid? Is a restored wilderness or natural entity just as valuable as the original?
2. Consider Hargrove's comments regarding the full-scale model cave with reproductions of the ancient paintings at Lascaux and that "the knowledge that the original still exists enhances the experiences afforded by the representation." Is this true? Does it have any bearing on the restoration thesis?
3. Consider the conservationist criticism that the preservationists, in appealing to the criterion of *originality* in nature against anthropocentric tampering, are themselves guilty of using an anthropocentric value. For example, while we would probably judge the copy to be less valuable than the original, this is a cultural judgment, not having to do with intrinsic or natural worth. Presumably, nature doesn't care in the least. Does this undermine the preservationists' argument for natural areas?
4. Consider the thought experiment mentioned in this chapter where God creates *Super Grand Canyon II*. How would a preservationist reply to this?
5. Consider Hettinger, Throop, and Wilson's claim that the need for the *wild* has been evolutionarily grounded in our nature. Can we change our nature to satisfy that need in other ways—for example, through religion, art, culture, virtual reality, or other creative activities? Just how plastic is our nature to adopt surrogate experiences for those of the wild? When I lived in an urban ghetto, where youth expressed fear of the wilderness, I noted that the tall buildings, fast automobiles, and other impressive artifacts impressed them more than they did me. They had a whole different repertoire of aesthetic values. How valid are these musings? Is the wild a primordial need so deeply rooted in us

that we reject or ignore it only at our peril? Are we oversocialized, or is further socialization the way of our salvation? Is urbanity our cure from nature or our curse, the instrument of alienation from our mother, nature?

6. Consider the quotation by Jose Lutzenberger where he argues that stewardship of the rain forest should be an international, not just a national concern: "This talk of 'we can do with our land what we want' is not true. If you set your house on fire it will threaten the homes of your neighbors." Do you agree? What does he mean? What should be done about the destruction of the rain forest?
7. Consider the question whether species have intrinsic value and/or rights. What about the statements by Rescher, Callicott, and Routley at the beginning of this section? Do you agree with them? Why or why not?
8. Examine Lilly-Marlene Russow's arguments against the thesis that species have values or rights. How might a preservationist or intrinsicalist like Rescher or Rolston respond to Russow? What do you think? Explain your judgment.
9. Here is a passage by David Quamman on *island biogeography,* the thesis that ecosystems function like island communities with special internal relations that can be upset by seemingly small human (or other animal) interventions. It relates to the idea of the web of life and the problem of species extinction. Read it over and discuss its relevance for the issues in this chapter:

> Let's start by imagining a fine Persian carpet and a hunting knife. The carpet is twelve feet by eighteen, say. That gives us 216 square feet of continuous woven material. Is the knife razor sharp? If not, we hone it. We set about cutting the carpet into 36 equal pieces, each one a rectangle, two feet by three. Never mind the hardwood floor. The severing fibers release small tweaky noises, like the muted yelps of outraged Persian weavers. Never mind the weavers. When we're finished cutting, we measure the individual pieces, total them up—find that, lo, there's still nearly 216 square feet of recognizably carpetlike stuff. But what does it amount to? Have we got 36 nice Persian throw rugs? No. All we're left with is three dozen ragged fragments, each one worthless and commencing to come apart.
>
> Now take the same logic outdoors and it begins to explain why the tiger has disappeared from the island of Bali. It casts light on the fact that the red fox is missing from Bryce Canyon National Park. It suggests why the jaguar, the puma, and 45 species of birds have been extirpated from a place called Barro Colorado Island—and why myriad other creatures are mysteriously absent from myriad other sites. An ecosystem is a tapestry of species and relationships. Chop away a section, isolate that section, and there arises the problem of unraveling. . . .
>
> [Scientists] have invented terms for this phenomenon of unraveling ecosystems. *Relaxation to equilibrium* is one, probably the most euphemistic. In a similar sense, your body, with its complicated organization, its apparent

defiance of entropy, will relax toward equilibrium in the grave. *Faunal collapse* is another. But that one fails to encompass the category of *floral* collapse, which is also at issue. Thomas E. Lovejoy, a tropical ecologist at the Smithsonian Institution, has earned the right to coin his own term. Lovejoy's is *ecosystem decay*. . . . What he means is that an ecosystem—under certain specifiable conditions—loses diversity the way a mass of uranium sheds neutrons. Plink, plink, plink, extinctions occur, steadily but without any evident cause. Species disappear. Whole categories of plants and animals vanish. . . .

Lovejoy's term is loaded with historical resonance. Think of radioactive decay back in the innocent early years of this century before Hiroshima, before Alamogordo, before Hahn and Strassmann discovered nuclear fission. Radioactive decay, in those years, was just an intriguing phenomenon known to a handful of physicists—the young Robert Oppenheimer, for one. Likewise, until recently, with ecosystem decay. While the scientists have murmured, the general public has heard almost nothing. Faunal collapse? Relaxation to equilibrium? Even well-informed people with some fondness for the natural world have remained unaware that any such dark new idea is forcing itself on the world.

What about you? Maybe you have read something, and maybe cared, about the extinction of species. Passenger pigeons, great auk, Steller's sea cow, Schomburgk's deer, sea mink, Antarctic wolf, Carolina parakeet: all gone. Maybe you know that human proliferation on this planet, and our voracious consumption of resources, and our large-scale transformations of landscape, are causing a cataclysm of extinction that bodes to be the worst such event since the fall of the dinosaurs. Maybe you are aware, with distant but genuine regret, of the destruction of tropical forests. Maybe you know that the mountain gorilla, the California condor, and the Florida panther are tottering on the threshold of extinction. Maybe you even know that the grizzly bear population of Yellowstone National Park faces a tenuous future. Maybe you stand among those well-informed people for whom the notion of catastrophic worldwide losses of biological diversity is a serious concern. Chances are, still, that you lack a few crucial pieces of the full picture.[42]

Suggested Readings

Botkin, Daniel. *Discordant Harmonies: A New Ecology for the Twenty-First Century*. Oxford: Oxford University Press, 1990.

Elliot, Robert. "Faking Nature." *Inquiry* 25:1 (March 1982). Reprinted in *Environmental Ethics,* ed. L. Pojman. Belmont, Calif.: Wadsworth, 1998.

Katz, Eric. *Nature as Subject: Human Obligation and Natural Community*. Lanham, Md.: Rowman & Littlefield, 1997.

Myers, Norman. *The Sinking Ark*. Oxford: Oxford University Press, 1980.

Nash, Roderick. *Wilderness and the American Mind*. New Haven: Yale University Press, 1967.

Norton, Bryan. *Why Preserve Natural Variety?* Princeton: Princeton University Press, 1987.

Oelschlaeger, Max. *The Idea of Wilderness: From Prehistory to the Age of Ecology*. New Haven: Yale University Press, 1991.

Pojman, Louis, ed. *Environmental Ethics,* 2nd ed. Belmont, Calif.: Wadsworth, 1998; Chapter 6.

Shoumatoff, Alex. *The World Is Burning*. Boston: Little, Brown, 1990.

Stone, Christopher. *Should Trees Have Standing?* Los Altos, Calif.: Kaufmann, 1974.

Wilson, Edward O. *Biophilia*. Cambridge: Harvard University Press, 1984.

———. *The Diversity of Life*. Cambridge: Harvard University Press, 1992.

Notes

1. Darrel Morrison, "Landscape Restoration in Response to Previous Disturbances," in *Landscape Heterogeneity and Disturbance,* ed. Monica Turner (New York: Springer, 1987), defines *restoration* as "returning a site to some previous state, with the species richness and diversity and physical, biological, and esthetic characteristics of that site before human settlement and the accompanying disturbances." Cited in William Throop, "The Rationale for Environmental Restoration," in *The Ecological Community,* ed. Roger Gottlieb (New York: Routledge, 1997).
2. The restoration thesis states that *in principle* we can restore all the values we've destroyed in the ecology.
3. William Throop believes that we should restore natural areas even though it is better to leave them alone in the first place. Wildness does impute value to environments, though that value is not absolute and can be overridden by other considerations. A more radical (*absolutist*) type of preservationist would reject such a compromise. Throop, "The Rationale for Environmental Restoration."
4. Martin Krieger, "What's Wrong with Plastic Trees?" *Science* 179 (3 February 1973). Reprinted in *Environmental Ethics,* 2nd ed., ed. L. Pojman (Belmont, Calif.: Wadsworth, 1998).
5. Robert Elliot, "Faking Nature," *Inquiry* 25:1 (March 1982), pp. 81–93. Reprinted in *Environmental Ethics,* ed. L. Pojman.
6. Eugene Hargrove, *Foundations of Environmental Ethics* (Englewood Cliffs, N.J.: Prentice-Hall, 1989), p. 169.
7. Baird Callicott, "The Wilderness Idea Revisited: The Sustainable Development Alternative," in *The Environmental Professional* 13 (1991), pp. 235–47.
8. Nabhan, quoted in Callicott, "The Wilderness Idea Revisited," p. 243.
9. Callicott, "The Wilderness Idea Revisited."
10. Holmes Rolston III, in "The Wilderness Idea Reaffirmed," *The Environmental Professional* 13 (1991), p. 375.
11. Rolston writes, "We can and ought to have rural nature, and we will be glad to have rural nature with a high bird count. But we can have a rural nature with a high species count and not have anywhere on the landscape the radical values of wild, pristine nature. The loss would not be compensated for by the stepped-up species count in agriculturally disturbed lands. In wilderness, we value the interactions as a fundamental component of biodiversity."
12. Ibid., p. 376.
13. Ibid., p. 377.
14. Rolston has written, "A biotic community is a dynamic web of interacting parts in which lives are supported and defended, where there is integrity (integration of the members) and health (niches and resources for the flourishing of species), stability and historical development (dependable regeneration, resilience, and evolution)." Holmes Rolston III, *Conserving Natural Value* (New York: Columbia University Press, 1994), p. 78.
15. Ned Hettinger and William Throop, "Refocusing Ecocentrism: De-emphasizing Stability and Defending Wildness," in *Environmental Ethics,* 2nd ed., ed. L. Pojman, 1998. Much of this section is beholden to this paper.
16. Michael Soulé, "The Social Siege of Nature," in *Reinventing Nature?* ed. Michael Soulé and Gary Lease (Washington, D.C.: Island Press, 1995, p. 143), quoted in Hettinger and Throop, "Refocusing Ecocentrism."
17. Baird Callicott, "Do Deconstructive Ecology and Sociobiology Undermine Leopold's Land Ethic?" *Environmental Ethics* 18 (Winter 1996); quoted in Hettinger and Throop, "Refocusing Ecocentrism."
18. Hettinger and Throop, "Refocusing Ecocentrism."

19. Eugene Hargrove, quoted in Hettinger and Throop, "Refocusing Ecocentrism."

20. William Throop, "Wilderness and the Preservation of the World," paper delivered at the U.S. Military Academy, West Point, New York, 20 February 1998.

21. Edward O. Wilson, *The Diversity of Life* (Cambridge: Harvard University Press, 1992), pp. 349–50. My thoughts are indebted to this excellent work.

22. Daniel Botkin, *Discordant Harmonies: A New Ecology for the Twenty-First Century* (New York: Oxford University Press, 1990).

23. Ibid., pp. 15–19. Botkin quotes Sheldrick's wife, who presented her husband's position on the matter: "Hasn't man always had a regrettable tendency to manipulate the natural order of things to suit himself. . . . With amazing arrogance we presume omniscience and an understanding of the complexities of nature, and with amazing impertinence we firmly believe that we can better it. . . . We have forgotten that we, ourselves, are just a part of nature, an animal which seems to have taken the wrong turning, bent on total destruction" (p. 18).

24. Ibid., p. 190.

25. Edward O. Wilson, *The Diversity of Life,* pp. 346–48.

26. Ibid., p. 280.

27. The Worldwatch Institute, "Life Support: Conserving Biological Diversity" (Washington, D.C., 1992).

28. Jonathan Schell, *The Fate of the Earth* (New York: Knopf, 1982), p. 65. After a nuclear holocaust, "the United States would be a republic of insects and grass." See also "The Planet of the Beetles," *National Geographic* 193:3 (March 1998) for a splendid description of the panoply and power of this species.

29. Donella H. Meadows, "Biodiversity: The Key to Saving Life on Earth," *The Land Steward Letter* (Summer 1990). Reprinted in *Environmental Ethics,* ed. L. Pojman, pp. 204–206.

30. Arnold Newman, *The Tropical Rainforest* (New York: Facts on File, 1990), p. 141.

31. Eugene Linden, "Playing with Fire," *Time* (18 September 1989). The story of Chico Mendez and the destruction of the rain forest is related in Lisa H. Newton and Catherine K. Dillingham's *Watersheds 2* (Belmont, Calif.: Wadsworth, 1997), chap. 4. My discussion is indebted to this source.

32. Wilson, *The Diversity of Life,* p. 351.

33. Nicholas Rescher, *Unpopular Essays on Technological Progress* (Pittsburgh: University of Pittsburgh Press, 1980), p. 79.

34. Baird Callicott, "Animal Liberation: A Triangular Affair," *Environmental Ethics* 2 (1980). Reprinted in *Environmental Ethics,* ed. L. Pojman.

35. Richard Routley, "Is There a Need for a New, an Environmental Ethics?" *Proceedings, 15th World Congress of Philosophy I* (1973), pp. 205–10.

36. Lilly-Marlene Russow, "Why Do Species Matter?" *Environmental Ethics* 3 (1981).

37. Ibid.

38. Ibid.

39. Ibid.

40. Ibid.

41. Stephen Jay Gould, "The Golden Rule—A Proper Scale for Our Environmental Crisis," *Natural History* (September 1990). Reprinted in *Environmental Ethics,* ed. L. Pojman, pp. 212–16.

42. David Quammen, *The Song of the Dodo: Island Biogeography in an Age of Extinctions* (New York: Scribner, 1996), pp. 11–12.

CHAPTER SEVENTEEN

Economics, Ethics, and the Environment

In April 1993 in his first Earth Day speech, President William Clinton solemnly promised to cut greenhouse gases so that by the year 2000 emissions would be at their 1990 level. It was a promise broken, for in 1998 the United States will have belched out 15% more CO_2 than it did in 1990. Why did President Clinton, someone who seemed environmentally enlightened, break his promise? Because keeping it, claims environmentalist Bill McKibben, would have very likely raised the price of gasoline to $2.50 a gallon. At that price, claims McKibben, "we'd drive smaller cars, we'd drive electric cars, we'd take buses—and we'd elect a new President. We can hardly blame Clinton, or any other politician. His real goal has been to speed the pace of economic growth, which has been the key to his popularity. If all the world's leaders could be gathered in a single room, the one thing that every last socialist, Republican, Tory, monarchist, and trade unionist could agree on would be the truth of Clinton's original campaign admonition: 'It's the economy, stupid.'"[1]

Economics is at the heart of the ecological crisis. Interestingly, both words, *economy* and *ecology,* derive from the Greek word for "home"—*economy* refers to the *laws* of managing a household; *ecology* refers to the *logic* of managing a household. One would think they are somehow related, and so they are. But unless one is a student of these subjects, the relationship between economics and the environment, especially environmental ethics, is elusive and clear explanations are not easily found. In the first part of this chapter I try to elucidate the meaning of the GNP (gross national product) as it affects broader aspects of national and, by extension, global health. Next, I introduce the topic of the relationship between economics and environmental ethics and analyze three problems. Then I outline an alternative approach to neoclassical economics—sustainable economics, based on environmental considerations rather than unlimited growth. Finally, I discuss two rival strategies for cutting down industrial pollution.

IS THE GROSS NATIONAL PRODUCT THE TRUE MEASURE OF NATIONAL HEALTH?

The United States is often compared favorably with the rest of the world because of its high gross national product (GNP), defined as the market value of all goods and services produced by an economy during a year. If the GNP is growing by more than 2%, we are making progress as a nation. If it is not, we are stagnating. But is this really true? Is the GNP a valid indicator of national health?

Suppose that two nations about the same size in land area and population had different economic systems. The two nations are the present United States and an imaginary futurist country called *Ectopia.* The United States has a flourishing economy, producing big cars, houses, skyscrapers, airplanes, and prisons, but it also uses a humongous amount of nonrenewable resources (such as fossil fuels and minerals) and pollutes its water, soil, and air, which contributes to a high cancer rate. Its way of life also encourages a high crime rate. However, a lot of money is transferred, so the GNP looks great! Pollution actually helps the GNP thrice over: first, the production that causes the pollution employs people; then society pays more people to clean the pollution up; then it pays health professionals to take care of the sick people who have been damaged by the pollution. Take the use of hazardous chemicals to make a product. People are paid wages in the production of the product, which adds to the total economy, but the pollution caused by the toxic waste dump, which seeps into an aquifer and contributes to cancer, is not counted against the GNP. If the government employs people to clean up the waste dump or aquifer, funds are credited to the national economy. But if people clean up the dump site or aquifer voluntarily, no credit is registered. The medical expenses paid to treat those harmed by the pollution also is credited as a gain to the economy. Cancer makes the economy look good!

In the United States forests are cut down to produce paper, furniture, and buildings; then monospecies of trees are planted to replace the clear-cutting. In Ectopia natural variety is maintained in the forests as only a minimal number of trees are cut down for homes and furniture, and paper is recycled.

In Ectopia people have few cars, preferring public transportation—electric trolleys—and low-cost bicycles. There is minimal pollution, so there is little to inspect or clean up; fewer doctors and related staff are needed. Most people have their own little gardens or farms and grow much of their own food. They voluntarily assist each other in building one another's homes and harvesting crops and cleaning up their neighborhood. Since little money is exchanged, their GNP compares unfavorably with that of the United States. They are an "impoverished country."

In the United States research scientists are always busy, making a lot of money by creating stronger pesticides to combat new strains of resistant pests and developing new treatments for diseases like cancer. In Ectopia most food is grown organically. Since Ectopians are predominantly vegetarians, they obtain as much as ten times the nutrition per acre of land as do their counterparts in the United States.

A refrigerator built in the United States lasts an average of 6 years. In Ectopia refrigerators are built to last about 60 years, so the resulting GNPs record that the United States has 10 times the productivity of Ectopia. Cars in the United States are assembled with built-in obsolescence, intended to last about 6 years, while those in Ectopia are made to run for 15 or 20 years. Shoes last about 1 year in the United States but 20 years in Ectopia so, once again, the United States shows a higher GNP. Ectopians think that the Western all-or-nothing option—being employed and working long hours or being unemployed—is foolish. Why not split the difference and have everyone work a shorter work week? Because of the obsolescence factor, the average worker in the United States works a 40-hour week, whereas Ectopians, because they don't need to replace their machines and goods as frequently, work an average of 25 hours per week, spending the rest of their time with family and friends, tending gardens, writing essays, creating art and music, continuing their education, hiking, and participating in local political forums. But of course, they must pay a price: Ectopian salaries are far lower—about 10% of salaries in the United States—but then, Ectopians don't need private medical insurance, a large police force, big cars, fancy clothes, minks, or expensive meat. But, of course, their lifestyle results in a miniscule GNP and makes them economically a real loser!

Movies and TV programs in the United States generate high revenues, as does advertising. Villages in Ectopia have their own local drama societies, which provide most of the entertainment at a fraction of the cost. Poetry reading, science and philosophy clubs, scouting, mountain climbing, and ecology clubs are huge successes in Ectopia—largely taking the place of movies and TV. Ectopians also like to tell stories and simply converse with one another. Wilderness fills them with rapture, and forest hiking and backpacking are their national pastime. While professional sports in the United States generate megabucks, no professional sports exist in Ectopia. Everyone who is able plays a recreational sport—baseball, hockey, lacrosse, basketball, tennis, soccer, volleyball, football—without pay. No one would think of paying to watch other people exercise. This deficiency, of course, categorizes Ectopians as an *undeveloped* country. Besides, Ectopians do not play golf, deeming golf courses wasteful uses of good land. They have transformed their golf courses into parks and gardens.

High crime rates in the United States mandate that a lot of police are employed, and criminal lawyers, judges, prisons, prison wardens, and guards are needed—all of which generate the exchange of money. Security and prison guards are one of the fastest-growing occupations in the United States, and some states spend more on their prison systems than on education.

Since Ectopia has a low crime rate, there are few police, lawyers, or prison guards—the money spent for crime prevention is only 1% per capita of what it is in the United States—a real loss for the economy. Crime makes the GNP look good!

Whereas tuition, room, and board at a good private college in the United States costs more than $30,000 per year, Ectopians enjoy inexpensive education, which often lasts a lifetime. Research and development go on in Ectopia, but the emphasis is on sustainable energy production, such as solar heating, electric cars, and preventive medicine.

Each year in the United States the GNP is enhanced by the estimated $150 billion in health-care expenses, by funeral costs for the approximately 350,000 people who die prematurely, and by other costs of damages caused by air pollution. Ectopia supports national health insurance, but if individuals demonstrably abuse the system (by smoking or alcohol abuse, for example) they lose their insurance for a period until they reform. This system seems to work. The cancer rate is a fraction of ours, as is the rate of atherosclerosis, liver, and kidney disease. Asthma and emphysema are rare. People jog regularly into their mid-80s. Whereas life expectancy in the United States is about 75 years (78 for women and 72 for men), it's well over 95 years in Ectopia—and equal for men and women.

Children in Ectopia are taught that being a morally good person is more important than being wealthy or famous; and since there are plenty of role models of unselfishness, courage, integrity, and love, there is less need to distrust people. The highest compliment one can receive is to be called "a morally excellent person." Altruism abounds and there are few mental health problems.

Ectopians seem happier than people in the United States. No one is hungry or homeless, as everyone cares about each other. When the Muir family home burned down two months ago, the neighbors all helped the Muirs to build a new home, even better than the old one. But again, since this neighborly assistance did not show up as new housing construction and an exchange of money, the GNP was not helped. In fact, while the United States is a financial success with its gigantic GNP, Ectopia appears to be a miserable failure. Americans have been touched by the apparent poverty of the Ectopians and have discussed ways to help them. The Americans are sending a team of government and business experts to Ectopia to help raise their standard of living to the U.S. level. What they need is prosperity, big cars, and industrial plants.

Now for a real-life situation: Although the South Indian state of Kerala has one of the lowest per capita incomes ($200 as compared with $810 for developing countries), it has a high life-expectancy rate of 70 years (compared with 59 years for India as a whole and 64 years for developing countries), a low infant mortality rate of 17 per 100,000 (compared with 79 for

India and 68 for developing countries), 100% literacy (compared with 50% for India). It maintains a high level of education, good schools, public housing, clean drinking water, sanitation, immunization, free or inexpensive medical care, and ration cards for rice and other basic commodities. Since 1960 land reform has distributed small plots of land to more than 3 million landless poor. Its 30 million people (the population of Canada) live in an area the size of Vancouver Island. People in Kerala seem to be among the most content in India, if not the world.[2] Lesson: a country with a low GNP can provide citizens with high standards of education, health, and hope. Different economic systems may produce different kinds of societies with different values. This thesis will be elaborated as the chapter continues.

ECONOMICS AND ENVIRONMENTAL ETHICS

While political leaders, ethicists, and socially concerned citizens have become increasingly aware of growing global environmental degradation, mainstream economics has taken relatively little note of this problem. Yet economics lies at the core of our malaise. Agricultural and industrial production, energy consumption, and pollution create enormous environmental stress. The question is whether mainstream economics has the resources to deal with these growing concerns or whether we need a whole new understanding of economics. This question is the subject of the present chapter.

Let us examine each of the terms in our subtitle: *economics, ethics,* and *environment* in order to understand their relationship to each other.

Economics is concerned with the production, distribution, and consumption of goods and services. It has to do with the allocation of scarce resources in a way that maximizes efficiency—that is, material social well-being. Economics arose out of a type of moral philosophy, classical utilitarianism, which we examined in Chapter 5. Classical utilitarianism stipulates that acts are morally right if and only if they maximize social welfare. Classical economics stipulates that a policy is economically right if and only if it maximizes material social welfare. But mainstream capitalist economics is also *libertarian*, calling for *laissez-faire* policies. People should be completely free to make market exchanges as they see fit. Government should not interfere with market transactions unless force or fraud are taking place. The theory, going back to Adam Smith (1723–90), is that an *invisible hand* guides individual self-interested transactions in such a way that they actually result in utilitarian outcomes. "Every individual intends only his own gain, and he is in this, as in so many other cases, led by an invisible hand to promote an end which was no part of his intention." This is almost too good to be true—it's OK to act selfishly, for it is the best way to promote universal welfare!

Mainstream economics claims to be a *science*, a pure neutral description of how the market, led by supply and demand, works. It is free from value commitments or moral prescriptions. But this is a myth. It presupposes a powerful ideology—that selfish behavior is good, that material growth is always good, and that the former leads inexorably by an invisible hand to the latter. It teaches us that poverty and the suffering of those who lose out in market competition are the lesser of evils—the price we must pay for growth and freedom. Whether or not or to what extent these claims are true, the point is that neoclassical economics is filled to the brim with values.

Ethics, as we saw in Chapter 2, is the study of right and wrong action, of good and evil. There are many moral theories, some being deontological, others being consequentialist (such as classical utilitarianism), and others being combinations of the two types. Depending on what ethical theory you hold, a given economic policy may or may not be morally right. The prescriptions of an economic policy and those of a moral theory may conflict. A particular allocation of resources may be economically good but morally bad, or economically wrong but morally right. For example, classical economics may approve a system that results in the rich getting richer and the poor, simply through market mechanisms, becoming so poor that they suffer terribly and even starve—while many ethical theories would condemn this state of affairs as grossly unjust and immoral. What is economically sound may not be morally so and, vice versa, what is morally sound may not be economically so.

Now add the third term, *environment.* This idea further complicates an already complicated relationship. There are several problems with treating environmental concerns from the perspective of classical economics. They all have to do with the economic notion of cost-benefit analysis (CBA) and how it affects decisions regarding optional courses of action.

Pareto-Optimality, Cost-Benefit Analysis, and Environmental Ethics

Economics tries to inform us how we can make trade-offs among material goods and services in the most efficient manner. This "most efficient manner" is referred to as *Pareto-optimal* (named after the Italian economist Wilfredo Pareto, who formalized the thesis). A state is Pareto-optimal when all economic resources are allocated and used efficiently, such that it is impossible to make anyone economically better off without making someone else economically worse off.[3]

The problem is that Pareto-optimal distributions can occur in an ethically and environmentally harmful way. A pattern of allocation can be efficient without being fair. Slavery could be efficient, even Pareto-optimal, but not fair. A system wherein the rich are very well off and the poor undeservingly miserable can be Pareto-optimal, but unjust. That is, Pareto-optimality has many of the weaknesses of its parent moral philosophy, utilitarianism.[4] Similarly, a pattern of using scarce environmental resources and creating pollution could be efficient in the short term without being morally good—or efficient in the long term. In our opening thought experiment we suggested that a nation's GNP may signify good economics but terrible environmental morality. This is especially so since economics usually evaluates local or national economies within short-term runs, whereas environmental concerns are international and intergenerational, dealing with distant and future people. A key feature of economic evaluation is cost-benefit analysis (CBA).

What is cost benefit analysis? It is a method we use almost every day of our lives. For example, you have $20 and two possibilities for this evening's activities. You can dine out with a friend and spend the whole $20 on a sumptuous meal or you can choose a simpler meal for $10 with another friend and then go to see a much-heralded movie. You estimate that you'll probably get more enjoyment from a simple meal and a movie with the second friend than a more sumptuous meal with the first friend, so you choose the second option. Students often say things like, "Well, I'd rather major in philosophy than business, but in the long run being a business major will bring more financial security." Normally, when we are faced with a choice of a major, a school, or commercial adventure, we compare the costs to the expected benefits to be derived from the costs. If the benefits significantly outweigh the costs, we proceed. If the costs are unacceptably high, we refrain from the adventure. Suppose we are a growing family, so that our house is getting crowded. We want to build a new larger house to suit our growing needs. We estimate the cost of the new house (minus profit from selling the old one) and compare it with the expected benefits of the new house. If constructing the new house will cost more money and time than we'd like to pay relative to the benefits, we content ourselves with our old house and try to accommodate ourselves to the situation. On the other hand, if the anticipated gains outweigh the anticipated cost, we proceed to build the house.

But some things are either not normally included in the total costs or not accurately measured. One of the costs of building a house may be cutting down some trees in the forest. Well, the forest will not miss a few trees, but if everyone starts to cut down trees, erosion may set in, depleting the topsoil. But the loss of the trees isn't usually factored in because it is an **external cost,** one not included in my cost-benefit assessment. I don't suffer from the cut-down trees, but my children may eventually have to live in a poorer environment—fewer trees, less topsoil, more fertilizer to substitute for the soil, more pollution from the petrochemical fertilizer as it leaches into the water supply, more cancer from the pollution, more medical bills, and shorter life spans from the cancer. These extra costs are left out of my cost-benefit analysis, but shouldn't they be included? We can categorize the issues involved in applying cost-benefit assessment to the environment as three different but related problems:

1. Classical economics holds that all costs and benefits can be expressed on a common scale (money) and thus can be compared with one another, even if they are not traded on the market. But environmental ethics, along with many standard ethical theories, holds that some things are *incommensurable*, not reducible to money or cash value. We say they are priceless or their price is infinite, not subject to economic comparison.
2. Economics typically has to do with short-term contracts between contracting parties, discounting the future; but environmental concerns are typically long term, reaching into the future and transcending the interests of the contracting parties to include those of future generations. That is, economics is mainly local in time and space, whereas environmental goals tend to be global and stretch far into time.
3. When doing a cost-benefit analysis, economics deals with actual anticipated costs and benefits between business transactors. We call these *internal* costs, because they are internal to the transaction. But some costs are not taken into account in the actual price, such as the pollution caused by an upstream company's dumping chemical waste into a river used by people who live downstream. The downstreamers pay *external* costs not included in the original transaction.

Let us analyze the three problems just mentioned and then discuss ways in which economics might take account of the claims of environmental ethics.

The Problem of Incommensurability

In 1513 the Roman Catholic Pope Leo X, needing money to complete the magnificent St. Peter's Cathedral in Rome, issued full indulgence, which included remission of sins as well as the elimination of punishment in purgatory, to those who made significant financial contributions to the Church. The priest in charge of collecting the indulgence fees in Germany was John Tetzel (d. 1519), a Dominican monk from Leipzig. Tetzel went from parish to parish proclaiming the good news of salvation for hire:

> As soon as the money in the coffer clings
> The soul from purgatory springs.

It was a big success. St. Peter's was built and paid for. However, one German monk protested. On October 31, 1517, Martin Luther posted 95 theses

"How much would you pay for all the secrets of the universe? Wait, don't answer yet. You also get this six-quart covered combination spaghetti pot and clam steamer. Now how much would you pay?"

on the door of the Wittenberg Cathedral, arguing that God's grace didn't have a market price. You can't buy forgiveness of sin, and eternal life cannot be reduced to monetary value.

An economic cost-benefit analysis treats all goods as commensurable, having a common denominator, money. It makes sense to ask the comparative monetary value of a four-bedroom house in the country or of a new sports car. But it seems impossible to put a price on life or love, freedom, clean air, or the wilderness. Just as Pope Leo X made a category mistake in treating forgiveness as a commercial commodity, so economists make a category mistake in treating life or environmental concerns as reducible to monetary evaluations.

This doesn't mean that we can't make ordinal rankings. We can say that it is better to save ten humans rather than one, or that the enhanced quality of urban life is worth the statistical risk of a few more deaths, all things being equal. But sometimes things are not equal. If I have to choose between killing my mother or my innocent child and saving ten or even a hundred other people, I am at a loss for what to do. Given a big enough number I might agree that I had an obligation to kill my loved ones (for instance, the rest of the world would perish if I didn't kill them). But I still wouldn't know how to put a monetary value on these people. If my children were subject to a cost-benefit analysis, a price could be fixed at which point I would be willing to sell my children into slavery. But I would not sell my children into slavery for any price—or if I did, I would consider myself to have done a terrible wrong. Nor do I think that intelligence or freedom can be reduced to monetary assessment. Otherwise, we could buy some people as slaves if they didn't value their freedom at a high enough price. Perhaps we could offer them a bargain: 4,000 hedons of enjoyment for every year they serve as Simon Legree's slave.

Mark Sagoff attributes the crass materialism of the cost-benefit analysis to the economist's fixation with *preference*. The individual's preferences constitute the standard by which we are to judge his or her well-being. They are neither right nor wrong in themselves, but the bases of right and wrong. It follows that we can simply evaluate the worth of any object by the relative preference for it. How much would you be willing to pay for it?

> How much do you value open space, a stand of trees, an "unspoiled" landscape? Fifty dollars? A hundred? A thousand? This is one way to measure value. You could compare the amount consumers would pay for a town house or coal or a landfill and the amount they would pay to preserve an area in its "natural" state. If users would pay more for the land with the house, the coal mine, or the landfill, than without—less construction and other costs of development—then the efficient thing to do is to improve the land and thus increase its value. That is why we have so many tract developments. And pizza stands. And gas stations. And strip mines. And landfills. How much did you spend last year to preserve open space? How much for pizza and gas? "In principle the ultimate measure of environmental quality," as one basic text assures us, "is the value people place on these . . . services or their *willingness to pay*."[5]

Sagoff argues that not all values can be reduced to mere preferences. We also make judgments about what is valuable, judgments that may conflict with our preferences. As *consumers* we have preferences and make choices based on them, but we also regard ourselves as *citizens*. As consumers we choose for ourselves, but as citizens we judge what we think good for the community. Our judgments and preferences may conflict. Sagoff illustrates the tension between the two categories:

> I am a schizophrenic. Last year, I fixed a couple of tickets, and was happy to do so since I saved fifty dollars. Yet, at election time, I helped to vote the corrupt judge out of office. I speed on the highway; yet I want the police to enforce laws against speeding. I used to buy mixers in returnable bottles—but who can bother to return them? I buy only disposables now, but to soothe my conscience, I urge my state senator to outlaw one-way containers. I love my car; I hate the bus. Yet I vote for candidates who promise to tax gasoline to pay for public transportation. I send my dues to the Sierra Club to protect areas in

> Alaska I shall never visit. . . . I support almost any political cause that I think will defeat my consumer interests. This is because I have contempt for—although I act upon—those interests. I have an "Ecology Now" sticker on a car that leaks oil everywhere it's parked.[6]

In January 1969 the U.S. Forest Service approved a plan by Walt Disney Enterprises to develop a ski resort in California's Mineral King Valley, a wilderness area next to Sequoia National Park. Disney Enterprises would have built a complex of hotels, restaurants, swimming pools, parking lots, ski lifts, and ski slopes to attract large crowds, offering recreation to an average of 14,000 visitors daily. To provide access to the resort, the state of California would have constructed a 20-mile-long highway, a section of which would have traversed Sequoia National Park. The Sierra Club filed a suit in a federal court, arguing that the Forest Service failed to take adequately into account the aesthetic and ecological factors, which outweighed the economic ones. In 1972 in *Sierra Club v. Morton,* the Supreme Court decided in favor of the Sierra Club.

Sagoff reports using this case in a class to illustrate the discrepancy between our consumer and citizen mind-sets. He first asked his students how many of them had visited or thought they would visit Mineral King as a wilderness area. A few hands went up. Then he asked them how many would like to visit it if it were developed as a Disney resort and ski park. A lot more hands went up. Then he asked them if they thought the government was right in giving Disney Enterprises a lease to develop Mineral King. "The response was nearly unanimous. The students believed that the Disney plan was loathsome and despicable, that the Forest Service had violated a public trust by approving it, and that the values for which we stand as a nation compel us to preserve the little wilderness we have for its own sake and as a heritage for future generations. On these ethical and cultural grounds, and in spite of their consumer preferences, the students opposed the Disney plan to develop Mineral King."[7]

If this analysis is accurate, it would seem that preferences themselves can be evaluated as good or bad, in which case the moral goal would be to acquire higher preferences, to allow our choices to be guided by our moral judgment. Our judgments as citizens may override our preferences as consumers. Hence, a simple preference-based cost-benefit analysis seems inadequate as a guide to political policy.

Economists often suppose that in deciding on environmental policy all we have to do is ask people how much they will pay to save the rain forest or the Hetch Hetchy or Mineral King. They call this the willingness-to-pay test (WTP). But WTP is not always a reliable indicator. The Israeli philosopher Avner de-Shalit performed the following experiment. He asked a group of 40 students to imagine that an ecological disaster had occurred, and that there was an urgent need to clean our country's coast. The first 20 students were given papers in which they were asked whether they would contribute 1% of their salaries that month to clean the coast. They all said yes. Then Shalit asked them whether they thought it would be possible to know the average amount that their fellow students would be willing to pay. Ninety percent replied that it would be possible; the average WTP predicted was 1.85% of one's salary. The second group of 20 students were told the same story but asked whether they would pay 4% of their pay. They acquiesced to this, and thought the average WTP would be 6.16% of one's salary. The conclusion is that the WTP reflects nothing about "individuals' autonomous wills," but rather it reflects the fact that the starting point affects the outcome.[8] Cultural practices and expectations are often decisive factors in determining individual preferences and willingness to pay for goods and services. So culture determines economic decisions. But if we assume a materialist, hedonistic culture from the outset, we're likely to get an economic system that serves those vacuous goals.

All sorts of other difficult questions arise regarding cost-benefit analysis (CBA). How should we use the surplus projected as a result of balancing our federal budget? Should we use all of it to protect social security or use it to pay our basic federal bills, so that we can lower taxes for the citizenry? And how useful is CBA for environmental planning? Classical economists argue that we can use CBA to guide our environmental planning, while many environmental economists argue that CBA is too crude a method to be of much assistance with environmental issues. Miller suggests that using "cost-benefit analysis to make decisions is somewhat like trying to detect a car speeding at 100 miles per hour with a radar device so unreliable that at best it can only tell us that the car's speed is somewhere between 50 mph and 5,000 mph."[9] On the neoclassical side, Robert Goodland and George Ledec admit that while CBA has the sort of deficiencies just discussed, it can be useful. "Even unreasonably low or highly inaccurate estimates of environmental benefits and costs are better than none, because the alternative is to assume implicitly that these benefits and costs are zero. Rather than abandoning CBA, environmentalists should insist that it take environmental and other social costs explicitly into account."[10] The exact value of CBA is still a highly controversial matter.

The Problem of the Discount Rate and Future Generations

A related problem is that of the **discount rate**—an estimate of a resource's future economic value compared to its present value. The size of the discount rate is a primary factor affecting the outcome of any cost-benefit analysis. If you are offered the choice between $1,000 today or $1,000 in 10 years, you will no doubt choose the $1,000 today, since you could invest it at interest and expect to reap a sizable profit in 10 years. Furthermore, you estimate that $1,000 today is equivalent to $2,000 in 10 years, accounting for inflation. Another reason to choose $1,000 today is that you can't count on the future; for example, you might be dead or independently wealthy or someone who doesn't need money—say, a hermit. A bird in the hand is worth two in the bush. We tend to value present over distant profit, discounting the future. Economists tend to discount the future at a high (5 to 10%) discount rate. The U.S. Office of Management and Budget uses a 10% annual rate. They calculate that technological innovation or changes in consumer preferences will affect the value of a given product.

When we apply this kind of thinking to the environment, however, we are faced with a problem. Suppose an area of a rain forest is worth $10 million today. Given the standard 10% annual discount rate, this area will be worth less than $4 million in 10 years and about $85,000 in 50 years. So it makes economic sense to cut down the rain forest and invest the profits.

Environmentalists reject this kind of thinking about natural resources. High discount rates tend to encourage immediate, short-term exploitation and destruction of the environment, its air, water, topsoil, forests, biodiversity, and minerals, as opposed to long-term utilization—all in the name of faith in technology. It is questionable whether we can put a money value on such resources as clean air and water or the value of human health or life and we have no idea what the future value of a species will be. Even if we agree to play the game of putting a dollar value on the wilderness or clean air, different people will assign different values to these things. Furthermore, self-interest tends to affect our estimates of even monetary assessments. For example, one industry-sponsored cost-benefit study estimated that compliance with a standard to protect workers from vinyl chloride would cost between $65 and $90 billion. When the standard came into effect, the cost of compliance came to $1 billion.[11]

The matter of valuing an environmental resource, such as the rain forest or a species, is very difficult because we are interested in long-term value—not what will be their value in 10 years, but what will they be worth in 500 or 1,000 years, which is impossible to predict. The concept of a discount rate doesn't seem even to apply to these kinds of situations. Nevertheless, our present tendency to discount the future at high rates is environmentally destructive. Reason would seem to suggest that different resources will have different discount rates and that unique and scarce resources should be protected by having a 0% or even a negative discount rate. Since future people don't exist, they can't vote, but we can apply the Golden Rule and ask ourselves, "What sort of environment would we want people of this generation to leave us, if we were living 100 (or 500) years from now?" Although we can't know everything about what they would want and need for a good life, we can be certain that clean air, clean water, and nourishing food are among those things, and we can be pretty sure that they include wilderness and the benefits of biodiversity. And this is what we are aiming for in conserving nature.

The Problem of External Costs

If the overall effects of environmental degradation are more harmful than beneficial, why does so much of it exist? People aren't simply maliciously wasting resources and energy and polluting for the fun of it. The answer is that those who—through their agricultural or industrial production or energy consumption or pollution—harm the environment gain personally more than they lose. The logic is the same as we saw in the tragedy of the commons (see Chapter 4). In producing a product which is both useful and degrading, the producer gains the whole unit of utility (minus internal costs), but pays only a small fraction of the degrading effects, those effects being dispersed widely over the whole population and on to distant and future people.

These additional costs are called *externalities*. All economic goods and services have both internal and external costs. For example, the price you pay for a car includes the firm's costs of raw material, labor, machinery, marketing, shipping, and the like. Once purchased, you have to pay for the car's gasoline, maintenance, and repair. All of these direct costs that the seller and buyer pay are called *internal costs*. They are paid directly by the agents to the transaction.

However, extracting and processing raw material to make and fuel cars depletes resources, produces hazardous wastes, pollutes the air and water, increases the buildup of harmful greenhouse gases, and contributes to illness in people who have nothing to do with the purchase of the car, some of whom are yet unborn. These negative effects are *external costs*, which are passed on to workers, the general public, and distant and future people.

Economists tend to dismiss external costs as a minor defect in the otherwise beneficial market system, a defect that can be easily rectified by selling pollution and water rights and so forth; but it is doubtful that air and water can be divided in this way. And this solution doesn't take into account the fact that it is impossible for future people to own property now. Because business has a need to continue to grow, because degradation is often a byproduct of economic growth, and because entrepreneurs pay only a fraction of the total cost of their degradation, they have no economic incentive to factor in external costs. In fact, the company that breaks ranks and does take account of them, voluntarily installing costly environmentally sensitive mechanisms, will be at a competitive disadvantage. It may go bankrupt as the polluters smile at such naïveté.

This is the lesson of the tragedy of the commons. Prudential reason dictates that if one can attain a unit of profit by adding a unit of degradation to the environment, which will be dispersed throughout the environment, one ought to do so, for the personal profit outweighs the personal cost. The solution to the problem of external costs is the same as it is in any tragedy of the commons: "mutually agreed upon, mutually coercive regulations" which cancel out any advantage that flows from the fruits of degrading production. Governments must intervene in the economy, passing laws to penalize polluters and tax resource use, so that companies are sufficiently deterred from wasteful and degrading practices. The goal is fairness: to get the business to *internalize externalities,* to see to it that the price of the product reflects its *full cost,* including its long-term effects on the environment. As Miller puts it, "The goals are *to have prices that tell the environmental truth* and *to have people and businesses pay the full costs of the harm they do to others and the environment.*"[12]

SUSTAINABLE ECONOMICS

What exactly is *sustainability* or *steady-state economics*? Two of its proponents, Robert Goodland and George Ledec, define *sustainability* this way: "a pattern of social and structural economic transformations (i.e., development) which optimizes the economic and other societal benefits available in the present, without jeopardizing the likely potential for similar benefits in the future."[13] That is, we need to reject, rethink, or revise mainstream economics to take into account long-term environmental welfare. What would implementing this policy involve? What can be done to change the way economics relates to the environment? Several things. First of all, we must recognize that the GNP is an inadequate measure for determining social success. It has some validity in systems like ours in terms of measuring a certain level of material wealth, but after basic needs are met, it tends to be less and less valid as an index of prosperity. It tends to measure everything in terms of material growth, but, the point is, we must move from a "growth" model to a more adequate model with a plurality of indices. As environmental economist Herman Daly has said, when the only tool you have in your toolbox is a hammer, everything tends to look like a nail. Daly and other environmental economists urge us to adopt a new nongrowth, steady-state model:

> The steady-state economy is basically a physical concept with important social and moral implications. It is defined as a constant stock of physical wealth and people. This wealth and population size are maintained at some desirable, chosen level by a low rate of throughput of matter and energy resources so as a result the longevity of people and goods is high. Throughput is roughly equivalent to GNP, the annual flow of new production. It is the cost of maintaining the stocks of final goods and services by continually importing high-quality

> matter and energy resources from the environment and exporting waste matter and low-quality heat back to the environment.
>
> Currently we try to maximize the growth of the GNP, but the reasoning just given suggests that we should relabel it "gross national cost," GNC. We should minimize it, subject to maintaining stocks of essential items. For example, if we can maintain a desired, sufficient stock of cars with a lower throughput of iron, coal, petroleum, and other resources, we are better off, not worse. To maximize GNP throughput for its own sake is absurd. Physical and ecological limits to the volume of throughput suggests that a steady-state economy will be necessary. Less recognizable but probably more stringent social and moral limits suggest that a steady-state economy will become desirable long before it becomes a physical necessity.[14]

Extracting from Daly's statement, *sustainable economics* can be described by three essential theses that are not found in mainstream (called *neoclassical*) economics: (1) It fits the paradigm of physical sciences with its notions of the laws of thermodynamics and closed systems, (2) it holds that economics is inherently value-laden and, as such, it should be so shaped as to serve moral goals, and (3) it has implications for restricting population growth and resource depletion and pollution that seem radical to those of us who have been indoctrinated in mainstream economics. Let us look at each of these points.

The New Paradigm: Economics and the Laws of Thermodynamics

Environmental economics changes the paradigm of mainstream economics from a self-contained system with infinite resources and infinite growth to one of limits, limited by the laws of physics, especially the two laws of thermodynamics. It supposes that economics is limited by the fact that Earth is a closed, not an open, system. What does this mean? A system is open if it exchanges both matter and energy with its environment. A system is closed if it exchanges energy but not matter from its environment. The Earth is a closed system, since it receives energy from the Sun, but virtually no matter is exchanged (we will ignore the tiny amounts of meteorites). Matter cannot be created or destroyed, so all production in Earth's system (the ecosphere) involves the transformation of a fixed amount of matter into other material forms according to the laws of physics. Unlike the Earth itself, its parts, such as humans and ecosystems, are open systems, receiving both matter and energy from their environment.

The first law of thermodynamics is that energy can be neither created nor destroyed. The second law (the entropy law) states that energy doesn't disappear, but is transformed into heat. In every energy transformation, some of the original energy is changed into heat energy and is no longer able to be transformed into other forms of energy. When coal burns, its chemical energy changes into heat. When a pendulum swings, kinetic energy is transformed to potential and back again, but with each swing, friction removes some energy as heat. Electric energy in a light filament is changed partly to light, but partly to heat. The light falls on the walls and floor and surroundings and is then converted to heat energy. In all energy transformations, some of the original energy becomes heat—that is, disordered molecular motion. Once it is heat, energy cannot be recovered in its original form.

The entropy law is vital to our understanding of economics. It tells us that we can't go on using the same energy forever. We will have to rely more on the sun's energy and use processes that are low in entropy.

This use of entropy to support the idea of sustainable or "stationary" economics goes back to John Stuart Mill, who wrote:

> But in contemplating any progressive movement, not in its nature unlimited, the mind is not satisfied with merely tracing the laws of its movement; it cannot but ask the further question, to what goal? . . .
>
> It must always have been seen, more or less distinctly, by political economists, that the increase in wealth is not boundless: that at the end of what they term the progressive state lies the *stationary state,* that any progress in wealth is but a postponement of this, and that each step in advance is an approach to it . . . if we have not reached it long ago, it is because the goal itself flies before us [as a result of technical progress].
>
> I cannot . . . regard the stationary state of capital and wealth with the unaffected aversion so generally manifested toward it by political economists of the old school. I am inclined to believe that it would be, on the whole, a very considerable improvement on our present condition. I confess I am not charmed with the ideal of life held out by those who think that the normal state of human beings is that of struggling to get on; that the trampling, crushing, elbowing, and treading on each other's heels which form the existing type of social life, are the most desirable lot of human kind, or anything but the disagreeable symptoms of one of the phases of industrial progress. The northern and middle states of America are a specimen of this stage of civilization in very favorable circumstances; . . . and all that these advantages seem to have yet done for them (notwithstanding some incipient signs of a better tendency) is that the life of the whole of one sex is devoted to dollar-hunting, and of the other to breeding dollar-hunters.

Mill thought it obvious that natural resources were limited, so that we would sooner or later have to modify our production processes in the light of this fact—even if technology could delay it somewhat. His point is that we should value a high quality of life. Development of our institutions, not unlimited material growth, should be our goal. He continues:

> Those who do not accept the present very early stage of human improvement as its ultimate type may be excused for being comparatively indifferent to the kind of economical progress which excites the congratulations of ordinary politicians; the mere increase of production and accumulation. . . . I know not why it should be a matter of congratulations that persons who are already richer than anyone needs to be, should have doubled their means of consuming things which give little or no pleasure except as representative of wealth. . . . It is only in the backward countries of the world that increased production is still an important object: in those most advanced, what is economically needed is better distribution, of which one indispensable means is a stricter restraint on population.[15]

Economics should not aim at maximizing material wealth, but be a servant to moral ends. Poor nations may need to concentrate on material goods, but rich countries have enough material goods and should be aiming at a high quality of life, which includes restricting population growth. We should aim at zero population growth.

> There is room in the world, no doubt, and even in old countries, for a great increase in population, supposing the arts of life to go on improving, and capital to increase. But even if innocuous, I confess I see very little reason for desiring it. The density of population necessary to enable mankind to obtain, in the greatest degree, all the advantages both of cooperation and of social intercourse, has, in all the most populous countries, been attained. A population may be too crowded, though all be amply supplied with food and raiment. It is not good for a man to be kept perforce at all times in the presence of his species. . . . Nor is there much satisfaction in contemplating the world with nothing left to the spontaneous activity of nature; with every rood of land brought into cultivation, which is capable of growing food for human beings; every flowery waste or natural pasture plowed up, all quadrupeds or birds which are not domesticated for man's use exterminated as his rivals for food, every hedgerow or superfluous tree rooted out, and scarcely a place left where a wild shrub or flower could grow without being eradicated as a weed in the name of improved agriculture. If the earth must lose that great portion of its pleasantness which it owes to things that the unlimited increase of wealth and population would extirpate from it, for the mere purpose of enabling it to support a larger, but not a happier or a better population, I sincerely hope, for the sake of posterity, that they will be content to be stationary, long before necessity compels them to it.

Then Mill reiterates his thesis that steady-state economics can free us to develop our culture, social institutions, and individual lives:

> It is scarcely necessary to remark that a stationary condition of capital and population implies no stationary state of human improvement. There would be as much scope as ever for all kinds of mental culture, and moral and social progress; as much room for improving the Art of Living and much more likeli-

> hood of its being improved, when minds cease to be engrossed by the art of getting on. Even the industrial arts might be as earnestly and as successfully cultivated, with this sole difference, that instead of serving no purpose but the increase of wealth, industrial improvements would produce their legitimate effect, that of abridging labor.

Mill's views on economics have been rejected by mainstream economists, but rediscovered by environmental economists like Kenneth Boulding, Herman Daly, Paul Hawken, Michael Jacobs, and others. Boulding contrasts mainstream economics (suitable to a "cowboy" mentality with wide-open frontiers) with "spaceship economics."[16] Earlier civilizations imagined themselves living on an illimitable plane. There was always more room further on, a seemingly endless frontier. Cowboys in the Wild West were profligate in their use of resources and pollution, for resources were plentiful and nature healed itself of their wounds and cleansed itself of their pollution. It was not until airplanes began circling the Earth that the global nature of our planet entered into our collective consciousness. The implications of this closed, spaceship-like world still haven't captured our psychological and political imagination. We still think like cowboys—there are abundant resources for everyone, illimitable space for billions more; and Earth will heal and cleanse itself of our damage and harm to its infrastructure. We identify with Julian Simon: The Earth's resources are infinite.

The new eco-economists call such thinking illusory—the Great Delusion. Rather than infinite resources on an illimitable plane, open to infinite growth, they argue that there are severe limits to growth. Earth is a closed system, so that the time has come to recognize the ethical imperative of a *no growth* philosophy.[17] Development replaces growth when you have only limited resources. The environmentalist Donella Meadows writes:

> What we need is smart development not dumb growth. When something grows, it gets quantitatively bigger. When something develops, it gets qualitatively better. Smart development invests in insulation, efficient cars, [bikes], and ever-renewed sources of energy. It ensures that forests and fields continue to produce wood, paper, and food, recharge wells, harbor wildlife, and attract tourists. Dumb growth crashes around looking for more oil. It clear-cuts forests to keep loggers and sawmills going just a few more years until the trees run out. It covers the landscape with the same kind of honkytonk ugliness tourists leave home to escape. We need to meet dumb growth with some questions. What really needs to grow? Who will benefit? Who will pay? What will last?[18]

Better not *bigger* signifies Mill's approach to environmental economics. We must learn to live off our interest, not our precious capital. So unlimited growth is an illusion. Granted, as Mill notes, *substitutability,* brought on by technology, can lengthen the life of our resources—but the cornucopian may put too much faith in human ingenuity to replace natural resources.

The pace of our depletion and pollution threatens to outrun our technology; and the sheer scale of the complexity of adjusting to all of the variegated problems involved in new processes (such as new drugs having unanticipated side effects that may be harmful) is overwhelming. Mill's idea of slowing down the birth rate as well as the growth rate in favor of sustainable development and high-quality living seems uncommon common sense. His ideas are even more relevant today than in his own time.

Economics and Morality: The Ultimate Ends

Cornucopian economist Julian Simon described humans as Earth's *ultimate resource,* with the ability, through science and technology, to extend our material resources forever, to infinity. Herman Daly, on the other hand, argues that our resources are essentially limited—even with human-devised substitutes for natural resources. He adds a feature to economics: an *ultimate end.* This means that economics is made for the human good, not the human good for economics.[19] That is, we must first figure out what our goals are and then use economics to help get us there. In understanding these goals, we must realize and incorporate moral principles. Daly argues that there are objective moral norms, which we simply must reaffirm as our guides at every step of development. Traditional free-enterprise economics has relied on an invisible hand, operating in the free market to work things out for the best. But critics point out that the free market, while valid to the point of providing incentive, has weaknesses. Besides the invisible benevolent hand, which leads self-interest unwittingly to serve the common good, there is the *invisible foot,* which "leads self-interest to kick the common good to pieces." While private ownership and private use under a competitive market give rise to the invisible hand, public ownership with unrestrained private use gives rise to the invisible foot. What is needed is public restraint—what Hardin calls "mutually agreed upon, mutually coercive rules"—to preserve our environmental capital and to distribute resources more equitably. Daly's idea of sustainable development holds that we must modify market processes so that they both take into account the Earth's capital and redistribute resources globally—in order that everyone has enough, but no one has more than his or her fair share. Daly would abolish unrestricted inheritance and restrict earnings to around $100,000 per annum. He would also have economics take into account the fact that we have obligations to future generations to leave nature's capital intact, so they can live lives as good as ours.

The Radical Implications of Sustainable Economics

From the point of view of mainstream economics, the revised version of Mill's zero-sum environmental economics has radical implications for restricting population growth and resource depletion and pollution. Daly would impose rigorous family planning restrictions and resource-depletion quotas onto society. A stationary population with low birth and death rates would lead to a greater percentage of old people than in the present growing population; the average age would likely change from 27 to 37. Here is the heart of his proposal:

> The economic and social implications of the steady state are enormous and revolutionary. The physical flows of production and consumption must be *minimized, not maximized,* subject to some desirable population and standard of living. The central concept must be the stock of wealth, not, as presently, the flow of income and consumption. Furthermore, the stock must not grow. For several reasons, the important issue of the steady state will be distribution, not production. The problem of relative shares can no longer be avoided by appeals to growth. The argument that everyone should be happy as long as his absolute share of the wealth increases, regardless of his relative share, will no longer be available. Absolute and relative shares will move together, and the division of physical wealth will be a zero-sum game. In addition, the arguments justifying inequality in wealth as necessary for savings, investment, and growth will lose their force. With production flows (which are really *costs* of maintaining stock) kept low, the focus will be on the distribution of the stock of wealth, not on the distribution of the flow of income. Marginal productivity theories and "justifications" pertain only to flow and therefore are not available to explain or "justify" the distribution of stock ownership. Also, even though physical stocks remain constant, increased income in the form of leisure will result from continued technological improvements. How will it be distributed, if not according to some ethical norm of equality? The steady state would make fewer demands on our environmental resources but much greater demands on our moral resources. In the past, a good case could be made that leaning too heavily on scarce moral resources, rather than relying on abundant self-interest, was the road to serfdom. But in an age of rockets, hydrogen bombs, cybernetics, and genetic control, there is simply no substitute for moral resources and no alternative to relying on them, whether they prove sufficient or not. . . .
>
> With constant physical stocks, economic growth must be in nonphysical goods: service and leisure. Taking the benefits of technological progress in the form of increased leisure is a reversal of the historical practice of taking the benefits mainly in the form of goods and has extensive social implications. In the past, economic development has increased the physical output of a day's work while the number of hours in a day has, of course, remained constant, with the result that the opportunity cost of a unit of time in terms of goods has risen. Time is worth more goods, a good is worth less time. As time becomes more expensive in terms of goods, fewer activities are "worth the time." We become goods-rich and time-poor.[20]

So steady-state economics would promote leisure growth, which would result in pushing up the price of material intensity relative to time intensity.

In addition pollution costs would be internalized by charging pollution taxes. Economic efficiency requires only that a price be placed on environmental benefits. It does not tell us who should pay the price. A free-marketer may pass these costs off onto the consumer. "Whoever wants air and water to be cleaner than it is should pay for it." But surely clean air and water ought to be a basic right of all citizens, so that the polluters, not the consumers, should pay these costs. Furthermore, it seems impractical to demand that we trace the causes of our diminished quality of life and health, including cancer, back to their ultimate sources, the ones who caused the pollution. Lag time and the mixing of molecules, as well as our inability to trace them, militate against such measures. Government regulations are not the complete answer to resource depletion and pollution, but they are part of it. Whether there is a place for pollution vouchers, whereby companies may buy and sell the right to pollute, is a difficult issue, which depends on the level of pollution in a given area and other economic needs.

Daly asks what political institutions will provide the control necessary to keep the stocks of wealth and people constant, with maximal individual freedom. He answers, "It would be far too simpleminded to blurt out, 'socialism' as the answer, since socialist states are as badly afflicted with growthmania as capitalist states." Marxism does not have an enlightened population policy. "However it is equally simpleminded to believe that the present big capital, big labor, big government, big military type of private profit capitalism is capable of the required foresight and restraint and that the additions of a few pollution and severance taxes here and there will solve the problem."[21]

Our concerns about developing sustainable global economics require taking into consideration the current enormous disparities between the rich and the poor, including the rich and poor nations. Statistics show that the richest fifth of the world receives 82.7% of its income, whereas the poorest fifth receives 1.4% of its income. While these statistics may be misleading, not taking into account different types of social systems, it is incontrovertible that millions of people live in extreme poverty. The current estimates put that number of people who are not able to meet their basic economic needs at 1.2 billion. Why that is so and what should be done about it, are difficult and highly controversial matters which we have touched on in Chapters 11 and 12. In addition, some have suggested forgiving the less developed countries a large part of the more than $2 trillion debt now owed to the richer nations. Biodiversity of flora and fauna and the rain forests should be viewed as global resources that require a shared responsibility. The wealthier nations should help countries like Brazil and Indonesia, where rain forests are in danger, to develop sustainable agricultural policies, concentrating on activities such as rubber tapping, fruit production, and medicinal uses of the myriads of plants and herbs located in these forests. Perhaps the rich nations need to compensate the poor nations for not cutting down the rain forests.

Here is a list of the kinds of broad policies that global economic sustainability would require:

- Reducing poverty in developing countries, including improvement of literacy for all
- Slowing population growth
- Encouraging and rewarding Earth-sustaining behavior
- Using penalties and taxes to discourage Earth-degrading behavior
- Using environmental indicators to measure progress toward environmental and economic sustainability and human well-being
- Using full-cost pricing, which includes the external costs of goods and services in their market prices
- Using a low discount rate for evaluating the future worth of irreplaceable or scarce resources
- Establishing global public agencies to manage and protect public lands and fisheries
- Revoking the government-granted charters of environmentally and socially irresponsible businesses
- Making environmental concerns a key part of all trade agreements and of all loans made by international lending agencies
- Reducing waste of energy, water, soil, and mineral resources
- Reducing future ecological damage and repairing past ecological damage

POLLUTION PERMITS AND PRIVATE PROPERTY

There is one other mechanism for reducing pollution that has been put into effect as a regulating procedure: issuing marketable pollution permits (MPP), allowing companies to sell them or trade them in for financial benefits. This may be the most controversial proposal of all. On the face of it, it seems contradictory to aim at reducing pollution by issuing permits to pollute, but such permits, tied as they are to financial incentives to reduce pollution, may be just the mechanism needed in a free-market economy to have the desired result.

The idea of marketable pollution permits has become an increasingly popular proposal for environmentalists and economists since J. H. Dales first set forth the idea in 1968.[22] In 1990 Congress adopted MPPs as part of its Clean Air Act Amendment. The government decides that a certain type of pollutant in a geographical area be reduced by a certain percentage and provides market incentives to do so. Firms that cannot reach their goal of reduced emissions may purchase a permit from the party whose reduction exceeds the mandated amount, thus providing an incentive for each party to reduce pollution below the required level.

Suppose that in a given area there are ten power plants spewing out 100 million tons of SO_2, at an average of 10 million tons per plant. The EPA decides that they need to reduce this to 30 million tons of the pollutant. Older and less efficient plants will have a harder time getting their emissions down to the average 3 million tons, while newer and more efficient plants can with some innovations go even lower than their quota. It is now in the interests of the newer plants to reduce their pollution to a very low level so that they can sell their pollution permits to the older plants. The government can adjust its goals and determine how many permits will be given out each year.

Pollution permits have received a mixed review. Environmentalists have praised the mechanism as a step forward, combining environmental concerns with free-market forces. MPPs have the advantage over Hardin's "mutually coercive, mutually agreed upon regulations" in that they substitute economic incentive for governmental coercion. Power plants have reduced SO_2 and particulates dramatically. CFCs and nitrogen oxides in the atmosphere have also been reduced. But free marketers like Martin Anderson, Robert McGee, and Walter Block have criticized MPPs as a strange amalgam of socialism and market incentive—"market socialism."[23] First, the trading of pollution permits create a bureaucratic command structure: Effectively regulating the numerous types of chemical pollutants in the air, land, and water will demand an unwieldy plethora of government-created and monitored markets. Second, they are a pervasive violation of property rights. In reality, they are licenses to pollute, payoffs to companies for not harming us. But it makes no more sense to pay people for not polluting than it does to pay criminals for not robbing us. Anderson has put the matter this way:

> Fortunately, there is a simple, effective approach available—long appreciated but underused. An approach based solidly on . . . private property rights. At its root all pollution is garbage disposal in one form or another. The essence of the problem is that our laws and the administration of justice have not kept up with the refuse produced by the exploding growth of industry, technology, and science.
>
> If you took a bag of garbage and dropped it on your neighbor's lawn, we all know what would happen. Your neighbor would call the police and you would soon find out that the disposal of your garbage is your responsibility, and that it must be done in a way that does not violate anyone else's property rights.
>
> But if you took that same bag of garbage and burned it in a backyard incinerator, letting the sooty ash drift over the neighborhood, the problem gets more complicated. The violation of property rights is clear, but protecting them is

> more difficult. And when the garbage is invisible to the naked eye, as much air and water pollution is, the problem often seems insurmountable.
>
> We have tried many remedies in the past. We have tried to dissuade polluters with fines, with government programs whereby all pay to clean up the garbage produced by the few, with a myriad of detailed regulations to control the degree of pollution. Now some even seriously propose that we should have economic incentives, to charge polluters a fee for polluting—and the more they pollute the more they pay. But that is just like taxing burglars as an economic incentive to deter people from stealing your property, and just as unconscionable.
>
> The only effective way to eliminate serious pollution is to treat it exactly for what it is—garbage. Just as one does not have the right to drop a bag of garbage on his neighbor's lawn, so does one not have the right to place any garbage in the air or the water or the earth, if it in any way violates the property rights of others.
>
> What we need are tougher, clearer environmental laws that are enforced—not with economic incentives but with jail terms.
>
> What the strict application of the idea of private property rights will do is to increase the cost of garbage disposal. That increased cost will be reflected in a higher cost for the products and services that resulted from the process that produced the garbage. And that is how it should be. Much of the cost of disposing of waste is already incorporated in the price of the goods and services produced. All of it should be. Then only those who benefit from the garbage made will pay for its disposal.[24]

Exactly how we go about monitoring air and water pollution is unclear. It's notoriously difficult to trace CO_2 or particulates in the air, or DDT or dieldrin in rivers and lakes, or even toxic waste in the underground water supply, as we noted in earlier chapters. Nor is it easy to see how property rights to air, lakes, oceans, and underground aquifers could be established. Would Bill Gates buy up one-half of the Pacific Ocean and Warren Buffett the other? Granted, allowing these areas to remain commons has drawbacks, but, at this point in history, the idea of turning the commons into private property seems both unworkable and, supposing we could extend the free market in this way, fraught with dangers of concentrating even more wealth in fewer hands. Free-market libertarians, with their unbounded faith in market mechanisms and technology to solve our environmental problems, would replace government regulations with consumer litigation. The burden is not on the firm to reduce pollution but on the citizen to prove harm. Whatever the wisdom of this shift in the burden of responsibility, we have reason to be wary of a process that increases litigation and puts more power in the hands of the legal system, judges, and juries. Anyone who has ever been sued, even by a frivolous suit, must shiver at the very thought of centering so much power in the courts. If the jurors in the O. J. Simpson case were illiterate with regard to the significance of

DNA findings, how much hope is there for the average citizen to understand the significance of biological magnification or the complexity of chemical pollution?

We asked earlier in this chapter whether mainstream economics could be extended to take environmental concerns adequately into account or whether a whole new way of doing economics is needed. The precise answer to that question is still in doubt, but, at the very least, new modes of thought have to supplement our traditional ways of viewing economics. We have to reevaluate the meaning of the discount rate and external costs in order to take into account future and distant people. Such resources as biological diversity, animal rights, and the ozone layer have to be considered in our calculations. We have to transcend our local perspective and think globally and with a view to the needs of posterity. If mainstream economics can reform in order to take account of these values, so much the better. If it cannot, we have to develop a new sustainable, Green economy. Paul Hawken has stated the basic Golden Rule for such an economy, "Leave the world better than you found it, take no more than you need, try not to harm life or the environment, and make amends if you do."[25]

UNCERTAINTY OVER ENVIRONMENTAL SAFETY

Many people, especially cornucopians, object that we have no idea what the future will bring, no idea of how to define safe minimal standards of CO_2 in the atmosphere, and no idea of the globe's carrying capacity for population. Almost all past forecasts about the future have been falsified, so why should we take the present doomsday reports seriously? Technology brings surprises. The future is simply unpredictable. So, we might ask, what does the principle of sustainability come to in the light of such overwhelming uncertainty?

The response is that we have learned something from experience, enough to cause us to exercise prudence about what we are doing, to conserve rather than spend all of our environmental capital. This may be illustrated by the idea of game theory. We have two possibilities: controlling potentially deleterious environmental effects and not controlling them. If we control, we will certainly have to pay higher costs for our energy, food, and pollution control; constrain our use of cars; and so forth. If we do not control our environmental spending, but adopt a *laissez-faire* policy, we are faced with gargantuan global uncertainty. If climatologists are right in projecting catastrophic changes in climate, then the cost of not controlling CO_2 emissions will be massive. So the benefits of controlling against catastrophe will be enormous. On the other hand, if the climatologists are wrong, and no great danger is in store, the cost of *laissez-faire* policy will be zero.

Table 17-1 The Payoffs of Environmental Policy Under Uncertainty

Policy	*If Catastrophe Occurs*	*If There Is No Catastrophe*	*Maximum Cost*
Sustainable development model: environmental control	Large benefit—control cost	Control cost	Control cost
Neoclassical model: laissez-faire	Catastrophe cost	0	Catastrophic cost

Based on Michael Jacobs, *The Green Economy* (London: Pluto Press, 1991), p. 100.

These possibilities are illustrated in the strategy of *minimizing the maximum cost,* developed by the British environmental economist Michael Jacobs. This Pascalian version of cost-benefit reasoning would direct us to adopt a policy of controlling pollution and resource degradation against the possibility of catastrophe.* If our doomsday predictions turned out to be wrong, we would pay the control costs but get no benefits—a bit like taking out an insurance policy on which we never collect. But if we adopt a laissez-faire policy and our projections are correct, we would experience a catastrophe (Table 17-1). Given the humongous costs of paying for a catastrophe—and it might be irreversible—reason would seem to dictate that we play it safe by adopting safe minimum standards for environmental pollution, hazardous waste, resource consumption, species preservation, and so forth. As Jacobs puts it, "Given scientific ignorance, prudent pessimism should be favored over hazardous optimism. . . . Uncertainty means that in many cases it may not be possible to set these sustainability targets with great accuracy or confidence. But in most circumstances we shall find that setting some is preferable to setting none at all."[26] In other words, given uncertainty, we should take out the environmental insurance policy.[27]

Review Questions

1. Review the thought experiment on the GNP. How accurate is this as a depiction of our using the GNP as a measure of success? Is the thought experiment hopelessly utopian? Discuss your ideas on the relationship of the GNP to environmental quality.

---

* Blaise Pascal (1632–62), French mathematician, physicist, and philosopher, argued that since we can prove neither that God exists nor doesn't exist and since the stakes are so high, we ought to bet that He does exist. The argument is known as "Pascal's Wager."

2. Discuss the relationship between economics and environmental ethics. What are the main problems with mainstream economic methods of assessment?
3. Discuss mainstream economics in relationship to future generations and animal rights. Does it account for these two concerns? Should it?
4. Analyze John Stuart Mill's idea of a stationary-state economy. How valid are his criticisms of mainstream economics? Based on your understanding of economics, do you think mainstream economics can accommodate environmental concerns, or do we need a whole new way of thinking about economics that more accurately takes environmental concerns into account?
5. Discuss Sagoff's distinction between preference and value judgments, which distinguish us as consumers and citizens. Is his characterization of the "schizophrenia" existing between these two categories accurate? Or can we reconcile the two categories into a common denominator?
6. Analyze Herman Daly's notion of *sustainable development*. Why does he think that "growth" economics is an immoral practice? Do you agree with him? How does the situation of the South Indian state of Kerala support his theses? Explain your answer.
7. Explain how the policy of granting pollution permits works. Is it a morally acceptable means of containing pollution? Will it work? Explain your answer.
8. Explain how free marketers (Anderson, McGee, and Block) would solve our pollution problems by creating property rights. Do you think this is a viable solution to our environmental problems? Explain your answer.

Suggested Readings

Daly, Herman, and John B. Cobb, Jr. *For the Common Good: Redirecting the Economy Toward Community, the Environment, and a Sustainable Future.* Boston: Beacon Press, 1994.

Daly, Herman, and Kenneth N. Townsend, eds. *Valuing the Earth: Economics, Ecology, Ethics.* Cambridge: MIT Press, 1993.

Field, Barry. *Environmental Economics: An Introduction.* New York: McGraw-Hill, 1997.

Haslett, David. *Capitalism with Morality.* New York: Oxford University Press, 1994.

Hawken, Paul. *The Ecology of Commerce.* New York: HarperCollins, 1993.

Jacobs, Michael. *The Green Economy.* London: Pluto Press, 1991.

Miller, G. Tyler. *Living in the Environment,* 10th ed. Belmont, Calif.: Wadsworth, 1998.

Perman, Roger, Yue Ma, and James McGilvray. *Natural Resources and Environmental Economics.* New York: Longman, 1996.

Sagoff, Mark. *The Economy of the Earth.* New York: Cambridge University Press, 1988.

Simon, Julian. *The Ultimate Resource.* Princeton: Princeton University Press, 1996.

Notes

1. Bill McKibben, "A Special Moment in History," *Atlantic Monthly* (May 1998), p. 76.
2. G. Tyler Miller, *Living in the Environment,* 10th ed. (Belmont, Calif.: Wadsworth, 1998), p. 709.

3. In a sense Pareto-optimality almost never obtains, since virtually any arrangement, however good, will leave someone worse off than he or she would have been.
4. Mane Hajdin reminded me that theoreticians prefer Pareto-optimality to utilitarianism because it does not depend on interpersonal utility comparisons.
5. Mark Sagoff, "At the Shrine of Our Lady of Fatima, or Why Political Questions Are Not All Economic," *Arizona Law Review* 23 (1981).
6. Ibid.
7. Mark Sagoff, *The Economy of the Earth* (New York: Cambridge University Press, 1988), p. 51.
8. Avner de-Shalit, "Is Liberalism Environment-Friendly?" *Social Theory and Practice* 21 (Summer 1995), p. 300.
9. Miller, p. 712.
10. Robert Goodland and George Ledec, "Neoclassical Economics and Principles of Sustainable Development," in *Environmental Ethics,* 2nd ed., ed. L. Pojman (Belmont, Calif.: Wadsworth, 1998), p. 515. See also Kristin Shrader-Frechette, "A Defense of Risk-Cost-Benefit Analysis," in the same volume, for a helpful discussion of this issue.
11. Miller, *Living in the Environment,* p. 712.
12. Ibid., p. 710. Miller writes, "Should we shift to full-cost pricing? As long as businesses receive subsidies and tax breaks for extracting and using virgin resources and are not taxed for the pollutants they produce, few will volunteer to reduce short-term profits by becoming more environmentally responsible. Suppose you own a company and believe it's wrong to subject your workers to hazardous conditions and to pollute the environment beyond what natural processes can handle. If you voluntarily improve safety conditions—but your competitors don't—your product will cost more than theirs, and you will be at a competitive disadvantage. Your profits will decline; you may eventually go bankrupt and have to lay off your employees. . . . One way of dealing with the problem of harmful external costs is for the government to levy taxes, pass laws, provide subsidies, or use other strategies that encourage or force producers to include all or most of these costs in the market prices of economic goods and services. Then that price would be the *full* cost of these goods and services: internal costs plus short- and long-term external costs. The goals are to have prices that tell the environmental truth and to have people and businesses pay the full costs of the harm they do to others and the environment."
13. Goodland and Ledec, "Neoclassical Economics," p. 519.
14. Herman Daly, "The Steady-State Economy in Outline," in G. Tyler Miller, *Living in the Environment,* p. 723.
15. John Stuart Mill, *Principles of Political Economy* (New York: Colonial Society, 1899). All quotes from Mill are from Vol. II, Book IV, Chapter VI, pp. 260–64.
16. Kenneth Boulding, "The Economics of the Coming Spaceship Earth," in *Environmental Quality in a Growing Economy,* ed. H. Jarrett (Baltimore: Johns Hopkins University Press, 1966).
17. One reader wondered whether my "no-growth philosophy" meant that we should cease to have children. We are permitted (even encouraged) to have children, for creating persons is one of the most important things we can do, but we can have too much of a good thing. Given the problems of population growth and all that goes with it, we need to restrict our procreative impulses. Aiming at fewer children but higher quality of lives for them should be our goal.
18. Quoted in G. Tyler Miller, *Living in the Environment,* p. 705.
19. Herman Daly, "Introduction to *Essays Toward a Steady-State Economy,"* in *Valuing the Earth: Economics, Ecology, Ethics,* ed. Herman E. Daly and Kenneth N. Townsend (Cambridge: MIT Press, 1993).

20. Ibid., pp. 34–36.
21. Ibid., p. 39.
22. John H. Dales, *Pollution, Property, and Prices* (Toronto: University of Toronto Press, 1968).
23. See Robert McGee and Walter Block, "Pollution Trading Permits as a Form of Market Socialism and the Search for a Real Market Solution to Environmental Pollution," in *Fordham Environmental Law Journal* 6:1 (1994).
24. Martin Anderson, "George Bush: Environmentalist," *The Christian Science Monitor*, 4 January 1989.
25. Paul Hawken, quoted in Miller, *Living in the Environment*, p. 719.
26. Michael Jacobs, *The Green Economy* (London: Pluto Press, 1991), p. 100.
27. I am greatly indebted to Dean Dudley, Mane Hajdin, Sterling Harwood, Stephanie Lofgren, and Michael Levin for critically and constructively analyzing an earlier version of this chapter.

CHAPTER EIGHTEEN

The Challenge of the Future: From Dysfunctional to Sustainable Society

In this concluding chapter I want to synthesize some material from previous chapters, developing points made earlier in this work. In Chapter 17, I argued against libertarians that the answer to pollution and environmental degradation is not to privatize the commons, but to develop more innovative incentives and regulations. In this chapter I want to elaborate the further implications of environmental stewardship on our notion of private property. Next I want to turn to the matter of the ecology of the city. Much of this book has been about the wilderness or natural habitats, as though cities were anti-environmental. I want to correct this impression, focusing on sustainable city life. Then I will bring together many of the themes in this book, pointing toward a universal environmental ethic and law that requires an international body to regulate. Finally, I want briefly to suggest some practical things we all can do to live responsibly toward the environment. First, I turn to the matter of the uses of private property.

PRIVATE PROPERTY AND ENVIRONMENTAL ETHICS

> *Nor shall private property be taken for public use without just compensation.*
>
> —From the fifth amendment of the Constitution of the United States of America, commonly known as the *Taking Clause*

In January 1986 David Lucas purchased two beachfront lots on South Carolina's Isle of Palm for $975,000; he planned to build a house on one lot and sell the other lot to a home builder. In 1988 the state passed the Beachfront Management Act, which prohibited building in the area, arguing that the proposed building projects would harm the dunes and hence were a nuisance. Lucas brought the case to a local court, which awarded him $1.2 million in compensation. However, the South Carolina Supreme Court reversed the decision, citing a rule that where property is taken to prevent public harm, no compensation is required. Lucas appealed to the U.S. Supreme Court, which heard his arguments on March 2, 1992. Justice John Scalia wrote the Court's decision, which reversed the South Carolina Supreme Court decision, remanding the case back to the state court to produce an equitable settlement. A strong component in the majority decision was the *Taking Clause* of the fifth amendment. Justice Scalia also quoted from Justice Oliver Wendell Holmes's exposition in *Pennsylvania Coal Co. v. Mahon* (1922) that "while property may be regulated to a certain extent, if regulation goes too far it will be recognized as a taking." And if a taking, then compensation is required. Justice Scalia conceded that the operative phrase "if regulation goes too far" was vague and relative to custom, but in *Agins v. Tiburon* the Court explained that a regulation goes "too far" if it "denies an owner [*all*] economically viable use of his land." Justice Scalia thought that this was occurring in Lucas's case.

Justice Harry Blackmun wrote a spirited minority dissent in which he defended the South Carolina Supreme Court's decision because it rested "on two premises that until today were unassailable—that the State has the power to prevent any use of property it finds to be harmful to its citizens, and that a state statute is entitled to a presumption of constitutionality."

> If the state legislature is correct that the prohibition on building in front of the setback line prevents serious harm, then, under this Court's prior cases, the Act is constitutional. "Long ago it was recognized that all property in this country is held under the implied obligation that the owner's use of it shall not be injurious to the community, and the Taking Clause did not transform that principle to one that requires compensation whenever the State asserts its power to enforce it." (*Keystone Bituminous Coal Assn. v. DeBenedictis*) The Court consistently has upheld regulations imposed to arrest a significant threat to the common welfare, whatever their economic effect on the owner.[1]

Blackmun's argument rests on the same reasoning as zoning laws (such as those that prohibit pornography from being sold openly in stores or taverns from being opened in certain residential communities). Just because I buy a piece of land with the intention of building a liquor store on it does not automatically mean that I'm entitled to compensation when a zoning law rules out my project. However, if no law exists and I start to build my liquor store, and then the state zones out liquor stores, I am entitled to compensation. The fact that Lucas had not actually started to construct his buildings before the law was passed seems to weigh against compensation. On the other hand, the state probably should offer to buy the property from him at a fair price. Or perhaps an environmental organization should purchase Lucas's land at a fair price.

But the larger issues are (1) whether the state has the right to prohibit what it considers harmful use of the land and (2) whether it has to compensate anyone when it prohibits ecological degradation from occurring on the land. If land is valuable for the common good and posterity, environmentalists argue, we are doing harm by depleting its quality. The ramifications of this idea are far-reaching. Many artificial lakes and reservoirs in the

Southwest, used mainly for recreation, cover precious bottom land, land with good topsoil. We are running out of good topsoil. So should we prohibit interest groups from constructing these artificial lakes? Dammed-up lakes also affect the natural environment in deleterious ways. Should this be a reason for prohibiting their construction? even if the land is privately owned? Golf courses often take up similarly good land and typically require enormous quantities of water and fertilizer, with the latter seeping into the underground water supply. Should we prevent private country clubs from creating golf courses? What about highways and parking lots and shopping malls constructed on land that will one day be needed for farming? And should owners be compensated for the restrictions we put upon them? Similarly, suppose I want to fill in wetlands I own in order to construct a shopping mall. The courts have ruled that the state may forbid my filling in these wetlands, since wetlands provide natural habitats for wildlife, help reduce flooding by slowing the runoff of heavy rains, and promote the cycling of nutrients. Should it be made to compensate me for not filling in the wetlands? This would be very costly and likely strain the state's coffers.

Taking the environment seriously into consideration certainly limits our freedom to do what we want with the land. Private property no longer has the same meaning it once had. Until recently we thought that the owner had virtually absolute control of his or her private property, absolute discretion on what uses he or she saw fit for the land—just as long as no one else was being harmed. But if environmentalists are right, there is no such absolute right to do what you want with your land. Rather, we have to extend the notion of harm to include future use—even to future people. Joseph Sax calls such protections of nature "public rights," since their infringement need not harm any identifiable person; rather, violation of these rights may bring diffuse harm and affect future people or existing people in the future. We now have to say that property rights are severely constricted since the land is precious, a common good that must remain in good condition for future use in perpetuity. We are not absolute owners of the land, but stewards of it.

This thinking is not new. Suppose a pharmaceutical company, at great cost, produces a powerful drug that it believes will be a cure for some malady. But then it turns out that the drug has unanticipated harmful side effects. The company has no claim to compensation from the state, which has prohibited the use of the drug, but must swallow the loss itself. Similarly, when we discover that a use of the land, hitherto thought to be benign, is harmful, the state has a right to prohibit such use without having to compensate the owner. Or take the fact that refrigerators and air conditioners used chlorofluorocarbons (CFCs). In our ignorance we thought that such use was benign. Now we have evidence that CFCs are depleting the ozone layer, contributing to the increased incidence of cancer. Aren't we within our rights to pass legislation restricting the use of CFCs and seeking environmentally safer substitutes for these harmful chemicals? Just because they were allowed in a time of ignorance is no reason for thinking that companies should now be compensated for being required to stop using such chemicals.[2] Whether the state should help finance research and development of substitutes is another matter. Perhaps it should.

Environmental considerations seem to require that we amend our understanding of *land* property rights. These were once thought to be absolute as long as we were not directly using our property to unjustly harm others; but we now see that some uses of the land, formerly thought benign, may well harm others—including those who will live after us. Or if these uses do not harm them directly, they prohibit others from using the land for optimal purposes (for example, if I fill in the wetlands, species may be lost and we may be unable to restore it to its original state). We must see the land not as our inalienable, absolute property, but as an entrustment—that is, the ownership model must be replaced by a stewardship model. The land is on loan to us to use and develop, as long as we do not make it significantly worse.

The Reaction to Property Restrictions: The "Wise Use" Movement

I hope the preceding discussion about the need for a more responsible attitude toward property sounds rational, even commonsensical, to you. One might say that what environmentalists are advocating is wise use of the land, as well as the wilderness, the forests, and the waterways. Unfortunately, words are cheap and names misleading. Anti-environmentalists have coopted the good phrase "wise use" in order to defeat many of the programs necessary to maintain and preserve the environment. In 1979 Ron Arnold wrote a series of articles for *Logging Management* in which he called for a coalition to counter the environmental movement. Alan Gottlieb followed up on his call and together they organized what has become known as *The Wise Use Movement*, a phrase borrowed from Teddy Roosevelt's chief of the Forestry Service, Gifford Pinchot (see Chapter 9). It has become an umbrella organization for more than 200 organizations dedicated to fighting the environmentalist agenda; the activities carried out by these groups include promoting the cutting of old growths in national forests, removing protection from endangered species, allowing private development of public lands and strip mining, and mandating compensation for *any takings* by the government—when they are taken for environmental causes.

The "Wise Use Movement" is a fearful reaction to the radical changes that are occurring in our understanding of humanity's relationship to its environment. It is hard to change old ways of thinking, and sometimes environmentalists are not patient, understanding, or ready to negotiate with those who haven't developed a sufficiently progressive attitude toward nature. Indeed, some environmentalists seem fanatical, resorting to eco-sabotage. While the "Wise Use Movement" must be exposed and resisted, more work probably needs to be done to find intermediate solutions for those who stand to lose their jobs, investments, and property as we pursue the quest for ecological wholeness.[3]

Let me now turn to another difficult issue on which more work needs to be done.

THE CITY

In 1800 about 2% of the world's population lived in urban areas; by 1900 the figure had doubled to 4%; in 1950 it had reached 30%—750 million people. The most recent figure (1997) put the world's urban population at 2.64 billion (46% of the world's population), projected to reach 3.3 billion by 2005. The 61 million people being added to cities each year come mainly through rural-to-urban migration. The United Nations estimates that in the next few years the percentage of people living in cities will reach 50%. By 2025 the projection is that almost two-thirds of humanity will live in cities. The most dramatic shifts are taking place in the developing nations, where urban population growth is 3.5% per annum, as opposed to 1% in the more developed nations. In 1800, 6% of the U.S. population lived in cities. By 1900 the figure was 40% and by 1997, 75%. Until a few generations ago most Americans lived on farms, produced their own food and clothes, fished in nearby streams and rivers, rode horses rather than cars, educated children at home, and enjoyed a simple life, far from the urban problems of overcrowding, pollution, and crime.

The number of urban areas is also increasing. In 1800 London was the only city with more than 1 million people. Today 326 cities have more than that figure, while 14 are "megacities," urban areas with populations over 10 million. Tokyo is currently the most populous city with 27 million people, about one-fourth of Japan's entire population. Lagos, Nigeria, growing at a rate of 5.4% each year, will probably increase from 10.3 million to 13.5 million from 1995 to 2000, heading for a 25-million mark by 2015.

The founders of the United States were committed to small-town, rural America. The first antiurban tract, *Notes on the State of Virginia* (1781), written by Thomas Jefferson, deplored urbanity as being opposed to good government: "The mobs of the great cities add just as much to the support of pure government as sores do to the strength of the human body."[4] In a letter to Benjamin Rush, Jefferson wrote, "I view great cities as pestilential to the morals, the health, and the liberties of man."[5] He envisioned the simple-living citizen-farmer, uncorrupted by urban luxury and sophistication, as the ideal democrat, the morally righteous man who had a stake in good government and who had no need for the unnecessary frills of urban society.

Nineteenth-century Americans generally shared Jefferson's faith in the farmer. Essayist Ralph Waldo Emerson thought that only farmers created wealth and that all trade depended on their endeavors. He shared Jefferson's views on the moral superiority of farmers: "The uncorrupted behavior which we admire in animals and in young children belongs to [the farmer], to the hunter, the sailor—the man who lives in the presence of nature. Cities force growth and make men talkative and entertaining, but they make them artificial."[6] Similar sentiments are found in the writings of Plato (in Book II of the *Republic*), Henry David Thoreau, Herman Melville, Nathaniel Hawthorne, and Edgar Allan Poe, who compared the city to a sewer of evil and wickedness. William James (1842–1910) deplored the "hollowness" and "brutality" of large cities and advocated their decentralization. James's colleague at Harvard, Josiah Royce (1855–1916), offered three criticisms of urbanity: (1) Cities were so overwhelmed with large numbers of alienated and unassimilated people that the essential fabric of society was stretched to the breaking point, (2) the centralization of culture produced mass conformity and intellectual stagnation, and (3) cities promoted the "spirit of the mob," which is the enemy of individualism and liberty.

These criticisms, though debatable, have continued throughout the twentieth century and seem likely to trouble us as we enter into the twenty-first century. Counterculture critic Ted Roszak sums up the present criticism—cities are decadent. According to Roszak, the city is seen as

> an imperialistic cultural force that carries the disease of colossalism in its most virulent form. . . . At the same time, the city is a compendium of our society's ecological bad habits. It is the most incorrigible of wasters and polluters; its economic style is the major burden weighing upon the planetary environment. Of all the hypertrophic institutions our society has inflicted upon both the person and the planet, the industrial city is the most oppressive.[7]

The case against cities goes like this: Cities are parasites on the agricultural base located in the country. They take in resources—water from the mountains, food from the farms, oil from other nations, coffee and tea and other products from the developing countries. While they take advantage of concentrated labor and produce important goods and services, which rural areas are unable to do, they also create expensive luxury items, which no one really needs and which may actually weaken society's moral fiber—items like indulgent department stores for the rich, neon-light districts, the compressed trees called the Sunday *New York Times,* and energy-inefficient buildings. Furthermore, they typically are filled with vehicles belching poisonous stench into our lungs. This emphasis on material goods is intensified by advertising, which creates false consumer consciousness. (American firms spent $468 per American on advertising in 1991.) As sociologist Robert Bellah noted, "That happiness is to be attained through limitless material acquisition is denied by every religion and philosophy known to humankind but is preached incessantly by every American television set."[8] The city is a cesspool of pollution, a sewer of vice, violence, crime, corruption, poor schools, poverty, unemployment, high taxes, suffering, and alienation. Dense with the anonymous homeless, panhandlers, muggers, and drug addicts; pungent with the smell of decay; fraught with the ugly sights of gaudy graffiti and garish advertisements and the noise of boomboxes, garbage trucks, and ambulances, the urban ulcer ubiquitously bombards our senses and crowds out our thoughts, alienating us from ourselves. The barrage of sensory stimulation overwhelms us, suffocating the inner voice within, so that we become alienated individuals in the lonely crowd. A common, superficial media-culture informs our ideas and dictates our tastes and fashions. These unnatural conditions close people off from the realities of the wilderness and agriculture—children and some adults actually suppose that food naturally comes wrapped in clean cellophane packages. Sheltered from the killings, from the blood and stench of the slaughterhouse, from the screams of the cattle and sheep and pigs, from the chicken *tortured* by our modern death chambers, the people of the city live in ignorance—not *blissful* ignorance, however, for decadence, disease, and death haunt their lives and render all too many of them meaningless. Urban poverty is typically more degrading than rural poverty, where food is usually available and a social support system is more intact than in the anonymous welfare hotels of the city.

Yet, in defense of the city, urban life offers civilization:[9] culture and convenience, commerce and industry, employment and job training, business headquarters and research centers, libraries and universities, music and theater, museums and art galleries, and a wide range of diverse ideas and attitudes. Urban existence offers enhanced freedoms and a wide range of friendships. Its standards of sanitation and health care are usually better than those of rural areas. For the prosperous, city life can be liberating. The concentration of people, wealth, culture, and business offers enormous opportunities for those equipped to take advantage of them.

The evils of cities tend to compete with and even outweigh these virtues, causing many people to fear urban existence, treating the city as a nice place to visit but a bad place to live. The question is, Since most of us want the benefits of city life, how can we restructure our cities so that they are environmentally sustainable centers of human flourishing?

This is the big question, and I can only point to some attempts that have been made. When we visit cities, we see children growing up without trees to climb or rivers in which to swim. In their place are urban jungles where skyscrapers replace redwoods on the near horizon. Comparing this scene with my own edge-of-the-town, semirural upbringing, in which I spent summers playing barefoot in the woods, swimming in ponds and rivers, a nostalgic sadness overwhelms me, for the simple joys urban children are missing. I see children who are oversocialized, programmed from the nursery school to the university, never feeling the call of the wild nor imbibing the wide open spaces of the prairie or the flow of the river, captives of too much repressive civilization, their watches mechanically dictating their schedules from their earliest years. When I worked with youth in Bedford-Stuyvesant, Brooklyn's urban ghetto, devoid of parks and greenery, someone offered us a large cabin by Lake Sebago in New York's Harriman State Park. Enthusiastically, I told the teenagers about this wonderful opportunity, only to be met with bewildering dread. The wilderness was so alien to them that I had to turn down the offer for want of takers. These young people had grown up without a sense of the wilderness in their consciousness. Yet I felt their lives suffered from the constrictions of urbanicity. If our historic, evolutionary roots are in the forests by the rivers and lakes, then, I hypothesize, we will generally flourish with an ample supply of our native environment. Or we must bring nature into our cities.

Cities like Minneapolis, Vancouver, Canada, and Melbourne, Australia, with their public parks, lakes, walking trails, bicycle paths, open spaces, low crime rates, and children's playgrounds stand out as models for the future. Every home in Minneapolis is within six blocks of green spaces. Melbourne reduced land taxes to attract the middle class to its environs, restricted the height of buildings to 131 feet (about 12 stories), and successfully renovated the decrepit structures of its inner city. Some 50,000 people now live in the central business district, a fivefold increase from a decade ago. But only time will tell what difference the recent addition of a mammoth casino will make.[10]

In 1993, Cajamarca, Peru, one of the poorest communities in the world, was racked by disease, unemployment, and water problems. Its infant mortality rate of 94.7 per 1,000 live births was 82% higher than the Peruvian national average. The Kilish River, a source of drinking water for the poor, had been contaminated by mining operations and untreated sewage. Overgrazing in nearby rural areas and clear-cutting of forests for fuel had caused severe soil erosion, exacerbating flooding problems and contributing to a depressed economy. In 1993 nongovernmental organizations (NGOs), in cooperation with businesses and local unions, organized communities in the urban and nearby rural areas into 76 "minor population centers," each with its own mayor and council. This dramatic decentralization of government power enabled the people to deal with local issues while communicating with an area-wide overarching authority. Together they set

up carpentry schools, an efficient water delivery system, refuse collection, health services, and park improvements. In the rural areas outside the city they terraced the steep hillsides and put into operation a plan to reduce mining pollution.[11]

In 1980 an NGO developed the Orangi Pilot Project (OPP) in the poverty-stricken, ethnically diverse city of Karachi, Pakistan. The residents were organized into groups of 20 to 40 families living along the same lane and taught to use appropriate technology to construct low-cost sanitation facilities. After this four-year project was successful, OPP developed basic health and family-planning programs, including immunization programs for children. Next it created a credit program to fund loans for small family enterprises, a low-cost housing-upgrade program, a program to assist in improving educational facilities, a women's work-center program, and a rural development program. Each house received a sanitary latrine. The Karachi government contributed to the construction costs of health and sanitation facilities, but by simplifying design and standardizing parts, these costs were greatly reduced, in some cases to one-fifth of similar improvements elsewhere in the city. Within a decade 95% of the children were being immunized, 44% of the families were practicing birth control, epidemic disease was under control, and hygiene and nutrition had improved. Infant mortality fell from 130 per 1,000 live births in 1982 to 37 in 1991. Through the work center, women learned to stitch clothing, enabling them to do piecework bound for international export at higher wages than they had ever earned, thus contributing to the overall wealth of the community.[12]

Such examples of sustainable urbanization, often in conjunction with nearby rural development, are impressive and offer models of a better future; but success stories are still too few and far between. Accessible clean water, sanitation systems, decentralized government, local empowerment to men and women, education, job training, and inexpensive basic health care—all seem necessary but not sufficient for sustainable city life. A moral consciousness must exist to energize and synthesize a community. People must believe in environmental goals and commit themselves to them.

But the task of rejuvenating our urban centers is herculean. First of all, there are too many people in the cities. Every week one million people are added to these urban centers. Poor people are lured from the countryside or other countries by the promise of a better life, only to add to the malaise of urban poverty. In too densely populated cities, the friction of our encounters ruffles our nerves and leaves us yearning for clean open spaces, for freedom of movement. We need more parks in our cities—following the European model—and fewer cars and trucks. New York City, with its ban on motor vehicles in Central Park on weekends, has taken a step in the right direction. The ban should be extended to the other five days of the week and to other parts of our cities, for motor vehicles are the main air polluters in New York City and elsewhere. Affordable public transportation should replace cars wherever feasible. Recycling of aluminum cans, glass, plastic, and paper is cumbersome but is an environmental necessity, since we need to husband our scarce resources. We must make it both natural and economical.

People in cities need more places to plant trees and gardens, where they may grow flowers and vegetables. Tall buildings should not dominate the skyline; zoning laws should require smaller ones that allow the sun's light to shine on urban inhabitants. Political decentralization is necessary in order to afford people a greater opportunity to participate in the political process. Efficient government, a streamlined court system, and a sense of fairness must bind people to each other—promoting the commonweal. But this sense of a common life and a common cause is difficult to create, especially where politicians and intellectuals emphasize difference rather than commonality, where ethnic and cultural diversity are allowed to divide people. Neither a nation, nor a city divided against itself can long survive, let alone flourish, and a political structure that allows unjust discrimination, including strong types of "affirmative action" (reverse discrimination), will sink in the quicksand of the swamps of exaggerated racial identity, ethnicity, and hate. Diversity may be enriching and has a legitimate place—especially diversity of ideas—as long as people adhere to a common core morality, an agreed-upon political process that brings us together as a moral community—*e pluribus unum*. But where we do approach that common culture, in TV programs and films, for example, it is often shallow, self-indulgent, and amoral.

Because we prize freedom so much and depend on an unplanned free-market economy to such a remarkable degree, it is difficult to solve these seemingly intractable environmental and social problems. Capitalism, our economic system, is like a powerful machine that is under no one's control, satisfying short-term wants and offering wealth, but threatening to uproot our traditions and all those spiritual bonds that tie us together. The truth is that we are not dealing with it successfully, but allowing it to proceed unchecked in a manner dangerous to our future—as the rich get richer and the poor get prison and toxic waste dumps.

Environmentalists have focused virtually all of their attention on the wilderness, on pristine nature. But in doing so, they may have missed something equally important, the urban environment. The challenge of the twenty-first century will be not only to preserve the wilderness, but also to reinvigorate our urban centers with simple dignity and natural beauty.

GLOBALISM: ONE WORLD, ONE ETHIC

We have just seen that part of the solution to urban crises lies in decentralized authority combined with a supporting, enlightened government. Reducing the locus of power to the smallest possible group, down to the individual, makes sense, since each individual or small group is a better authority on where the shoe pinches than is a distant bureaucracy. On the other hand, individuals and communities often lack resources to lift themselves from poverty or environmental degradation. So an overarching umbrella authority is necessary to distribute goods and services. Moreover, there is the problem of the tragedy of the commons to contend with, which leads to the necessity of an overarching regulatory system. My business or community or country is likely to reason that it is in our interest to use CFCs (or burn fossil fuels) since the benefits we reap are solely ours, whereas we share the harms, a depleted ozone layer (or enhanced greenhouse effect) with others. But if everyone thinks this way, the ozone layer is likely to be destroyed and everyone will suffer a global cataclysmic disaster. Similarly, with regard to emitting greenhouse gases, if we all act in our perceived immediate interest, we are bound to reap total and global ruin. So it is in all our interest autonomously to give up some of our autonomy and accept "mutually agreed upon, mutually coercive regulations" which, if followed by the majority, will result in mutual benefit. Recently, I was in North Carolina and heard about the problem of disposing of pig waste (a pig, I was told, discharges enormous waste—five times the amount of an average human). The sanitation facilities were inadequate and pig waste was seeping into the water supply in parts of the state, but the state government was reluctant to force the pig industry to invest in better waste-disposal systems lest it move out of North Carolina to a state with more relaxed regulations. The solution in such cases is for a federal standard, nationally enforced. Similarly, I was told that Switzerland had imposed strict safety regulations on the pig industry. The result? All pig industries have moved out of Switzerland to less demanding countries. The Swiss still eat the same amount of pork but pay more for it. The solution is obvious: For the health of all people, have an international regulatory commission monitoring and enforcing safety standards.

Many of our most intractable environmental problems are international in nature. Radiation from Chernobyl was experienced as far west as Sweden and Switzerland; air pollution from Poland's factories drifts over to neighboring countries; greenhouse gases in the United States will affect climate patterns all over the globe; the depletion of the ozone layer will affect the health of people in many nations; and we all, as well as posterity, will suffer from the loss of biodiversity. Rivers and underground water tables—aquifers—do not respect national boundaries, so that if country A depletes its water table, country B, frugal though it be, will also experience the loss of water. The recent conferences on the environment in Stockholm (1972), Rio de Janeiro (1992), and Kyoto, Japan (December 1997), fragmented and seemingly fraught with controversy and national self-interest though they were, are a fledgling step in the right direction. At least, we're talking with each other about global environmental degradation and solutions to that degradation, seeking to work out a set of universal rights from which will arise global environmental law. Principles 7 and 8 of the Rio Declaration put the matter this way:

> *Principle 7:* States shall cooperate in a spirit of global partnership to conserve, protect and restore the health and integrity of the Earth's ecosystem. In view of the different contributions to global environmental degradation, States have common but differentiated responsibilities. The developed countries acknowledge the responsibility that they bear in the international pursuit of sustainable development in view of the pressures their societies place on the global environment and of the technologies and financial resources they command.
>
> *Principle 8:* To achieve sustainable development and a higher quality of life for all people, States should reduce and eliminate unsustainable patterns of production and consumption and promote appropriate demographic policies.

Other principles call for compensation of victims of pollution (13), a prohibition of reallocation of toxic substances to poorer countries (14), the internalization of the environmental costs of pollution (16), and the ecological protection of weaker countries from oppression and domination by the wealthier corporations and nations (23). The Charter of the United Nations will be the "appropriate means" for resolving "all their environmental disputes."[13]

An international body, such as the United Nations, will be needed to regulate and enforce these environmental laws. This will not be easy for nationalists to swallow, but gradually we are moving toward universal government to complement and qualify national autonomy. The world is shrinking. We live in a global economy, where a drop in the Japanese market sends reverberations as far as Argentina and New York. Already, several multinational corporations are among the wealthiest bodies in the world, richer than most nations.[14] Even as the capitalist economy has become global, the regulation of the environment must also. The road to an enforceable global environmental law will be fraught with obstacles, but in the end we must realize our common humanity, a common objective morality, and a common commitment to ecological wholeness and sustainable living. Box 18-1 details Americans' disproportionate share in consumption and pollution—a factor that must be considered when we see ourselves as global citizens.

WORKING FOR A SUSTAINABLE SOCIETY

On May 14, 1998, Marjory Stoneman Douglas died at the age of 108. The author of many short stories, novels, and works of nonfiction, she is best known for her influential 1947 call to arms, *The Everglades: River of Grass,* a natural and political history of the wetlands of southern Florida. Mrs. Douglas protested against the poor land management that was imperiling the Everglades' ecosystem, opposed state and local policies that encouraged overdevelopment, and led the campaign to have the central core of the Everglades preserved as a national park. The Everglades has shrunk from more than 4,000 square miles to less than half that size, the result of overdrainage, urban sprawl, and pollution from government-supported sugar cane and dairy farming. Many environmentalists believe that its fate is still in doubt. Regarding the apathy of the people of southern Florida to the plight of the Everglades, Mrs. Douglas said, "They could not get it through their heads that they had produced some of the worst conditions themselves, by their lack of cooperation, their selfishness, their mutual distrust, and their willful refusal to consider the truth of the whole situation." Unless people act responsibly "over-drainage will go on . . . and the soil will shrink and burn and be wasted and destroyed, in a continuing ruin." In 1969 she helped found Friends of the Everglades, a conservation organization that now has 5,000 members. Joe Podgor, the former executive director, called her "the giant on whose shoulders we all stand." In 1990, on her one-hundredth birthday, blind, hearing-impaired, and frail, she continued to speak out against

BOX 18-1

Americans: The Biggest Consumers in the World

- An American in one lifetime (70 years) consumes and wastes:

Resource Consumption	*Waste*
623 tons of fossil fuel	840 tons of agricultural waste
613 tons of sand, gravel, stone	823 tons of garbage, industrial waste
26 million gallons of water	7 million gallons of polluted water
21,000 gallons of gasoline	70 tons of air pollution
50 tons of food	19,250 bottles
48 tons of wood	7 automobiles
19 tons of paper	

- An American uses 70 times as much energy as a Bangladeshi, 50 times as much as a Malagasy, 20 times as much as a Costa Rican.
- Since we typically live longer, the effect of each of us is further multiplied. In a year an American uses 300 times as much energy as a Malian; over a lifetime, 500 times.

Even if all such effects as the clearing of forests and burning of grasslands are factored in and attributed to poor people, those who live in the poor parts of the world are typically responsible for the annual release of a tenth of a ton of carbon each, whereas the average for residents of the Western nations is 3.5 tons. The richest tenth of Americans annually emit 11 tons of carbon apiece.

those who plundered the Everglades. Roderick J. Jude, a long-time leader of the Florida chapter of the Sierra Club, said, "The Everglades wouldn't be there for us to continue to save if not for her work through the years." Finally in 1996, after decades of struggle, the voters of Florida approved a constitutional amendment for cleaning up of the Everglades. In 1997, hoping to rescue the endangered ecosystem from polluted runoff from the sugar cane industry, the Clinton Administration and the state of Florida agreed to buy more than 50,000 acres of sugar cane fields on the outskirts of the Everglades National Park. In 1993 President Clinton awarded Marjory Douglas the Presidential Medal of Freedom and said, "Long before there was an Earth Day, Mrs. Douglas was a passionate steward of our nation's natural resources, and particularly her Florida Everglades."[15]

Marjory Douglas deserves to be ranked with Henry Thoreau, John Muir, Theodore Roosevelt, Aldo Leopold, Herman Daly, Rachel Carson, Chico Mendes, and Lois Gibbs, all mentioned earlier in this work, as one of the friends of the Earth—people who, by their integrity, courage, and commitment, made important contributions toward preserving and promoting ecological well-being. They all attest to the fact that citizens can make a difference in making this a better world. These are our present-day heroes, our much-needed role models for simple living, local acting, and global thinking.

The fate of the Earth is still in doubt. Many questions about the state of the environment are still in doubt. Good and honest people can differ on their reading of the evidence regarding the best energy policy, the best ways to limit pollution, the prognosis of the greenhouse effect, the implications of population growth and so forth. Some of you, reading this book, will opt for radical action to save the planet, others for more conservative policy, and still others for mixed strategies.

On October 18, 1998, an underground environmental organization, Earth Liberation Front, burned down the Two Elks Restaurant and Ski Lodge and two other buildings on Vail Mountain in Colorado, causing estimated damage of at least $12 million. The fire came after Vail had won a major court battle against environmental groups, who opposed expansion of the ski facilities. The environmental groups, led by the Earth Liberation Front, claimed that the clear-cutting of 885 acres by Vail, Inc., will destroy the best lynx habitat in Colorado. The following e-mail was received by radio station KCFR:

> On behalf of the lynx, five buildings and four ski lifts at Vail were reduced to ashes on the night of Sunday, October 18. Vail, Inc., is already the largest ski operation in North America and now wants to expand even further. The 12 miles of roads and 885 acres of clear-cuts will ruin the last, best lynx habitat in the state. Putting profits ahead of Colorado's wildlife will not be tolerated. This action is just a warning. We will be back if this greedy corporation continues to trespass into wild and unroaded areas. For your safety and convenience, we strongly advise skiers to choose other destinations until Vail cancels its inexcusable plans for expansion. (Earth Liberation Front)[16]

The Vail Mountain burnings were just the latest and most destructive incidents in a growing trend of eco-terrorism. In the past 20 years more than 1,500 attacks have been reported. Oregon loggers and mill workers have reported serious injuries from cutting trees containing spikes; pipe bombs have been used to blow up feed trucks in an attack on a fur breeder cooperative in Sandy, Utah; and ski lifts in Arizona have been destroyed by eco-sabotage.

If these acts draw attention to the importance of preserving our environment, good may yet come out of them. Katie Fedor, a spokeswoman for the Animal Liberation Front, announced that this "is a nonviolent revolution. Unfortunately, the traditional routes to societal change such as lobbying haven't worked. Constituents are not being heard. We are forced to take nonviolent action." This is clearly false. Burning down property is an act of violence and endangers people who might be in the buildings. It is not a big step from burning down the restaurants and homes of your opponents to the acts of Unabomber Ted Kaczynski. Kaczynski, whose bombings killed three people and wounded 29 others, selected some of his victims on the basis of their opposition to ecological well-being.

We live in a democracy, which sometimes is dull and sluggardly in promoting the common good, but which affords opportunity for open debate about these important and difficult environmental issues. But though democratic processes are often painfully slow, they seem the most moral—or least dangerous—process at our disposal. Those who become impatient with this process may engage in nonviolent protest to get their point across; but certainly, if history has taught us anything, it is that violence is counterproductive, whether it be perpetrated in the name of the Palestinian cause on the West Bank or that of Catholic freedom in Northern Ireland—or whether, as eco-sabotage, it be done in the name of the environment. Violence begets more violence, destroying even the good that exists. Concerned citizens, then, must engage in the peaceful political process, working for a raised consciousness about environmental concerns in the public domain. We must also live out our ecological philosophy—for, to a remarkable degree, the personal is the political. Our actions speak louder than our words.

Much has been accomplished since the 1960s, but much has yet to be done. On the plus side, a growing number of citizens have become conscious of environmental concerns, as the membership in environmental organizations such as the Sierra Club, Wilderness Society, Nature Conservancy, and others indicates. The celebration of Earth Day each April 22 since 1990 represents a heightened awareness of the environmental crisis. Many school systems, such as those of Wisconsin, incorporate environmental education into the curriculum. In the United States we've seen the passing of the Federal Water Pollution Control Act and the Clean Air Acts; the Wilderness Act, setting aside or protecting several ecosystems; the Endangered Species Act, protecting species from harm; and the Toxic Substance Control Act, requiring the screening of new substances before they are widely used. These and the recent international conferences on environmental concerns such as global warming and biodiversity, already discussed, are steps in right direction.

On the negative side, the greenhouse effect is getting worse, carbon dioxide in the atmosphere has increased from 280 ppm at the beginning of the industrial revolution to 360 ppm and threatens to reach 500 ppm by the middle of the twenty-first century.[17] The great glaciers on Antarctica are breaking up, and climate patterns may be dangerously changing. The

BOX 18-2

Why Recycle Paper?

1. To save forests: recycling 1 ton of office paper saves 17 trees.
2. To save energy: it takes 60% less energy to manufacture paper from recycled stock than from virgin materials. Every ton of recycled paper saves 4,200 kilowatts of energy, enough to meet the energy needs of at least 4,000 people.
3. To save water: making paper from recycled paper stock uses 15% less water than making paper "from scratch." Recycling 1 ton of paper saves 7,000 gallons of water, enough to supply the daily water needs of almost 30 households.
4. To reduce garbage overload: Every ton of paper not landfilled saves 3 cubic yards of landfill space.

ozone layer continues to be depleted and acid rain and other pollutants continue their destructive effect on lakes and forests. The world's rivers and underground aquifers are increasingly polluted, and rich topsoil continues to be eroded. The destruction of the rain forests and the forests everywhere continues at a menacing pace. The future of the Earth is in jeopardy.

The Earth's population, which has passed 6 billion, continues to grow exponentially. People in the developing countries seek to increase their living standards and consumption to match those of the developed countries, depleting resources and producing enormous pollution. Add to this the fact that we're losing much of our topsoil and our food production is declining. In 1981 Julian Simon in his book *The Ultimate Resource* showed how global food production had continued to rise, and so become cheaper, for several decades. He wrote, "The obvious implication of this historical trend toward cheaper food—a trend that probably extends back to the beginning of agriculture—is that real prices for food will continue to drop. . . . It is a fact that portends more drops in price and even less scarcity in the future."[18]

A few years later, however, the sharp growth rates in food production began to level off. Now the gains in grain production are coming in smaller increments, too small to keep pace with the world's population growth. As Bill McKibben points out, "The world reaped its largest harvest of grain per capita in 1984; since then the amount of corn and wheat and rice per person has fallen by 6%. Grain stockpiles have shrunk to less than two months' supply."[19]

What can we do? If the thoughts set forth in this work have any validity, we can and ought to live more simply. We in the West must lower our consumption levels and reduce the pollution we cause, while encouraging people everywhere to deal with exponential population growth and resource consumption. We can use less and more efficient electricity; recycle paper (Box 18-2), plastics, glass, and metal; use fluorescent lights; incline toward a vegetarian diet; walk and cycle for short distances and use public transportation wherever possible, instead of using a car. We can keep in good physical condition and decrease energy use by walking up stairs instead of using an elevator. Instead of turning up the thermostat, put on an extra sweater. Wherever possible, we can install solar panels in our buildings. We can strive to make our cities more environmentally wholesome and, at the same time, promote organic farming and local home gardens. We can plant trees to add beauty and oxygen to our communities. We can increase our appreciation of the wilderness and spend time camping and hiking, observing wildlife and appreciating the breathtaking beauty and sacred stillness of mountains, lakes, forests, and canyonlands. We can join and support an environmental organization that best identifies our values and concerns. We can support political leaders who promote environmental integrity. We can share our ideas and vision of a better world with others, encouraging them to join the environmental movement for a better world. We can become informed citizens and then educate the media, newspapers, radio, and television personnel to the significance of environmental concerns. Our hope is in the young. If we can instill an environmental consciousness in children in home, church, and school, we may be able to save our global home, our planet.

Although much ecological damage has been done, there is still time to save the planet and make this a better world for posterity.

In sum: *Live simply so that others may simply live.*

Review Questions

1. What should be done in cases like David Lucas's, where property rights are at issue? Was the Supreme Court wrong in stating that he was entitled to compensation? Could such a policy bankrupt governments?
2. What is the justification for private property? Should property be redistributed to the poor and for environmental purposes?
3. Why are cities crucial to environmental ethics? Do you agree that a major challenge of the twenty-first century will be to produce sustainable cities? Explain your answer.
4. What are the most important features in a sustainable city?
5. Assess the argument that the development of a global environmental ethic requires international environmental law. Will we need a global regulating body to enforce such law? Or is there a more effective way of dealing with environmental problems? Explain your answer.

6. What else can be done to produce a sustainable society?
7. What can you do to promote environmental well-being?

Suggested Readings

Berry, Wendell. *The Unsettling of America: Culture and Agriculture.* San Francisco: Sierra Club, 1977.

Brown, Lester. *Tough Choices: Facing the Challenge of Food Scarcity.* New York: Norton, 1996.

Brown, Lester R., Christopher Flavin, and Hilary French. *State of the World, 1998.* New York: Norton, 1998.

Durning, Alan Thein. *How Much Is Enough?* New York: Norton, 1992.

Ferre, Frederick, and Peter Hartel, eds. *Ethics and Environmental Policy.* Athens, Ga.: University of Georgia Press, 1994.

Greider, William. *One World, Ready or Not.* New York: Simon & Schuster, 1997.

Meyer, William B. *Human Impact on the Earth.* New York: Cambridge University Press, 1996.

Newton, Lisa, and Catherine Dillingham. *Watersheds 2: Cases in Environmental Ethics.* Belmont, Calif.: Wadsworth, 1997.

Sadler, A. E. *The Environment: Opposing Viewpoints.* San Diego: Greenhaven Press, 1996.

Saign, Geoffrey C. *Green Essentials: What You Need to Know About the Environment.* San Francisco: Mercury House, 1994.

Schumacher, E. F. *Small Is Beautiful: Economics As If People Mattered.* New York: Harper & Row, 1977.

World Resources: The Urban Environment 1996–97. New York: Oxford University Press, 1996.

Notes

1. *Lucas v. South Carolina Coastal Council* (Blackmun, J., dissenting).
2. I am indebted for this illustration to Gary Varner, "The Eclipse of Land as Private Property," in *Ethics and Environmental Policy,* ed. F. Ferre and P. Hartel (Athens, Ga.: University of Georgia Press, 1994). Varner's article contains a helpful discussion of these matters.
3. For a good discussion of the "Wise Use Movement," see Lisa Newton and Catherine Dillingham, *Watersheds 2* (Belmont, Calif.: Wadsworth, 1997), chap. 10.
4. Thomas Jefferson, *Notes on the State of Virginia* (New York: Harper & Row, 1964), p. 158.
5. Thomas Jefferson, *Works of Thomas Jefferson,* vol. 4, ed. P. Ford (New York: G. P. Putnam, 1905), pp. 146–47.
6. Ralph Waldo Emerson, *Society and Solitude* (Boston: Houghton Mifflin, 1883), p. 148.
7. Theodore Roszak, *Person/Planet* (London: Gollancz, 1979), pp. 253–54.
8. Robert Bellah, quoted in Alan Thein Durning, "An Ecological Critique of Global Advertising," *Worldwatch* 6:3 (May–June 1993).
9. *Webster's Dictionary* defines *civilization* as "(1) a relatively high level of culture and technological development; specifically, the stage of cultural development at which writing and the keeping of written records is attained; (2) refinement of thought; (3) a situation of urban comforts."
10. Brendan I. Koerner, "Cities at Work," *U.S. News & World Report,* 8 June 1998.
11. Jeb Brugmann, "Cities Take Action: Local Environmental Initiatives," in *World Resources: The Urban Environment 1996–97* (New York: Oxford University Press, 1996), pp. 128–29.
12. Akhtar Badshah, "The Orangi Pilot Project, Karachi, Pakistan," in *World Resources: The Urban Environment 1996–97,* pp. 132–33.

13. *The Rio Declaration,* approved by the United Nations Conference on Environment and Development (Rio de Janeiro, Brazil, June 3–14, 1992) and endorsed by the forty-seventh session of the United Nations General Assembly on December 22, 1992. Reprinted in *Environmental Ethics,* 2nd ed., ed. L. Pojman (Belmont, Calif.: Wadsworth, 1998).
14. For a good discussion of the coming global economy, see William Greider, *One World, Ready or Not* (New York: Simon & Schuster, 1997).
15. "Marjory Douglas, Champion of Everglades, Dies at 108," *New York Times,* 15 May 1998, p. A23.
16. Steven K. Paulson (Associated Press), "Animal Rights Group Steps Up Eco-Terrorism with Vail Fire," *Daily Herald* (Provo, Utah), 23 October 1998.
17. "Climate is an angry beast, and we are poling it with sticks." Wallace Broecker, in Bill McKibben, "A Special Moment in History," *Atlantic Monthly,* May 1998, p. 70.
18. Julian Simon, *The Ultimate Resource* (Princeton: Princeton University Press, 1981).
19. Bill McKibben, "A Special Moment in History," p. 62.

Glossary

absolutism The thesis that moral principles are not overridable by other moral principles.

altruism Unselfish regard or concern for others; disinterested, other-regarding action.

anaerobic digestion The process of transforming garbage and waste into methane gas and soil fertilizer by decomposing it in a warmed airtight container.

anthropocentrism The theory that humans alone or primarily have objective value or should be accorded special status. A difference exists between **strong anthropocentrism,** which asserts that only humans have objective value, and **weak anthropocentrism,** which asserts that humans have higher value or have higher moral status than other organisms, which also have moral status. Kant holds to the first type. The split-level theory is an example of the second. See Chapter 7.

anthropogenic Having a human origin without necessarily being human-centered. For example, the land ethic is a theory of human origin, but not human-centered.

autonomy (from the Greek for "self-rule") Self-directed freedom. The autonomous individual arrives at his or her moral judgments through reason rather than through simply accepting authority. The *autonomy thesis* states that ethical truths can be known and justified on the basis of human reason without the need for divine revelation. See also **heteronomy.**

bacteria Microscopic single-celled organisms that lack nuclear membranes around the genes or do not contain a well-defined nucleus, such as prokaryotes.

benign demographic transition theory (BDT) The theory that states that population change goes through stages and that improved social conditions will inevitably lead to population stabilization. See Chapter 12.

biodiversity The variety of organisms in both their intraspecies and interspecies aspects. It usually refers to the variety of communities of organisms living within an ecosystem.

biological magnification The increased concentration of pollutants in living organisms. Many synthetic chemicals, such as DDT and PCB, are soluble in fat but insoluble in water and are slowly degradable by natural processes. That is, when the organism takes in DDT or PCB, the chemical becomes concentrated in the fatty tissue of the animal and is degraded very slowly. When a large fish consumes plankton with these pesticides, the chemicals become more concentrated still. Sufficient levels of these chemicals in an organism can harm it. Biological magnification of certain chemicals in human fatty tissue can cause cancer.

biotic community "A biotic community is a dynamic web of interacting parts in which lives are supported and defended, where there is integrity (integration of the members) and health (niches and resources for the flourishing of species), stability and historical development (dependable regeneration, resilience, and evolution)." From Holmes Rolston III, *Conserving Natural Value* (New York: Columbia University Press, 1994), p. 78.

categorical imperative The categorical imperative commands actions that are necessary in themselves without reference to other ends. They are contrasted with **hypothetical imperatives,** which command actions not for their own sakes but for some other good. For Kant, moral duties command categorically; they represent the injunctions of reason, which endows them with universal validity and objective necessity.

cultural relativism The theory that different cultures have different moral rules. It makes no judgment on the validity of those rules and, as such, is neutral between ethical objectivism and ethical relativism. Sometimes the moral systems of a culture are referred to as *positive morality:* any existing moral code as distinguished from an adequate or justified moral code. For example, Nazi morality is a moral code, but most objectivists would deny that it is an adequate or justified moral code. It contains invalid principles such as "Always kill Jews and gypsies and Poles." See Chapter 1; see also **objectivism** and **ethical relativism.**

demarcation Identification of principles or properties that describe which objects have value and which do not.

demographic momentum The process which shows that when there is a positive population growth, the population continues to grow even though the percentage of new births has declined.

demographic transition The theory that countries or classes of people tend to stabilize their populations as they become affluent. Bringing down the death rate of a group with enhanced social benefits usually results in lower birth rates.

demography The science of human population, its size and changes in a geographical area or the world.

deontological ethics (from the Greek *deon,* duty or obligation) Ethical systems that consider certain features in the moral act itself to have intrinsic value. These are contrasted with **teleological ethics** (below), which holds that the ultimate criterion of morality lies in some nonmoral value that results from actions. For example, for the deontologist, there is something right about truth telling, even when it may cause pain or harm, and there is something wrong about lying, even when it may produce good consequences.

discount rate A time preference concept. We generally value benefits today rather than tomorrow. A bird in the hand is worth two in the bush. Economists generally value imminent profit more than future profit, so that a benefit postponed five, ten, or a hundred years will get a lower value (often discounted at 5 or 10%). Discounted at 10%, something worth $1,000 today will be worth $900 next year and $810 the following and $729 the one after that. A high discount rate favors using up resources now rather than later.

ecosystem The organisms (animals and plants) living in a particular environment, such as a prairie, lake, forest, or desert—and the physical geography in which they dwell. The organisms are called the community.

egoism *Psychological egoism* is a *descriptive* theory about human motivation that holds that people always act to satisfy their perceived best interests. *Ethical egoism* is a *prescriptive* or normative theory about how people *ought* to act; they ought to act according to their perceived best interests.

ethical relativism The theory which holds that the validity of moral judgments depends on cultural acceptance. It is opposed to **objectivism.**

eutrophication The process of creating excessive plant nutrients in aquatic ecosystems, such as lakes, rivers, and ponds. This happens naturally through erosion and runoff of nitrogen and phosphorus compounds into the aquatic system. This natural process can be exacerbated by **cultural eutrophication,** the artificial enrichment from human activities, such as synthetic fertilizers and manure from agriculture and discharge of sewage. The excessive nutrients can cause gargantuan growth of algae. The algae dies and sinks to the bottom of the lake or stream. Oxygen-consuming bacteria decompose dead algae, depleting oxygen from the water. Fish, dependent as they are on oxygen, die; and bacteria decompose the dead fish, further depleting oxygen. If this process is not stopped, the chemical cycling system of the lake can be destroyed and the lake can become devoid of life.

external costs Costs not borne by the parties to the market transaction, buyer and seller, but by third parties. For example, if Company X freely pollutes a river, it doesn't suffer the consequences, but people living downstream do. They "pay" the external costs in sickness and hospital bills. Environmentalists want the external costs internalized into the cost of the product itself.

food chain The food web of a particular community of organisms, consisting of predators and prey, from the top predators (such as lions, eagles and human beings) to the decomposers that consume dead organisms.

genetic drift Evolution in the genetic endowment of a species which takes place by chance.

habitat A particular environment, such as a rain forest or tall-grass prairie.

hedon (from the Greek *hedone,* pleasure) Possessing a pleasurable or painful quality. Sometimes *hedon* is used to stand for a quantity of pleasure.

hedonism *Psychological hedonism* is the theory that motivation is to be explained exclusively in terms of desire for pleasure and aversion of pain. *Ethical hedonism* is the theory that pleasure is the only intrinsic positive value and that pain or "unpleasant consciousness" is the only thing that has negative intrinsic value (or intrinsic disvalue). All other values are derived from these two.

heteronomy Rule by another; the opposite of autonomy; a heteronomous person is one who lacks personal self-directedness.

hypothetical imperative A command that enjoins actions because they are useful for the attainment of some end that one desires to obtain. Ethicists who regard moral duties as dependent on consequences would view moral principles as hypothetical imperatives. They have the form: "If you want X, do action A" (for example, "If you want to live in peace, do all in your power to prevent violence"). This is contrasted with the **categorical imperative.**

intuitionism The ethical theory that the good or the right thing to do can be known directly via the intuition. G. E. Moore and W. D. Ross hold different versions of this view. Moore is an intuitionist about the Good, defining it as a simple, unanalyzable property; Ross is an intuitionist about what is right.

keystone species A species that affects the survival and flourishing of many other species in the community in which it lives. Its removal results in a significant shift in the composition of the community. The sea otter off the coast of California is an example of such a species.

naturalism The theory that ethical terms are defined through factual terms, in that ethical terms refer to natural properties. *Ethical hedonism* is one version of ethical naturalism, for it states that the Good, which is at the basis of all ethical judgment, refers to the experience of pleasure. Naturalists such as Geoffrey Warnock speak of the content of morality in terms of promoting human flourishing or ameliorating the human predicament.

natural law The theory that an eternal, absolute moral law can be discovered by reason. First set forth by the Stoics but developed by Thomas Aquinas in the thirteenth century.

niche The place occupied by a species in its ecosystem: where it lives, what it eats, and the general role it plays in the system.

objectivism (or **ethical objectivism**) The view that moral principles have objective validity whether or not people recognize them as such; that is, moral rightness or wrongness does not depend on social approval, but on such independent considerations as whether the act or principle promotes human flourishing or ameliorates human suffering. Objectivism differs from absolutism in that it allows that all or many of our principles are overridable in given situations.

panentheism The doctrine that God is *in* all things, but not to be identified with all things. He still enjoys independent existence as a person.

pantheism The doctrine that everything there is constitutes a unity and that this unity is divine.

photovoltaic cells A type of active solar energy. Thin films of sensitive material in which a flow of electrons, caused by exposure to sunlight, produces direct electric current. Collecting ten together increases the output tenfold; intensifying the brilliance of the sunlight on the cells also magnifies the power. The cells pose no threat to the environment. They can be installed anywhere—on roofs, windows, or walls.

prima facie Latin term that means "at first glance." It signifies an initial status of an idea or principle. In ethics, beginning with W. D. Ross, it stands for a duty that has a presumption in its favor but may be overridden by another duty. *Prima facie* duties are contrasted with *actual duties* or *all-things-considered duties*.

relativism See **cultural relativism.**

slippery slope fallacy The fallacy of objecting to a proposition on the erroneous grounds that the proposition, if accepted, will lead to a chain of other propositions that will eventually result in an absurdity. For example, I might object in the following manner to the statement that some people are rich: You will agree that owning only one cent does not make one rich and that adding one cent to whatever we own will not in itself make anyone rich. So imagine that I have only one cent and then imagine giving me an additional penny. I still am not rich. You can give me as many pennies as you like but at no point will you change my status from being poor to being rich. Even though I might eventually end up with a million dollars' worth of pennies, there is no point where the transition from poverty to wealth takes place. Therefore neither I nor anyone else can be rich. This conclusion, of course, is false.

solipsism The view that only I exist (asserted by the speaker); everyone else merely exists in my mind. *Moral solipsism* is the view that only I am worthy of moral consideration; it is an extreme form of egoism.

species The basic unit of biological classification, consisting of a population of closely related organisms; a population of organisms that freely interbreed with one another in natural conditions.

split-level theory The theory first set forth by Martin Benjamin and Donald VanDeVeer which accords higher moral status to human beings on the basis of their rational self-consciousness but accords moral status to animals on the basis of their sentience. It stipulates that with regard to basic needs humans should be given prior consideration, but with regard to human trivial needs (or luxuries) vis à vis animal basic needs, animals should be given priority. See Chapter 7.

subjectivism The view that morality is in the eyes of the beholder. Each person creates or invents his or her own morality.

supererogatory (from the Latin *supererogatus,* beyond the call of duty) A supererogatory act is one that is not required by moral principles but contains enormous value; it is "beyond the call of duty," such as risking one's life to save a stranger. Although most moral systems allow for the possibility of supererogatory acts, some theories (most versions of classical utilitarianism) deny that there can be such acts. See Chapter 1.

sustainable development The use of natural resources, such as land, air, minerals, and water, to sustain production indefinitely without environmental deterioration, ideally without loss of native biodiversity. Ideally, a policy of social and economic development, which optimizes the economic and other societal benefits to the present generation, without jeopardizing the potential for similar benefits for future generations.

symbiosis The living together of two or more species in a prolonged and intimate ecological relationship.

teleological ethics Teleological ethical theories (also called **consequentialist** ethical theories) place the ultimate criterion of morality in some nonmoral value (for example, happiness or welfare) that results from acts. Whereas **deontological** ethics ascribes intrinsic value to features of the acts themselves, teleological theories see only instrumental value in the acts but intrinsic value in the consequences of those acts. Both **ethical egoism** and **utilitarianism** are teleological theories. See Chapter 2.

trophic level A group of organisms that obtain their energy from the same part of the food web or biological community.

trophic pyramid The levels of the food chain from less complex organisms to secondary and tertiary consumers.

ungulate A hoofed, typically herbivoric, quadruped mammal, such as a ruminant, deer, elk, swine, sheep, horse, hippopotamus, or camel.

universalizability The principle, found explicitly in Kant's and R. M. Hare's philosophy and implicitly in most ethicists' work, that states that if some act is right (or wrong) for one person in a situation, then it is right (or wrong) for any relevantly similar person in that kind of situation. It is a principle of consistency that aims to eliminate irrelevant considerations from ethical assessment.

usufruct In Roman and civil law, the right of enjoying all the advantages derivable from the use of something which belongs to another or the community. New England Native Americans reportedly practiced usufruct with regard to land. See Chapter 6.

utilitarianism The theory that the right action is that which maximizes utility. Sometimes *utility* is defined in terms of *pleasure* (Jeremy Bentham), *happiness* (J. S. Mill), *ideals* (G. E. Moore and H. Rashdall), or *interests* (R. B. Perry). Its motto, which characterizes one version of utilitarianism, is "The greatest happiness for the greatest number." Utilitarians further divide into *act* and *rule utilitarians*. Act utilitarians hold that the right act in a situation is that which results (or is most likely to result) in the best consequences, whereas rule utilitarians hold that the right act is that which conforms to the set of rules that in turn will result in the best consequences (relative to other sets of rules). See Chapter 2.

zoological sentientism The view of animal liberationists and utilitarians that all animals, insofar as they are sentient, capable of experiencing pain and pleasure, are morally considerable. See Chapter 7.

Credits

Chapter One: p. 1 Excerpt from *Antigone* in *Sophocles, The Oedipus Cycle: An English Version,* by Dudley Fitts and Robert Fitzgerald. Copyright © 1939 by Harcourt Brace and Company and renewed 1967 by Dudley Fitts and Robert Fitzgerald. Reprinted by permission of the publisher.

Chapter Six: p. 99 St. Francis of Assisi, *Canticle of Brother Sun*, translated by R. J. Armstrong and I. Brandy. Copyright © 1982 by Paulist Press.

Chapter Seven: p. 124 The Far Side. Copyright © 1992 by Farworks, Inc. Used by permission. All rights reserved.

Chapter Eight: pp. 146, 147 Holmes Rolston, from "Values at Stake: Does Anything Matter: A Response to Ernest Partridge." Reprinted by permission of the author.

Chapter Nine: p. 160 G. Tyler Miller, *Living in the Environment*, Third Edition, Wadsworth Publishing Company. Reprinted by permission of the publisher. p. 167 From *The Ages of Gaia: A Biography of Our Living Earth,* by James E. Lovelock. Copyright © 1988 by The Commonwealth Fund Book Program of Memorial Sloan-Kettering Cancer Center. Reprinted by permission of W.W. Norton & Company, Inc.

Chapter Ten: p. 177 Reprinted from Arne Naess, "The Shallow and the Deep, Long-Range Ecological Movement," *Inquiry* 16 (Spring 1973). By permission of Scandinavian University Press, Oslo, Norway, and the author. p. 186 Paul Taylor, "The Ethics of Respect for Nature," from *Respect for Nature.* Copyright © 1986 by Princeton University Press. Reprinted by permission of Princeton University Press. p. 196 Bill Devall and George Sessions, *Deep Ecology: Living As If Nature Mattered.* Salt Lake City: Gibbs Smith, Publisher, 1985.

Chapter Eleven: p. 203 Foto Marburg/Art Resource, New York. p. 214 By permission of Myles Duffy.

Chapter Twelve: pp. 229, 230 Reprinted by permission of Garrett Hardin.

Chapter Fourteen: p. 272 G. Tyler Miller, *Living in the Environment*, Third Edition, Wadsworth Publishing Company. Reprinted with permission of Wadsworth Publishing Company.

Chapter Sixteen: p. 302 Reprinted from Robert Elliot, "Faking Nature" *Inquiry* 25, no. 1, pp. 81–93. By permission of Scandinavian University Press, Oslo, Norway, and the author. pp. 309, 314, 315 Reprinted by permission of the publisher from *The Diversity of Life,* by E. O. Wilson. Cambridge, MA: Harvard University Press. Copyright © 1992 by Edward O. Wilson. p. 323 Reprinted by permission of David Quammen. All rights reserved. Copyright © 1996 by David Quammen.

Chapter Seventeen: p. 335 The New Yorker Collection. Copyright © 1981 by Michael Maslin from cartoonbank.com. All rights reserved. p. 341 "Introduction to Essays Towards a Steady-State Economy," by Herman Daly, in *Valuing the Earth: Economics, Ecology, Ethics*, eds. Herman Daly and Kenneth Townsend. MIT Press, 1993. Reprinted by permission of MIT Press.

Chapter Eighteen: p. 365 Copyright © 1992 by Sidney Harris. p. 375 By permission. From Merriam-Webster's Collegiate® Dictionary, Tenth Edition. © 1998 by Merriam-Webster, Incorporated.

Index